A Mathematica Companion for Differential Equations

Selwyn Hollis

Armstrong Atlantic State University

Prentice Hall

Upper Saddle River, NJ 07458

Editor-in-Chief: Sally Yagan
Executive Editor: George Lobell
Supplement Editor: Jennifer Brady
Assistant Managing Editor: John Matthews
Production Editor: Donna Crilly
Supplement Cover Manager: Paul Gourhan
Supplement Cover Designer: Joanne Alexandris
Manufacturing Buyer: Ilene Kahn

Pearson Education, Inc.
Upper Saddle River, NJ 07458

Printed in the United States of America

10 9 8 7 6 5 4 3 2 1

ISBN 0-13-046329-9

Pearson Education Ltd., *London*
Pearson Education Australia Pty. Ltd., *Sydney*
Pearson Education Singapore, Pte. Ltd.
Pearson Education North Asia Ltd., *Hong Kong*
Pearson Education Canada, Inc., *Toronto*
Pearson Educación de Mexico, S.A. de C.V.
Pearson Education—Japan, *Tokyo*
Pearson Education Malaysia, Pte. Ltd.
Pearson Education, *Upper Saddle River, New Jersey*

Contents

4. Approximation of Solutions

5. Coupled Pairs of First-Order Equations

6. Second-Order Equations and Applications

7. Linear Differential Systems

8. Nonlinear Autonomous Systems in the Plane

9. A Gallery of Nonlinear Systems

10. Fourier Series and Sturm-Liouville Eigenvalue Problems

11. Linear Partial Differential Equations

Preface

Mathematica® is a powerful and elegant tool with which one can explore essentially every branch of mathematics. But in particular, its ease of use and its extraordinary graphics capabilities make it an ideal learning laboratory for the elementary differential equations course. This manual is intended to provide students with a guide along their journey through this course.

The material here is organized along the lines of the standard, modern differential equations course—modern in the sense that numerical solutions, graphics, and systems are emphasized and introduced early, standard in the sense that exact solutions are given their deserved place when appropriate. With coverage of topics through nonlinear systems, Sturm-Liouville eigenvalue problems, and the diffusion and wave equations, this manual will be useful in both semesters of the typical two-semester sequence.

Several standard applications play an important role here: projectile motion with resistance, spring-mass systems, the simple pendulum, the logistic population model, and predator-prey and competition models. In addition, the spruce budworm problem and a half-wave rectifier circuit each appear in the context of both single equations and coupled systems. Chapter 9, A Gallery of Nonlinear Systems, contains numerous other applications, including the chemostat, the Fitzhugh-Nagumo equations, the Brusselator, the Oregonator, the double pendulum, and the three-body problem.

In numerous instances, we take advantage of *Mathematica*'s ability to animate graphics. In addition to animating graphs of solutions, we create "physical" animations of a simple pendulum and a spring-mass system. (Some of these animations, as well as dozens of others, can be found in the form of QuickTime® movies at www.prenhall.com/hollis.)

This entire manual was created with *Mathematica* 4.1. Each chapter is a *Mathematica* notebook.

Learning *Mathematica*

Even though most students called upon to use *Mathematica* in the differential equations course will have used *Mathematica* to some extent in calculus, we do not assume that the student has any prior experience with *Mathematica.* Yet we do assume that a student at the level of the differential equations course has sufficient technical savvy to pick up on the basic ideas fairly readily. The introductory chapter (Chapter 0) begins with a discussion of *how* one learns *Mathematica*, with special emphasis on using the wealth of information easily available in the Help Browser. This is followed by brief discussions of several of the more fundamental ideas in *Mathematica* related to functions, equations, lists, and rules. The chapter concludes with a section on troubleshooting, which is a distillation of years of experience helping students resolve difficulties (as well as solving our own problems) with *Mathematica.*

It is impossible to overstate the importance and usefulness of the Help Browser. The answer to virtually any question about *Mathematica* can be found there. For example, if one is unsure of the syntax of the `DSolve` function, then all one has to do is type and highlight "`DSolve`" and select Find Selected Function from the Help menu (or press the *help* key).

Throughout this manual we strive to present examples in a *style* of using *Mathematica* that is both transparent and efficient. We avoid the use of pure functions, for example, as well as most of the short "operator forms" such as `/@` for `Map`, etc., in order not to introduce unnecessary confusion or distract from our focus on differential equations. However, there will no doubt frequently arise questions about what some option is for or why a command is given a certain way. The way to find answers to such questions is to *change it and see what happens*. This cannot be emphasized enough: the user of this manual should actively experiment! Indeed, it can be argued that experimentation is precisely the purpose for which *Mathematica* exists.

The DiffEqs Packages

An important part of this manual is a collection of robust, fully documented, easy-to-install *Mathematica* packages, which can be found at www.prenhall.com/hollis. After an easy two-step installation process, the documentation will even appear in the Help Browser. An Appendix in this manual also provides documentation of these packages.

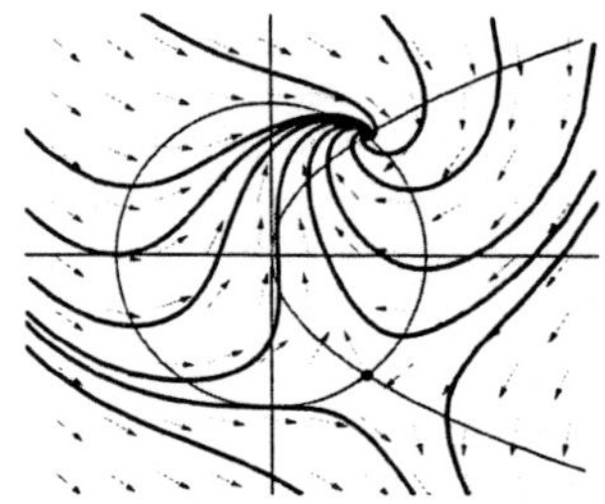

The `DEGraphics` package provides a number of functions that greatly simplify the process of plotting solutions of differential equations. In particular, `DEPlot` and `PhasePlot` make it easy to create very attractive plots of direction fields and solution curves with a single command. For example, the picture on the right was created with one command using `PhasePlot`. (For anyone who has used the standard *Mathematica* functions to create phase portraits, these routines are guaranteed to impress. To our knowledge, *no comparable package exists*.) Other functions in the package are `NDPlot`, `TimeStatePlot`, `ViewProjections`, `PlotImplicit`, and `PoincareTimeSection`.

The `DETools` package provides several functions that solve various types of equations encountered in the differential equations course. These include `FOLDSolve` (*First-Order-Linear-DSolve*), `BernoulliDSolve`, `RiccatiDSolve`, `SeparableDSolve`, and `SeriesDSolve`. Each of these returns solutions as expressions rather than rules, which makes them simpler for the novice to use than `DSolve`. `SeriesDSolve` is unique (as far as we know) in that will automatically compute partial sums of Frobenius-series solutions as well as ordinary power-series solutions.

The `MatrixTools` and `FourierTools` packages provide several handy functions for working with matrices and Fourier series.

Final Thoughts

Writing ancillary such as this offers unique challenges to an author. Since the student will have available both a textbook and *The Mathematica Book* (as hard copy or through the Help Browser), one must avoid being pedantic about *both* of the things the manual is about: *Mathematica* and differential equations. So it is necessary to make many statements concerning both with little if any explanation. The key, I believe, is to give good examples and let the student take it from there. I hope I have achieved at least some success in that respect.

Chapter 0

An Introduction to *Mathematica*

This chapter begins with a brief section in which we introduce the true beginner to the *Mathematica* front end and Help Browser. We also give some specific suggestions for how and where to learn the basics of *Mathematica*. (We don't attempt to give a thorough tutorial here.) In the next two sections we discuss four of the most important *Mathematica* objects: functions, equations, lists, and rules. The last section is about troubleshooting. There we discuss a few of the more common technical issues and potential pitfalls related to using *Mathematica*. Since our primary goal is to use *Mathematica* to help us understand differential equations, you should not initially spend a great amount of time pouring over the details of this chapter. Familiarize yourself with what's here, and refer back to it as needed.

0.1 Getting Started

■ The *Mathematica* Front End

When you start up *Mathematica*, the first thing you see is a window displaying the contents of a "**notebook.**" This window is displayed by *Mathematica*'s **front end**. The front end is the interface between you and the *Mathematica* **kernel**, which does the computations. The following is a notebook in a front end window (in Mac OS X).

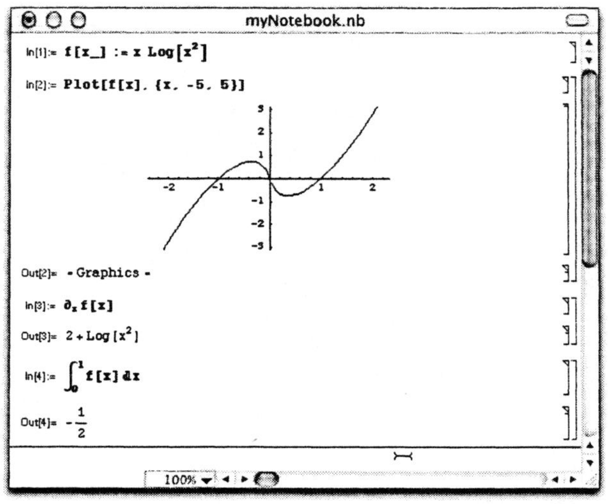

A *Mathematica* notebook is composed of **cells**. On the right side of the window you see the **cell brackets**. Each cell in the notebook shown above is either an *input cell*, an *output cell*, or a *graphics cell*. There are several other kinds of cells. Some of these are *text*, *title*, and *section*.

Also notice the horizontal line near the bottom of the window. This indicates the insertion point for the next new cell. To enter a command into a notebook, simply begin typing. The default cell-type is *input*. When you're done typing, just press **shift-return** (on a Macintosh, you can also use the "**enter**" key.) To evaluate an existing input cell, simply click anywhere inside the cell (or on the cell bracket) and press **shift-return** (or **enter**.)

To create a cell *between* two existing cells, move your cursor over one of the cells toward the other until the "I-beam" ( ) becomes horizontal ( ). Then click, and a horizontal line will appear, indicating the desired insertion point.

To delete a cell, click on its bracket and then choose Clear from the Edit menu.

Palettes. You can enter mathematical expressions so that they appear essentially the same as you would write them on paper or see them in your textbook. For example, to define the function $f(x) = \sqrt{x^2 + 1}$, we could use the "Input Form" that is necessary in versions of *Mathematica* prior to 3.0:

```
f[x_] := Sqrt[x^2 + 1]
```

or we could use "Standard Form":

$$f[x\_] := \sqrt{x^2 + 1}$$

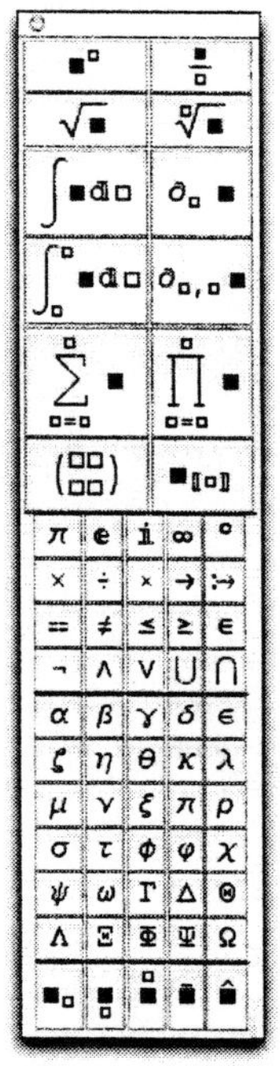

There is a vast set of keystroke combinations for typing such expressions. However, at first, you will probably want to take advantage of one or more or the standard *palettes* that are available. The image to the right shows the `BasicInput` palette. Clicking on one of the palette's buttons places the corresponding character/expression at the current input location. This particular palette probably appears by default when you start *Mathematica*, but if not, you can access it or any of the other palettes through your File menu as indicated below.

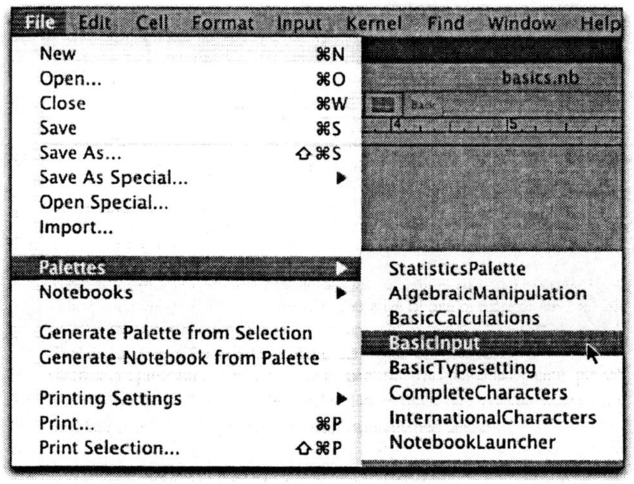

■ The Help Browser

Mathematica's Help Browser is truly one of its most extraordinary features. It gives easy access to an encyclopædic collection of information about *Mathematica*. To access the Help Browser, select Help Browser... from the Help menu.

The Help Browser provides a great deal of tutorial material. If you're a beginner, you should take a thorough look at the information under Getting Started/Demos: Tour of *Mathematica* ...

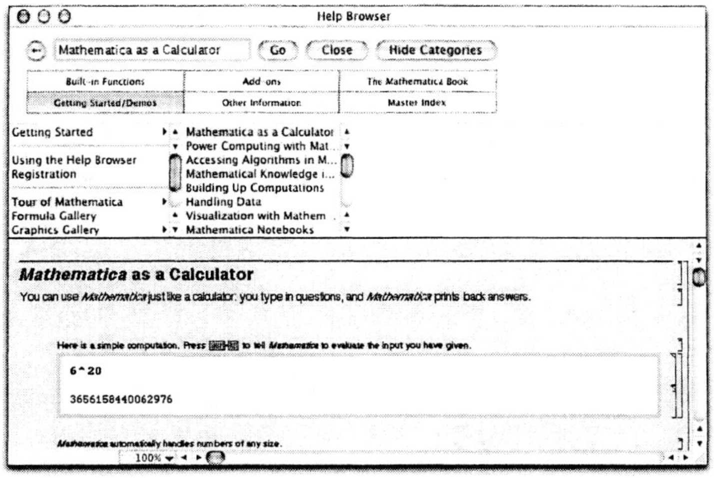

Also among the wealth of information available through the Help Browser are descriptions of all of *Mathematica*'s built-in functions, including "*Further Examples*" of their use. You can also enter commands from within the Help Browser.

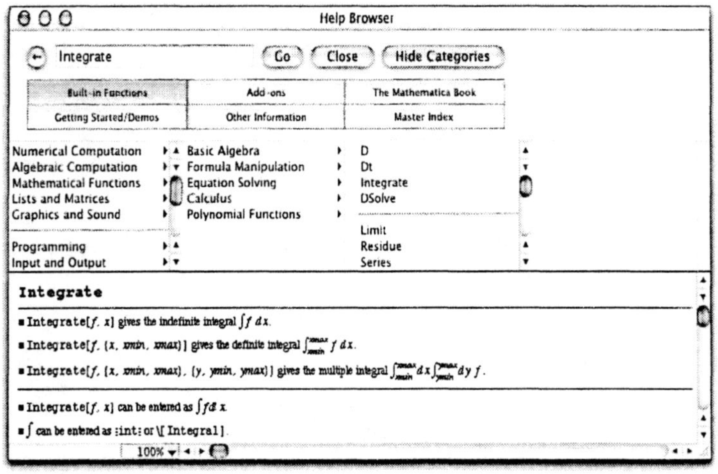

Under Other Information, you will find help in typing and editing mathematical expressions.

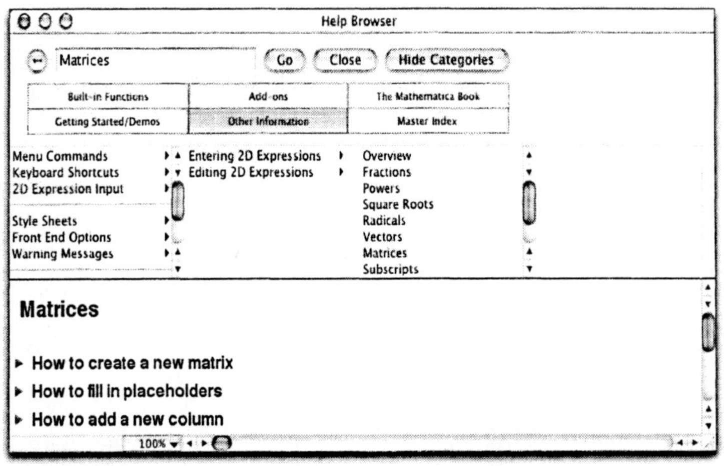

■ Learning the Basics

After you've perused the tutorial material found in the Help Browser under Getting Started/Demos ▷ Tour of *Mathematica*, you should click on the button labeled **The Mathematica Book** near the top of the Help Browser window. Then, under Practical Introduction to *Mathematica*, you'll find a thorough description of the basic usage of *Mathematica*.

Once you've got the basics down, look under Advanced Mathematics in *Mathematica* ▷Calculus, and go through the material there on Differentiation, Indefinite Integrals, Integrals That Can and Cannot Be Done, and Definite Integrals. After you've done all this, you should be ready to tackle the remainder of this manual.

You can further increase your *Mathematica* knowledge through the Resource Library located at `http://library.wolfram.com`. There you'll find a plethora of tutorials, demos, examples, FAQs, tips, and "how-tos." You'll also find there a Student Support Forum, where you can post your own questions or your answers to other students' questions.

■ *Mathematica*, the Word Processor

You will likely need to put the work you do in *Mathematica* into a form that will be presentable enough to submit to your professor. Fortunately, the *Mathematica* front end is a very versatile tool for creating technical documents. In fact, this manual was written entirely with *Mathematica*.

Cell Styles

You should always provide comments and narrative along with your calculations (whether you're using *Mathematica* or pencil and calculator). Any cell that contains text should be given Text Style by selecting Text from the Style submenu of the Format menu before you begin typing text into the cell. You can also give an existing cell Text Style by highlighting the cell bracket and selecting Text from the same menu.

You should also use Title, Section, Subsection cells, etc., to organize your notebook. These items are also in the Style submenu of the Format menu.

Aside from resulting in much nicer looking work, here is the main reason why this is so important: Making sure that all cells not containing *Mathematica* input are given a style other than Input Style allows you to evaluate all the input cells in your notebook sequentially by choosing Evaluate Notebook from the Kernel menu, without getting all kinds of errors and garbage that result from trying to evaluate invalid input.

```
For example notice what happens when I enter this
```

```
i enter example For happens notice this what when
```

```
or when I enter this.
```

```
Syntax::tsntxi : "this." is incomplete; more input is needed.
```

```
or when I enter this.
```

Page Breaks

Bad page breaks usually involve a large amount of blank space at the bottom of a page. Frequently this is caused by a graphic being just a little too large to fit on the current page. The default size of *Mathematica* graphics is larger than it usually needs to be. So by resizing (i.e., shrinking) graphics, you can avoid a lot of bad page breaks. A graphic can be resized by clicking on it and dragging a corner.

In order to tell where page breaks will occur before you print, select Show Page Breaks from the Format menu.

You can also *force* a page break above or below a given cell. This is done through the **Option Inspector**, which is accessed by choosing Option Inspector... from the Format menu. After highlighting a cell bracket, click on the Option Inspector window, show option values for Selection, and look under Cell Options – Page Breaking. There you can set PageBreakAbove or PageBreakBelow to True.

Warning: Be *very* careful about fiddling with options in the Options Inspector. Although—and probably *because*—it provides an enormous amount of freedom and flexibility, the Option Inspector is highly complex and can truly be a *Pandora's box* for the beginner.

Other Tips

In the Kernel menu you'll see Show In/Out Names. You might want to use this to hide the `In[ ]:=` and `Out[ ]:=` labels on input and output cells.

In the Printing Settings submenu of the File menu, you'll see Printing Options... In the resulting dialog box, you can set margins and specify whether or not to print cell brackets.

In the Style Sheet submenu of the Format menu, you can choose from among several standard style sheets. The choice of style sheet affects the appearance of title and section cells, background color, etc. Experiment to find one that you like.

0.2 Functions and Equations

■ 0.2.1 Functions

Blank

When defining a function, it is essential to follow each argument by a `Blank` (or "underscore"). Also, recall that the arguments of a function are enclosed by brackets. For example, we would define the function $f(x) = x^3 - 2$ by entering

```
f[x_] := x³ - 2 x
```

Then we can evaluate the function at any number

```
f[3]
```

```
21
```

or plot its graph:

```
Plot[f[x], {x, -2, 2}];
```

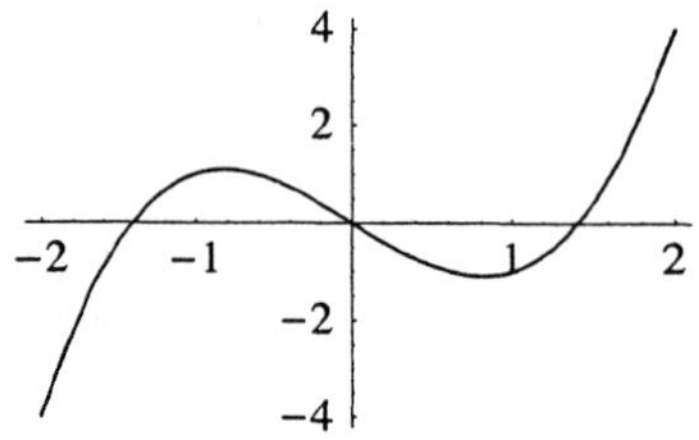

Note that the `Blank` appears next to `x` *only on the left side of the expression*. Also, when defining a function with more than one argument, a `Blank` must follow each one.

```
g[x_, y_, z_] := x + y + z
g[1, 2, 3]
6
```

`Set` (=) versus `SetDelayed` (:=)

Definitions of functions—and assignments of expressions to variables in general—can be made using either "equal" or "colon-equal." The difference between these two ways is described by the full *Mathematica* names of the = and `:=` expressions, which are `Set` and `SetDelayed`. When an assignment is made using `Set`, any calculations that are indicated on the right side are done when the assignment is entered. When an assignment is made using `SetDelayed`, any calculations that are indicated on the right side are delayed until the defined expression is used.

In many cases, such as in the definitions of `f` and `g` above, it makes no difference which is used. To see a simple example that indicates the importance of using `Set` rather than `SetDelayed`, let's suppose we want to define $f(x)$ to be the derivative of $(x+1)\cos x$. If we enter

```
f[x_] := ∂x ((x + 1) Cos[x])
```

notice what happens when we try to evaluate $f(2)$:

```
f[2]
General::ivar : 2 is not a valid variable.
∂2 (3 Cos[2])
```

However, if we enter

```
f[x_] = ∂x ((x + 1) Cos[x])
Cos[x] - (1 + x) Sin[x]
```

then f works the way we want it to:

```
f[2]
Cos[2] - 3 Sin[2]
```

When is it important to use `SetDelayed` rather than `Set`? Here's an example: Suppose we want to define a function (of a) to compute the limit of $f(x) = \frac{\sin x}{x}$ as $x \to a$. Let's first attempt this using `Set` (and suppress the output by ending the command with a semicolon):

```
lim[a_] = Limit[Sin[x]/x, x → a];
```

If we try out our function by entering

```
lim[π / 4]
```

$$\frac{2\sqrt{2}}{\pi}$$

then everything seems to be okay. But when we enter

```
lim[0]
```

```
Power::infy : Infinite expression 1/0 encountered.

∞::indet : Indeterminate expression 0 ComplexInfinity encountered.

Indeterminate
```

we do *not* get the correct result, as evidenced by

```
Limit[Sin[x] / x, x → 0]
```

```
1
```

The problem is that *Mathematica* evaluated the limit, assuming that $a \neq 0$ in order to give a "general" result, as soon as we entered the definition. To avoid that, we should use `SetDelayed` instead of `Set`. So let's clear the bad definition

```
Clear[lim]
```

and enter a good one:

```
lim[a_] := Limit[Sin[x] / x, x → a];
```

Now we obtain the correct result from `lim`:

```
lim[0]
```

```
1
```

Applying Functions with //

Suppose that we define a simple function such as

```
f[x_] := x (x - 1)
```

Naturally, we could evaluate this function at, say, $x = 3$, by entering

```
f[3]
```

```
6
```

Mathematica provides an alternative syntax for applying f to $x = 3$ as follows.

```
3 // f
```

```
6
```

We will use this *postfix* method of function application frequently, often for the purpose of applying either `Simplify` to some expression or `N` to some numerical calculation. For example, we will use the following style when doing a symbolic calculation:

```
2 x + x (5 x + 1) // Simplify
```

```
x (3 + 5 x)
```

The percent sign represents the most recent output. So the following style displays an exact value followed by its numerical value:

```
√585 / 33
% // N
```

$$\frac{\sqrt{65}}{11}$$

```
0.732933
```

Piecewise-Defined Functions

Mathematica has two logical functions that we can use to enter definitions of piecewise-defined functions. These are `If` and `Which`. `If` usually works best for functions with two pieces, such as

$$f(x) = \begin{cases} x, & \text{if } x \leq 1; \\ x-2, & \text{if } x > 1. \end{cases}$$

This function can be entered and plotted as follows.

```
f[x_] := If[x ≤ 1, x, x - 2]
Plot[f[x], {x, -1, 3}];
```

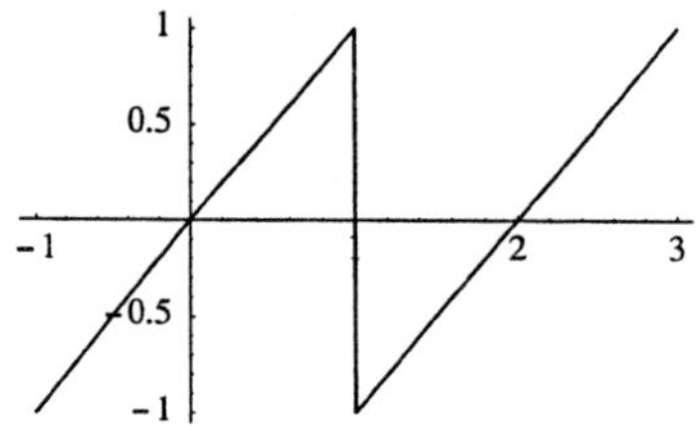

More complicated functions are best handled with `Which`. For example,

$$f(x) = \begin{cases} 1, & \text{if } x \leq -1 \\ -x, & \text{if } -1 < x \leq 1 \\ -1, & \text{if } x > 1 \end{cases}$$

can be entered and plotted as follows.

```
f[x_] := Which[x ≤ -1, 1, -1 < x ≤ 1, -x, x > 1, -1]

Plot[f[x], {x, -3, 3}];
```

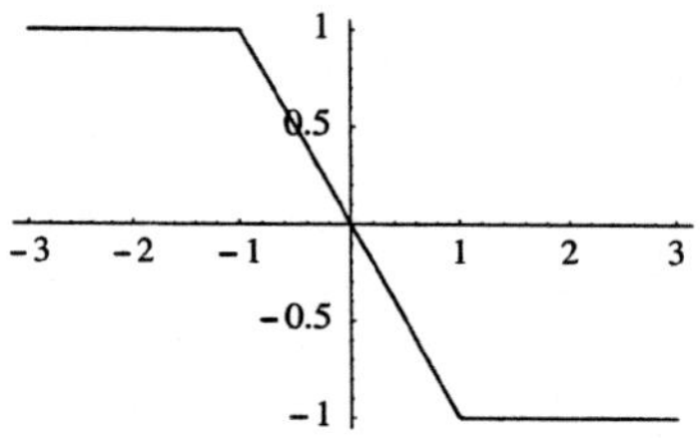

There are a handful of standard, built-in, piecewise-defined functions, including `Abs`, `Floor`, and `Mod`.

```
Plot[Abs[x], {x, -1, 1}];
Plot[Floor[x], {x, -2, 3}];
Plot[Mod[x, 1], {x, 0, 3}];
```

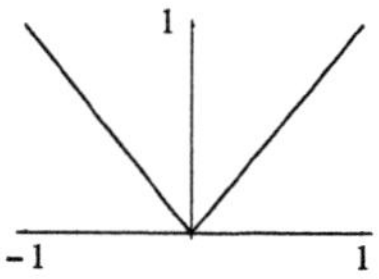

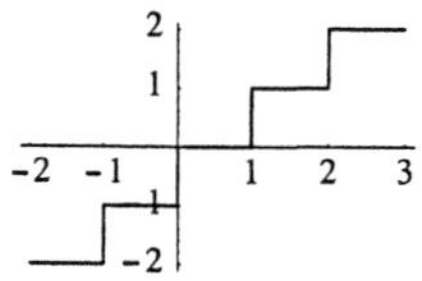

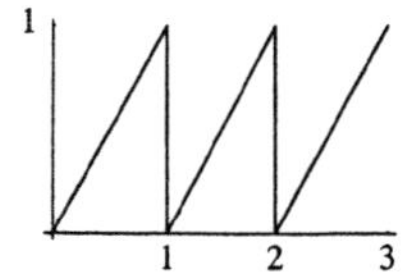

The following functions are built with `Abs`.

```
Plot[Abs[Cos[x] - .5], {x, 0, 4 π}];
Plot[Abs[Abs[x] - 1], {x, -3, 3}];
```

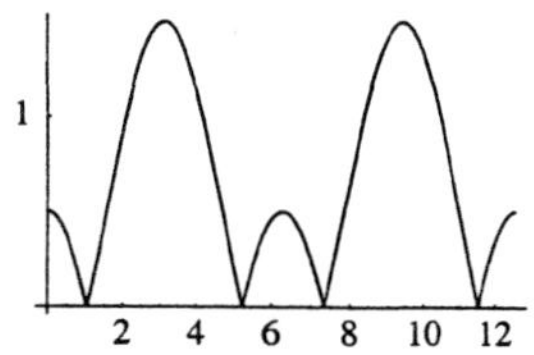

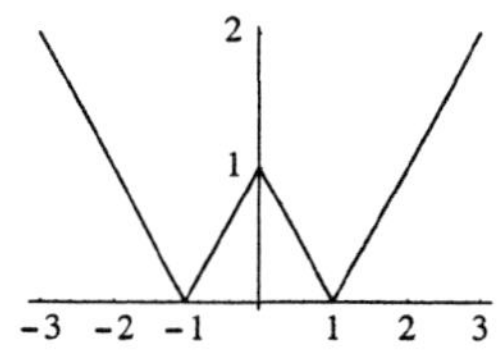

`Floor` helped create each of these *step functions*:

```
Plot[(-1)^Floor[x], {x, -2, 3}];
Plot[Sin[π/6 Floor[6 x]], {x, 0, 2}];
```

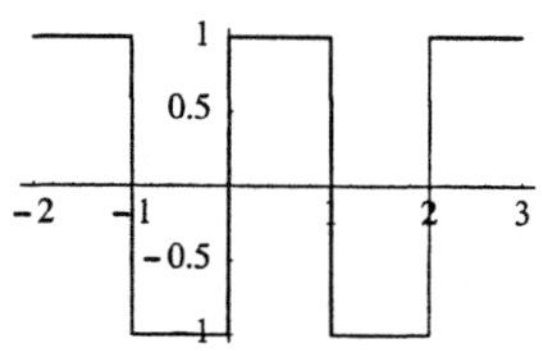

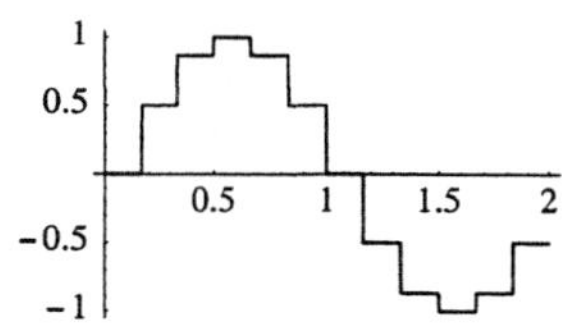

A *periodic function* based on a piece of any graph can be constructed with `Mod`.

```
Plot[Sin[π Mod[x, 1] / 2], {x, 0, 3}];
Plot[Abs[Mod[x, 2] - 1], {x, -4, 4}];
```

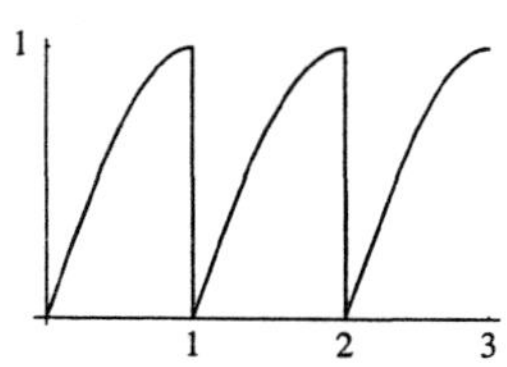

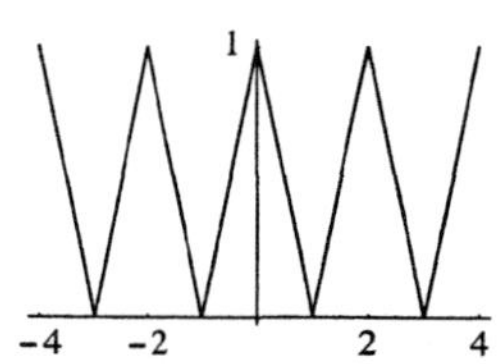

The Heaviside unit step function, `UnitStep`, is useful in defining piecewise-defined functions.

```
Plot[UnitStep[x] + UnitStep[x - 1] - UnitStep[x - 2] - UnitStep[x - 3],
  {x, -1, 4}, AspectRatio → Automatic, Ticks → {Range[-1, 4], {1, 2}}];
```

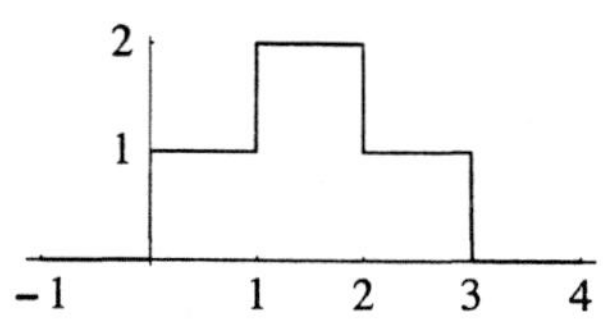

While its use can be cumbersome, one advantage that `UnitStep` has is that `Integrate` can handle expressions involving `UnitStep`. For example, on the interval [0, 3] the functions `Mod[x,1]` and `x-UnitStep[x-1]-UnitStep[x-2]` have the same values, except at $x = 1, 2$, and 3.

```
Plot[x - UnitStep[x - 1] - UnitStep[x - 2], {x, 0, 3},
  AspectRatio → Automatic, Ticks → {{1, 2, 3}, {1}}];
```

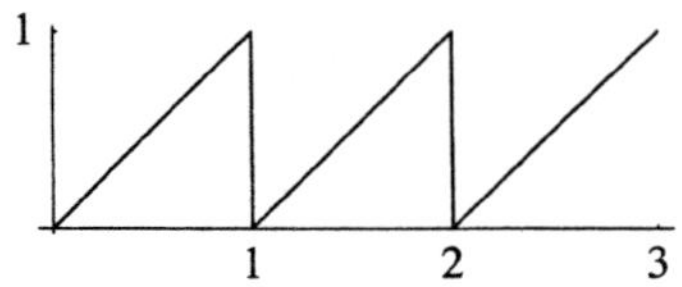

However, notice that the first of the following fails, while the second succeeds.

$$\int_0^3 \texttt{Mod[x, 1]}\, dx$$

$$\int_0^3 \texttt{Mod[x, 1]}\, dx$$

$$\int_0^3 (\texttt{x - UnitStep[x - 1] - UnitStep[x - 2]})\, dx$$

$$\frac{3}{2}$$

Exercises

1. Enter definitions for each of $f(x) = x(x-2)^2$, $g(x) = x-3$, and $h(x) = x^2 - 1$, and then compute and simplify each of the following:

   (a) $f(g(h(x)))$ (b) $h(g(f(x)))$ (c) $f(h(g(x)))$

2. Enter a definition for $f(x) = (1+x^2)^{-1}$. Then create a plot (over $-3 \le x \le 3$) containing the graphs of:

   (a) $f(x)$, $f(x-1)$, and $f(x+1)$ (b) $f(x)$, $f(2x)$, and $f(3x)$

3. Plot: (a) $f(x) = \begin{cases} -x, & \text{if } x < 0 \\ x^2 - 1, & \text{if } x \ge 0 \end{cases}$ (b) $f(x) = \begin{cases} 1, & \text{if } x \le \pi/2 \\ \sin x, & \text{if } x > \pi/2 \end{cases}$

4. Plot: (a) $f(x) = \begin{cases} 1, & \text{if } x < 0 \\ 1 - x^2, & \text{if } 0 \le x \le 2 \\ -3, & \text{if } x > 2 \end{cases}$ (b) $f(x) = \begin{cases} x+1, & \text{if } x < 0 \\ 1 - 2x, & \text{if } 0 \le x \le 1 \\ x - 2, & \text{if } x > 1 \end{cases}$

5. Let $\lfloor x \rfloor$ denote the floor function. Plot: (a) $f(x) = (x - \lfloor x \rfloor)^2$ (b) $f(x) = x\,(-1)^{\lfloor x \rfloor}$

6. For brevity, let $H(x)$ denote the Heaviside unit step function. Plot each of the following functions over $0 \le x \le 3$, and then express each one in a form (valid for $0 \le x \le 3$) that does not involve `UnitStep`.

   (a) $f(x) = (H(x) - 2\,H(x-1) + H(x-2))\sin(\pi x)$

   (b) $f(x) = x\,H(x) - (x-1)\,H(x-1) - H(x-2)$

7. Plot each of the following functions over $-1 \le x \le 1$, and then express each one (for $-1 \le x \le 1$) in terms of `UnitStep`. Make a plot to check your work.

   (a) $f(x) = \sqrt{|x|}$ (b) $f(x) = (-1)^{\lfloor x \rfloor}$

■ 0.2.2 Equations

Certainly one of the most frequent mathematical tasks that we need to do is to solve an equation. In order to be able to solve equations with *Mathematica*, we first need to understand how equations are formed. The important thing to remember is that *double equal signs* are used to form an equation.

`Equal` (double equal signs) is a *logical test* that returns either `True`, `False`, or the equation itself.

```
2 * 17 - 34 == 0
True
```

```
3 == 4
False
```

$x^2 - 4 == 0$

$-4 + x^2 == 0$

Notice, however, that *Mathematica* returns `True` only when the expressions on each side are *identical.* Only the most superficial simplification is done prior to the test, as in:

```
2 x + x - 2 == 3 x + 5 - 7
True
```

Notice that for this equation:

$x^2 - 4 == (x + 2)\ (x - 2)$

$-4 + x^2 == (-2 + x)\ (2 + x)$

a nontrivial operation must be done to one side or the other before the expressions become truly identical.

Notice that *Mathematica* returns an error message if we use a single equal sign improperly.

```
3 = 4
Set::setraw : Cannot assign to raw object 3.
```

Mathematica has three functions for solving ordinary equations. These are `Solve`, `NSolve`, and `Find-Root`. `Solve` works very well on polynomial equations and many other algebraic equations.

$\text{Solve}[x^4 - 2\,x^3 - x^2 + 6\,x - 6 == 0,\ x]$

$\{\{x \to 1 - i\},\ \{x \to 1 + i\},\ \{x \to -\sqrt{3}\},\ \{x \to \sqrt{3}\}\}$

$\text{Solve}[\sqrt{x - 1} + x == 4 + \sqrt{x + 4},\ x]$

$\{\{x \to 5\}\}$

`Solve` will also give solutions to trigonometric (or exponential/logarithmic) equations, but frequently gives a warning.

$\text{Solve}[2\,\text{Sin}[2\,x] == \text{Cos}[x]^2 - 1,\ x]$

```
Solve::ifun : Inverse functions are being
   used by Solve, so some solutions may not be found.
```

$\{\{x \to 0\},\ \{x \to -\pi\},\ \{x \to \pi\},\ \{x \to \text{ArcCos}[-\frac{1}{\sqrt{17}}]\},\ \{x \to -\text{ArcCos}[\frac{1}{\sqrt{17}}]\}\}$

`Solve` will also find solutions of a system of equations. The equations must be given as elements of a list, *i.e.*, separated by commas and enclosed in braces.

```
Solve[{x^2 - y == 1, -x + y == 1}, {x, y}]
```

```
{{y → 0, x → -1}, {y → 3, x → 2}}
```

The solutions of polynomial equations of degree five or greater generally cannot be found in any exact form. Notice how *Mathematica* "avoids" the problem. The expression `1 + 5 #1 - 10 #1^2 + #1^5 &` is the polynomial expressed as a *pure function*.

```
Solve[x^5 - 10 x^2 + 5 x + 1 == 0, x]
```

```
{{x → Root[1 + 5 #1 - 10 #1^2 + #1^5 &, 1]},
 {x → Root[1 + 5 #1 - 10 #1^2 + #1^5 &, 2]}, {x → Root[1 + 5 #1 - 10 #1^2 + #1^5 &, 3]},
 {x → Root[1 + 5 #1 - 10 #1^2 + #1^5 &, 4]}, {x → Root[1 + 5 #1 - 10 #1^2 + #1^5 &, 5]}}
```

In such situations, we can always resort to numerical solutions. `NSolve` finds a numerical approximation to each solution of a polynomial equation, including complex solutions.

```
NSolve[x^5 - 10 x^2 + 5 x + 1 == 0, x]
```

```
{{x → -1.22065 - 1.89169 i}, {x → -1.22065 + 1.89169 i},
 {x → -0.153102}, {x → 0.66946}, {x → 1.92494}}
```

In many situations where `Solve` is successful, such as

```
Solve[x^3 - 10 x^2 + 5 x + 1 == 0, x]
```

$$\left\{\left\{x \to \frac{10}{3} + \frac{85}{3\left(\frac{1}{2}\left(1523 + 9\,i\,\sqrt{1691}\right)\right)^{1/3}} + \frac{1}{3}\left(\frac{1}{2}\left(1523 + 9\,i\,\sqrt{1691}\right)\right)^{1/3}\right\},\right.$$

$$\left\{x \to \frac{10}{3} - \frac{1}{6}\left(1 + i\sqrt{3}\right)\left(\frac{1}{2}\left(1523 + 9\,i\,\sqrt{1691}\right)\right)^{1/3} - \frac{85\left(1 - i\sqrt{3}\right)}{3\;2^{2/3}\left(1523 + 9\,i\,\sqrt{1691}\right)^{1/3}}\right\},$$

$$\left.\left\{x \to \frac{10}{3} - \frac{1}{6}\left(1 - i\sqrt{3}\right)\left(\frac{1}{2}\left(1523 + 9\,i\,\sqrt{1691}\right)\right)^{1/3} - \frac{85\left(1 + i\sqrt{3}\right)}{3\;2^{2/3}\left(1523 + 9\,i\,\sqrt{1691}\right)^{1/3}}\right\}\right\}$$

it may still be preferable to use `NSolve`:

```
NSolve[x^3 - 10 x^2 + 5 x + 1 == 0, x]
```

```
{{x → -0.152671}, {x → 0.692369}, {x → 9.4603}}
```

Numerical solution of nonpolynomial equations requires the use of `FindRoot`. `FindRoot` can find only one solution at a time and requires us to supply an initial guess at the solution we're looking for. (An appropriate initial guess can usually be determined by examining a graph.) For example, to find the positive solution of the equation $x^2 = \cos x$, one can enter

```
FindRoot[x^2 == Cos[x], {x, .75}]
```

```
{x → 0.824132}
```

Exercises

1. The equation $x^3 = x + 1$ has one real solution. Find its exact value with `Solve` and its numerical value with `NSolve`.

2. Use `Solve` on the *underdetermined* system

$$x + y + 2z = 2, \quad x - 2y + z = 1,$$

and interpret the result. Try each combination of *solve variables*: `{x,y,z}`, `{x,y}`, `{x,z}`, and `{y,z}`. Which gives the "cleanest" solution?

3. For each of the following functions, plot the graph to determine the approximate location of each of its zeros. Then find each of the zeros with `FindRoot`.

   (a) $f(x) = x^2 e^{-x/2}$ (b) $f(x) = x - 9\cos x$ (c) $f(x) = x^2 - \tan^{-1} x$

4. For each of the following equations, plot both sides of the equation to determine the approximate location of each of its solutions in the specified interval. Then find each of the solutions with `FindRoot`.

   (a) $\sin x \cos 2x = \cos x \sin 3x,\ 0 \le x \le 2\pi$

   (b) $\sin x^2 = \sin^2 x,\ 0 \le x \le \pi$

   (c) $\tan x = x,\ 0 \le x \le 3\pi$

0.3 Lists and Rules

0.3.1 Lists

Lists are ubiquitous in *Mathematica.* A list is anything that takes the form of a series of objects separated by commas and enclosed in braces. Each of the following is a list:

```
{a, b, c}
{{1, 3}, {2, 5}}
{x, 1, 2}
{x^2 + y == 2, 2 x - y == 0}
```

Many built-in commands expect lists for certain arguments. For example, notice in the following plot command that the second argument is a list that specifies the name of the variable and the interval over which to plot.

```
Plot[Cos[π Sin[x]], {x, 0, 2π}];
```

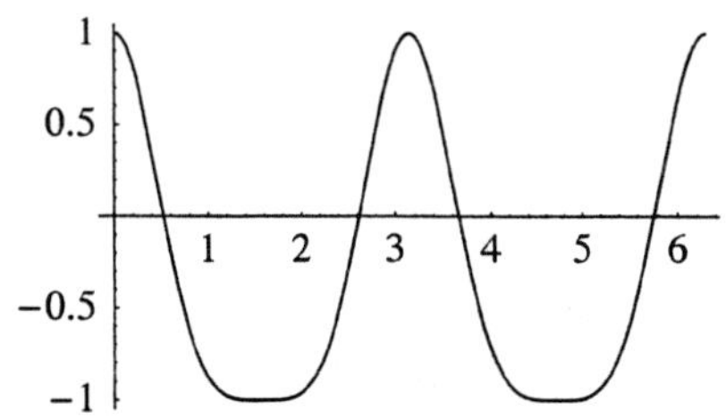

In the following `Solve` command each of the two arguments is a list. The result is also a list (of lists).

```
Solve[{x^2 + y == 2, 2 x - y == 0}, {x, y}]
{{y → 2 (-1 - √3), x → -1 - √3}, {y → 2 (-1 + √3), x → -1 + √3}}
```

Listable Functions

Most built-in *Mathematica* functions and operations are *listable.* When a listable function is applied to a list, it is applied to each element of the list and returns the result in the form of a list. For example:

```
{1, 2, 3, 4, 5, 6, 7, 8, 9}^2
{1, 4, 9, 16, 25, 36, 49, 64, 81}
```

$\sqrt{\{\{\{2, 3\}, \{4, 5\}\}, \{\{6, 7\}, \{8, 9\}\}\}}$

$\{\{\{\sqrt{2}, \sqrt{3}\}, \{2, \sqrt{5}\}\}, \{\{\sqrt{6}, \sqrt{7}\}, \{2\sqrt{2}, 3\}\}\}$

$\frac{1}{\{1, 2, 3, 4, 5\}}$

$\{1, \frac{1}{2}, \frac{1}{3}, \frac{1}{4}, \frac{1}{5}\}$

$2^{\{0,1,2,3,4,5,6,7,8,9,10\}}$

```
{1, 2, 4, 8, 16, 32, 64, 128, 256, 512, 1024}
Cos[{0, π / 4, π / 2, 3 π / 4, π}]
```

$\{1, \frac{1}{\sqrt{2}}, 0, -\frac{1}{\sqrt{2}}, -1\}$

While `Plus` and `Times` are listable,

```
{1, 2, 3} + {4, 5, 6}
{5, 7, 9}
{1, 2, 3} {4, 5, 6}
{4, 10, 18}
```

`Equal` is a notable exception:

```
{x, y} == {1, 2}
{x, y} == {1, 2}
```

It will occasionally be convenient to *make* `Equal` listable with the `SetAttributes` command.

```
SetAttributes[Equal, Listable]
```

Now notice the following result:

```
{x, y} == {1, 2}
{x == 1, y == 2}
```

□ Creating Lists

Mathematica provides three functions that are especially useful for creating lists. These are `Range`, `Table` and `NestList`.

Range

`Range` can be used with one, two, or three arguments. With one argument, it returns a list of consecutive natural numbers beginning with 1.

```
Range[10]
{1, 2, 3, 4, 5, 6, 7, 8, 9, 10}
```

`Range[`a, b`]` returns a list containing a, $a+1$, $a+2$, …, $a+n$, where $a+n \le b < a+n+1$.

```
Range[4.5, 15.1]
{4.5, 5.5, 6.5, 7.5, 8.5, 9.5, 10.5, 11.5, 12.5, 13.5, 14.5}
```

A third argument specifies the increment. (The default is 1.)

```
Range[0, 1, 1/10]
```

$\{0, \frac{1}{10}, \frac{1}{5}, \frac{3}{10}, \frac{2}{5}, \frac{1}{2}, \frac{3}{5}, \frac{7}{10}, \frac{4}{5}, \frac{9}{10}, 1\}$

Table

`Table` provides an easy way of constructing many kinds of lists. (It is similar to the seq() function on many calculators.) The following computations illustrate its use.

```
Table[k^2, {k, -3, 5}]
{9, 4, 1, 0, 1, 4, 9, 16, 25}
Table[Sin[k π x], {k, 4}]
{Sin[π x], Sin[2 π x], Sin[3 π x], Sin[4 π x]}
Table[i - j, {i, 4}, {j, 4}]
{{0, -1, -2, -3}, {1, 0, -1, -2}, {2, 1, 0, -1}, {3, 2, 1, 0}}
```

NestList

`NestList` creates lists whose elements are members of a *recursive sequence*; that is, given a function f and a starting point a_1, it creates a list containing members of the sequence $a_1, a_2, a_3, \ldots,$ where $a_{k+1} = f(a_k)$. For example, ten terms of the arithmetic sequence defined by $a_{k+1} = 2\,a_k - 1$ with $a_1 = 3$ can be computed by first defining

```
f[x_] := 2 x - 1
```

and then entering

```
NestList[f, 3, 9]
{3, 5, 9, 17, 33, 65, 129, 257, 513, 1025}
```

Notice that `NestList` has three arguments. The first is the name of the function, the second is the first member of the list, and the third is the number of "steps" to be computed (which is one less than the length of the resulting list).

□ Manipulating Lists

Flatten

There are occasions when we need to simplify lists by "merging" smaller lists that it contains. The `Flatten` command does this. For example:

```
Flatten[{{1, 2, 3}, {4, 5}}]
{1, 2, 3, 4, 5}
```

```
Flatten[{{{x, y}, {3}}, {4, q}}]
{x, y, 3, 4, q}
```

Append and Prepend

We will often need to add elements to the end or beginning of an existing list. These tasks can be done with `Append` and `Prepend`. Here are two examples:

```
Append[{1, 2}, 3]
{1, 2, 3}
Prepend[{1, 2}, 0]
{0, 1, 2}
```

Union and Join

Merging two or more lists into one can be done with `Union` or `Join`. `Union` does *not* maintain the order of elements:

```
Union[{a, 2}, {v, 8}, {h, s}]
{2, 8, a, h, s, v}
```

but `Join` does:

```
Join[{a, 2}, {v, 8}, {h, s}]
{a, 2, v, 8, h, s}
```

`Union` also provides a simple way of eliminating repetitions from a list.

```
Union[{3, 2, 1, 2, 2, 3, 3, 1, 4, 5, 5}]
{1, 2, 3, 4, 5}
```

Map

A very useful method for applying a nonlistable function to each element of a list is provided by `Map`. Suppose that we have a list of ordered pairs of numbers such as

```
points = Table[{2 i, 5 - i}, {i, 8}]
{{2, 4}, {4, 3}, {6, 2}, {8, 1}, {10, 0}, {12, -1}, {14, -2}, {16, -3}}
```

and that we want to create a list containing the sums of the numbers in each ordered pair in the list. To do this, we can create the function

```
addpairs[{x_, y_}] := x + y
```

and "map" it through the list of ordered pairs:

```
Map[addpairs, points]
{6, 7, 8, 9, 10, 11, 12, 13}
```

As an exercise, explain what goes on in the following:

```
Map[Flatten, {{3, {5, 6}}, {a, {b, c}}}]
{{3, 5, 6}, {a, b, c}}
```

Exercises

1. (a) Use `Table` and `Prime` to generate a list of the first 100 prime numbers.

   (b) Generate the same list using only `Prime` and `Range`.

2. Generate a list of values of the function $f(x) = \frac{\sin x}{x}$ for $x = .1, .2, \ldots, 1$, first using `Table`, then without `Table`.

3. Generate a list of the first 20 odd natural numbers, using:

   (a) `Table`; (b) `Range`; (c) `NestList`

4. Generate a list of the first 21 powers of 2 (beginning with 2^0), using:

   (a) `Table`; (b) `Range`; (c) `NestList`

5. Use `Table` to generate a list of ordered pairs $(x, f(x))$ for $x = 0, \frac{\pi}{12}, \frac{2\pi}{12} \ldots, \pi$, where $f(x) = \sin x$. Can you think of a way to do this without `Table`?

6. Create a list named `waves` that contains $\frac{1}{k} \sin kx$ for $k = 1, 2, 3, 4, 5$. Then plot the expressions in `waves` by entering

```
Plot[Evaluate[waves], {x, 0, 2 π}];
```

   Now enter

```
colors = Map[Hue, Range[.4, 1, .15]];
Plot[Evaluate[waves], {x, 0, 2 π}, PlotStyle → colors];
```

   Then enter

```
grays = Map[GrayLevel, Range[.8, 0, -.2]];
Plot[Evaluate[waves], {x, 0, 2 π}, PlotStyle → grays];
```

7. (a) The function $f(x) = x^5 - 2x^2 - 3x + 3$ has three real zeros. Plot the graph of f, and create a list named `guesses` that contains a rough estimate of each of its zeros.

   (b) Define the function

```
getZero[guess_] := FindRoot[f[x], {x, guess}]
```

   Find all three zeros of f with one command:

```
Map[getZero, guesses]
```

■ 0.3.2 Rules

Understanding rules is essential to making efficient use of *Mathematica*. For example, note that the `Solve` command returns its result as a list of rules:

```
soln = NSolve[{x^2 + x + y^2 == 2, 2 x - y == 1}, {x, y}]
{{y → 0.677033, x → 0.838516}, {y → -1.47703, x → -0.238516}}
```

To convert this result to a list of pairs of numbers, we apply the rules to the list `{x,y}` as follows:

```
{x, y} /. soln
{{0.838516, 0.677033}, {-0.238516, -1.47703}}
```

Anticipating this in advance, we might have combined these steps by entering

```
{x, y} /. NSolve[{x^2 + x + y^2 == 2, 2 x - y == 1}, {x, y}]
{{0.838516, 0.677033}, {-0.238516, -1.47703}}
```

The name of the object "`/.`" that we use to apply rules is `ReplaceAll`. The following are some simple examples that illustrate its use.

```
√(x² - x + 1) /. x → 3
√7
x + y /. y → x
2 x
x y + y z + x z /. {x → a, y → b + c, z → 5}
5 a + 5 (b + c) + a (b + c)
```

Patterns can be used in rules. For example, suppose we want to replace each order pair in the following list with the sum of its "coordinates."

```
points := {{2, 4}, {4, 3}, {6, 2}, {8, 1}, {10, 0}, {12, -1}, {14, -2}}
```

This can be done with a simple rule as follows:

```
points /. {x_, y_} → x + y
{6, 7, 8, 9, 10, 11, 12}
```

Exercises

1. Trigonometric identities provide a good context in which to learn about rules and gain a bit of insight into symbolic computation in general. For example, the sine addition formula can be applied via the rule

```
sinAdd := Sin[x_ + y_] → Sin[x] Cos[y] + Cos[x] Sin[y]
```

Notice what happens when the rule is applied to $\sin(3x + 5\,y)$:

```
Sin[3 x + 5 y] /. sinAdd
Cos[5 y] Sin[3 x] + Cos[3 x] Sin[5 y]
```

The same rule provides the sine difference formula as well.

```
Sin[t - ϕ] /. sinAdd
Cos[ϕ] Sin[t] - Cos[t] Sin[ϕ]
```

This rule also handles expressions with three or more summands, provided we use `//.` (`ReplaceRepeated`) instead of `/.`:

```
Sin[a + b + c] //. sinAdd
Cos[b + c] Sin[a] + Cos[a] (Cos[c] Sin[b] + Cos[b] Sin[c])
```

(a) Enter the definition of `sinAdd` and construct a similar rule, `cosAdd`, for the cosine addition formula. Test both rules on several different expressions.

(b) Enter the following multiple-angle expansion formula for sine:

```
sinMult := Sin[n_Integer x_] → Sin[(n - 1) x + x] /. sinAdd
```

Check that this rule works properly by entering

```
{Sin[2 x], Sin[3 x]} //. sinMult
```

```
{2 Cos[x] Sin[x], 2 Cos[x]^2 Sin[x] + Cos[2 x] Sin[x]}
```

(c) Construct a similar rule, `cosMult`, for the multiple-angle expansion formula for cosine. Test it on a few expressions.

(d) Notice the result of repeatedly applying all four rules (followed by `Expand`) by entering

```
Sin[2 x + y] //. {sinAdd, cosAdd, sinMult, cosMult} // Expand
```

Then enter `Simplify[%]` to verify that the expansion is correct.

(e) Define the function

```
trigExpand[expr_] := expr //.
   {sinAdd, cosAdd, sinMult, cosMult} // Expand
```

and test it by entering

```
Cos[x + 2 y] + Sin[3 x - y] // trigExpand
% // Simplify
```

Finally, create a table of multiple angle formulas for sine by entering

```
Table[{Sin[k x], Sin[k x] // trigExpand}, {k, 1, 5}] // TableForm
```

and create a similar table of multiple angle formulas for cosine.

2. (a) Use `NSolve` to find the zeros of the polynomial $f(x) = x^5 - 4x^4 + 12x^2 - 9x + 1$. Convert the result to a list of numbers.

(b) Compute the sum of the zeros of f using `Apply` and `Plus`. (See the previous section.) Combine the entire process into a single command.

(c) Repeat the process in parts (a) and (b) after changing the coefficient of x^4 to 3, and then once again after changing the coefficient of x^4 to 1. Try changing the other coefficients to see if they affect the result. What do you conjecture about the sum of the zeros of a fifth-degree polynomial?

(d) Compute the *product* of the zeros of f using `NSolve`, `Apply` and `Times`. Experiment with the coefficients to determine which affect the result. What do you conjecture about the product of the zeros of a fifth-degree polynomial?

(e) Experiment with a few polynomials of other degrees. Do your conjectures depend on degree? Also, are your conjectures consistent with *linear* polynomials?

0.4 Troubleshooting

■ Interpreting *Mathematica* Output When Things Don't Work

In many circumstances, *Mathematica* will give you a useful error message or warning when a bad command is entered. Here are three examples:

```
Plot[x^2, {0, 1}]
```

```
Plot::pllim : Limit specification {0, 1} is not of the form {x, xmin, xmax}.
```

```
Plot[x^2, {0, 1}]
```

```
Solve[x^2 + 5 x = 2, x]

Set::write : Tag Plus in 5 x + x^2 is Protected.

Solve::eqf : 2 is not a well-formed equation.

Solve[2, x]

sin[π/6]

General::spell1 : Possible spelling error: new
   symbol name "sin" is similar to existing symbol "Sin".

sin[π/6]
```

However, it is very common for *Mathematica* simply to give a problematic command back to you with no message. *Mathematica* does this whenever the syntax is correct, but there is something in the command that is unrecognizable. For example,

```
Ln[1]

Ln[1]
```

This also happens when we enter a command from a package that hasn't been loaded.

```
FilledPlot[Cos[x], {x, 0, π}]

FilledPlot[Cos[x], {x, 0, π}]
```

Common Causes of *Mathematica* Problems

Syntax Problems

- **Spelling errors (including capitalization).** Remember the basic rule: Built-in *Mathematica* objects are (almost) always spelled out completely, and each component word is capitalized.(For instance: `InverseLaplaceTransform`.) Most of the exceptions to this rule involve standard mathematical abbreviations such as `Abs`, `D`, `Min`, `Max`, `Re`, `Im`, `Mod`, `Sin`, `Cos`, etc.
- **Forgetting to type a space between multiplied expressions.** For example, if you accidentally type `xSin[x]`, *Mathematica* assumes that you are referring to a function nameee:d `xSin`.
- **Using parentheses instead of brackets or braces (or vice-versa.)** Parentheses, brackets, and braces have very specific and different uses. Parentheses are used only for grouping within expressions, brackets enclose function arguments, and braces enclose members of a list.
- **Using the wrong name or symbol for an object.** Remember that the natural log function is `Log` (not Ln) and that an ordinary "e" is not Euler's *e*. To get Euler's *e*, type [ESC]ee[ESC], or use a capital `E`. Another convention to remember: Inverse trig functions are `ArcSin`, `ArcTan`, etc.
- **Typing an equation with one equal sign (=) instead of two (==).** A single equal sign is used only for assignments; an equation requires two. See Section 0.2.2.

Naming Problems

- **Forgetting to use a `Blank` when defining a function.** See Section 0.2.1.
- **Making multiple definitions for a function name.** Suppose that we enter

```
f[x] = x^2 + 3 x
3 x + x^2
```

and we then realize that we forgot the `Blank` that we need to put beside the variable. So we then enter

```
f[x_] = x^2 + 3 x
3 x + x^2
```

and everything seems fine. *Later...* when working on a different problem, we redefine `f` as

```
f[x_] = x - 2
-2 + x
```

This function behaves as we expect; its graph is the expected straight line, etc. But then we enter

```
g[x_] = f[x]^2
(3 x + x^2)^2
```

which does not give the function `g` that we expect. So what's going on here? *Mathematica* remembers our original, "erroneous" definition of the expression `f[x]`.

💡 A simple way to reveal all the definitions associated with f (or any object) is to type its name after a question mark.

```
?f
Global`f
f[x] = 3*x + x^2
f[x_] = -2 + x
```

💡 *In general it is a good idea to* `Clear` *any variable name after making the* "forgotten `Blank`" *error.*

Package Problems

- **Forgetting to load a package before referencing something in it.** To correct the "shadowing" problem that results from this, you can enter `Remove[`*object*`]` where *object* is what you tried to use prior to loading the package. If multiple objects are involved, the simplest fix may be to quit and restart the kernel.

Computational Problems

- **Entering a command that relies on a previous definition that has not been entered during the current session.** Whenever you resume work from a previous session, be sure that you re-enter commands in order from the top of your *Mathematica* notebook.
- **Doing an enormous *symbolic* computation instead of a simple *numerical* computation.** This is most likely to happen as a result of doing some kind of recursive calculation. For example, suppose we want to calculate several terms in the sequence defined by

$$x_0 = 1 \text{ and } x_{k+1} = 3 \sin x_k - x_k \text{ for } k = 0, 1, 2, \ldots$$

Here is a typical *Mathematica* approach: We'll define the function

```
f[x_] := 3 Sin[x] - x
```

and use `NestList` to compute terms in the sequence. This computes the first five terms:

```
NestList[f, 1, 4]
```

```
{1, -1 + 3 Sin[1], 1 - 3 Sin[1] - 3 Sin[1 - 3 Sin[1]],
 -1 + 3 Sin[1] + 3 Sin[1 - 3 Sin[1]] + 3 Sin[1 - 3 Sin[1] - 3 Sin[1 - 3 Sin[1]]],
 1 - 3 Sin[1] - 3 Sin[1 - 3 Sin[1]] - 3 Sin[1 - 3 Sin[1] - 3 Sin[1 - 3 Sin[1]]] -
  3 Sin[1 - 3 Sin[1] - 3 Sin[1 - 3 Sin[1]] - 3 Sin[1 - 3 Sin[1] - 3 Sin[1 - 3 Sin[1]]]]}
```

If we had asked for ten terms instead of five, the result would have filled several pages. (Try it.) If we had asked for thirty terms, and *if* the computation had *eventually* succeeded, the result would have contained more than 1/2 *billion* copies of the expression `Sin`! (Do yourself a favor; *don't* try it.) To avoid this problem—and get the kind of result we wanted to begin with—we simply need to coerce *Mathematica* into doing the calculation *numerically* instead of *symbolically*. One simple way to do this is to start the sequence with the real number `1.` instead of the integer `1`. (Can you think of two other ways to accomplish the same thing?)

```
f[x_] := 3 Sin[x] - x; NestList[f, 1., 30]
```

```
{1., 1.52441, 1.47236, 1.51312, 1.48189, 1.50626, 1.4875, 1.5021,
 1.49082, 1.49959, 1.49281, 1.49807, 1.494, 1.49716, 1.49471, 1.49661,
 1.49514, 1.49628, 1.4954, 1.49608, 1.49555, 1.49596, 1.49564,
 1.49589, 1.4957, 1.49585, 1.49573, 1.49582, 1.49575, 1.49581, 1.49576}
```

When attempting a complicated computation, *start small!* In other words, see what happens when you do three steps before you try to do thirty.

■ What to Do When You Run into Trouble

- **Check for spelling mistakes, typos, and other syntax errors.**
- **Look for online help.** You can access the help browser through the Help menu or by simply pressing your "help" key. If you want help on a particular command, option, etc., highlight that item first.
- **Clear variable names.** Remember that entering `Clear[`*var1*, *var2*,...`]` clears variables.
- **Clear *everything*.** Here's a quick way to clear *all* previous definitions: `ClearAll["Global`*"]`
- **Quit and restart the kernel**. Do this by choosing Quit Kernel: Local from your Kernel menu. You can then choose Start Kernel: Local from your Kernel menu or simply enter a command to start a new kernel session.
- **Quit and restart *Mathematica*.** Be sure to save your work first.
- **Quit *Mathematica* and restart your computer**. Again, be sure to save your work.
- **Quit and restart your day.** (*Just kidding*.) Seriously though, if you get frustrated, *take a break!*

■ Miscellaneous Issues

The Cube Root Function

When you ask *Mathematica* for the cube root (or any root) of a negative number, it returns a complex number. This complex number is indeed the *principal value* the cube root function defined on the complex numbers.

```
∛-8.
1. + 1.73205 i
```

However, this is not what we want when we talk about the cube root function defined on the real numbers. One simple remedy is to define your own cube root function as follows:

```
Cbrt[x_] := Sign[x] ∛Abs[x]

Plot[Cbrt[x], {x, -4, 4}];
```

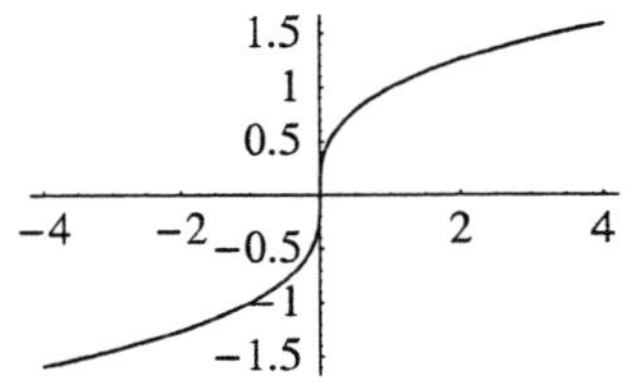

Suppressing Warning Messages

Occasionally *Mathematica*'s warning messages can become annoying. For example, suppose we wanted to define a function named Cost.

```
Cost[x_] = 10 x - x^2

General::spell :
 Possible spelling error: new symbol name "Cost" is
   similar to existing symbols {Cos, Cosh, Cot}.
```

Warnings about possible spelling errors are often helpful. Nevertheless, if you find spelling warnings more annoying than helpful, you can stop *Mathematica* from giving them by entering

```
Off[General::spell, General::spell1]
```

Here is *Mathematica*'s usage information on `Off`:

```
?Off

Off[symbol::tag] switches off a message, so that it is no longer
   printed. Off[s] switches off tracing messages associated
   with the symbol s. Off[m1, m2, ... ] switches off several
   messages. Off[ ] switches off all tracing messages. More...
```

A source of endless warnings is the `ParametricPlot` command (and `ParametricPlot3D`).

```
r[t_] := {Cos[t], Sin[t]};
ParametricPlot[r[t], {t, 0, π}, Axes → False];

ParametricPlot::ppcom : Function r[t] cannot be
  compiled; plotting will proceed with the uncompiled function.
```

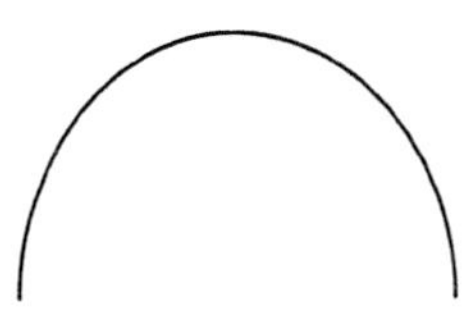

💡 Rather than entering `Off[ParametricPlot::ppcom]`, a better way to avoid the warning generated by `ParametricPlot` is to use `Evaluate` as follows. This allows the function to be compiled; thus plotting is faster.

```
r[t_] := {Sin[t^3], Cos[t^4]};
ParametricPlot[Evaluate[r[t]], {t, -1.65, 1.65}, Axes → False];
```

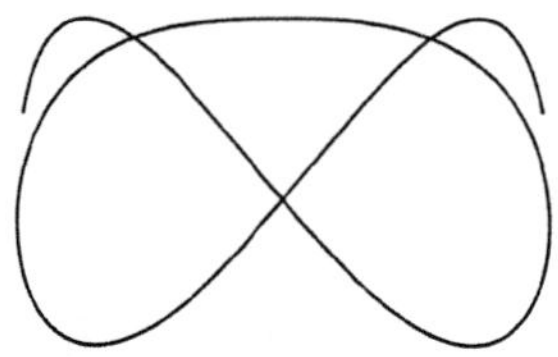

Capitalization and Protected Symbols

Since all built-in *Mathematica* objects have names that begin with an upper-case (capital) letter (or $ in *system variables*), it is a *good idea* always to begin user-defined symbols with a lower-case letter. We violate this tenet many times in this manual, but only because special care has been used to avoid conflicts with built-in *Mathematica* objects.

In fact, *most* built-in *Mathematica* objects have the `Protected` attribute, which prevents us from changing them or accidentally using their names. For example, suppose we decide to name a variable `D`:

```
D := 5

SetDelayed::wrsym : Symbol D is Protected.

$Failed
```

This happened, of course, because `D` is the name of *Mathematica*'s derivative operator. A list of all built-in, single-letter symbol names is generated as follows:

```
ClearAll["Global`*"]

Select[Names["*"], (StringLength[#] == 1) && LetterQ[#] &]

{C, D, E, I, K, N, O}
```

With the exception of `K`, these are the only single-letter, upper-case names that we cannot use. You should use the question mark to get information on each of these, for example,

```
?C

C[i] is the default form for the ith constant of
  integration produced in solving a differential equation with DSolve.
```

Similarly, one can get a list of all two-letter symbol names, and so on, as follows:

```
Select[Names["*"], (StringLength[#] == 2) &]

{Do, Dt, If, Im, In, On, Or, Pi, Re, Tr, Up}

Select[Names["*"], (StringLength[#] == 3) &]

{Abs, All, And, Arg, Box, Cos, Cot, Csc, Det, Dot, End, Erf, Exp,
 Fit, For, GCD, Get, Hue, LCM, Log, Map, Max, Min, Mod, Nor, Not, Off,
 Out, Put, Raw, Row, Run, Sec, Set, Sin, Sum, Tab, Tan, Top, URL, Xor}
```

It is also interesting to look at the output from the following command. It displays a long list of *all* currently defined symbols and their respective attributes. Notice that most are protected.

```
Map[{#, Attributes[#]} &, Names["*"]] // TableForm
```

Chapter 1

An Introduction to Differential Equations

1.1 Verifying Solutions

Let's begin by looking at how *Mathematica* can be used to verify solutions of differential equations.

Example 1.1.1

Suppose we want to verify that

$$y = (1 - e^t)^{-1}$$

is a solution of the equation

$$y' + y = y^2.$$

We define the function by entering

```
y[t_] = 1 / (1 - e^t)
```

$$\frac{1}{1 - e^t}$$

and then, noting that the differential equation is equivalent to $y' + y - y^2 = 0$, we enter

```
y'[t] + y[t] - y[t]^2
```

$$-\frac{1}{(1 - e^t)^2} + \frac{e^t}{(1 - e^t)^2} + \frac{1}{1 - e^t}$$

followed by

```
% // Simplify
```

0

So we conclude that the function is indeed a solution of the differential equation. Note that the process could have been done all at once by entering either of the following:

```
Simplify[y'[t] + y[t] - y[t]^2]
```

0

```
y'[t] + y[t] - y[t]^2 // Simplify
```

0

Another way of doing the same thing is to enter

```
y'[t] + y[t] == y[t]^2 // Simplify
True
```

Example 1.1.2

It also turns out that, for any number C,

$$y = (1 - C e^t)^{-1}$$

is a solution of the same equation as in Example 1.1.1,

$$y' + y = y^2 .$$

We can check this just as easily. After entering the function with a parameter c,

```
y[t_] = 1 / (1 - c e^t)
  1
------
1 - c e^t
```

we then enter either of the following:

```
y'[t] + y[t] - y[t]^2 // Simplify
0
```

```
y'[t] + y[t] == y[t]^2 // Simplify
True
```

Example 1.1.3

The initial-value problem

$$y' + y = y^2, \quad y(0) = y_0$$

is satisfied by

$$y = \frac{y_0}{y_0 - (y_0 - 1)\, e^t},$$

as is verified by the following calculations. First we enter the function:

```
                 y0
y[t_] := -------------------
         y0 - (y0 - 1) e^t
```

Then to verify that the function satisfies the differential equation, we enter

```
y'[t] + y[t] == y[t]^2 // Simplify
True
```

and to check the initial value we enter

```
y[0] == y0
True
```

■ Exercises

1. Verify that each of the following functions is a solution of the differential equation beside it.

a) $y = \sqrt{1 + e^{-2t}}$ $\quad y' + y = 1/y$

b) $y = \sqrt{1 - 2t}$ $\quad y' + y = -2t/y$

c) $y = \sqrt{t^4 + 7}$ $\quad y' y = 2t^3$

d) $y = \sqrt{3 - 2\cos t}$ $\quad y' y = \sin t$

e) $y = e^{-t} + \cos t + \sin t$ $\quad y' + y = 2\cos t$

f) $y = e^{t} - \cos t + \sin t$ $\quad y' - y = 2\cos t$

g) $y = 7\sin t - \cos 2t$ $\quad y'' + y = 3\cos 2t$

2. Verify that each member of the indicated family of functions is a solution of the differential equation beside it.

a) $y = \sqrt{1 + c\,e^{-2t}}$ $\quad y' + y = 1/y$

b) $y = \sqrt{1 - 2t + c\,e^{-2t}}$ $\quad y' + y = -2t/y$

c) $y = \sqrt{t^4 + c}$ $\quad y' y = 2t^3$

d) $y = \sqrt{c - 2\cos t}$ $\quad y' y = \sin t$

e) $y = c\,e^{-t} + \cos t + \sin t$ $\quad y' + y = 2\cos t$

f) $y = c\,e^{t} - \cos t + \sin t$ $\quad y' - y = 2\cos t$

g) $y = c\sin t - \cos 2t$ $\quad y'' + y = 3\cos 2t$

3. Verify that each of the following functions satisfies the initial-value problem beside it.

a) $y = \sqrt{1 + (y_0^2 - 1)\,e^{-2t}}$ $\quad y' + y = 1/y,\ y(0) = y_0$

b) $y = \sqrt{1 - 2t + (y_0^2 - 1)\,e^{-2t}}$ $\quad y' + y = -2t/y,\ y(0) = y_0$

c) $y = \sqrt{t^4 + y_0^2}$ $\quad y' y = 2t^3,\ y(0) = y_0$

d) $y = \sqrt{y_0^2 + 2(1 - \cos t)}$ $\quad y' y = \sin t,\ y(0) = y_0$

e) $y = (y_0 - 1)\,e^{-t} + \cos t + \sin t$ $\quad y' + y = 2\cos t,\ y(0) = y_0$

f) $y = (y_0 + 1)\,e^{t} - \cos t + \sin t$ $\quad y' - y = 2\cos t,\ y(0) = y_0$

g) $y = (1 + y_0)\cos t + y_1 \sin t - \cos 2t$ $\quad y'' + y = 3\cos 2t,\ y(0) = y_0,\ y'(0) = y_1$

1.2 Using DSolve

Mathematica has a built-in command named `DSolve` that can solve many types of differential equations.

```
? DSolve
```

```
DSolve[eqn, y, x] solves a differential equation
  for the function y, with independent variable x.
```

Basic use of `DSolve` is illustrated in the following examples.

Example 1.2.1

Let's find the general solution the following differential equation:

$$y' - \frac{1}{3+t}\,y = t.$$

First we'll `Clear` any previous definition of `y` and define the equation as `eqn`.

```
Clear[y]
eqn = y'[t] - y[t] / (3 + t) == t
```

$$-\frac{y[t]}{3+t} + y'[t] == t$$

Now we use `DSolve` to find the solution.

```
soln = DSolve[eqn, y[t], t]
{{y[t] → (3 + t) C[1] + (3 + t) (t - 3 Log[3 + t])}}
```

Notice that the result is given as a nested list containing a *rule* for the solution `y[t]`. Also notice that `C[1]` represents the arbitrary constant in the general solution. To convert the output of `DSolve` to a function we can work with, let's apply the rule produced by `DSolve` as follows:

```
y[t_] = y[t] /. soln[[1]]
(3 + t) C[1] + (3 + t) (t - 3 Log[3 + t])
```

Now, for example, we can check the solution by entering

```
eqn // Simplify
True
```

and plot its graph for a particular value of `C[1]` as follows:

```
Plot[y[t] /. C[1] → 3, {t, -3, 4}, PlotRange → All];
```

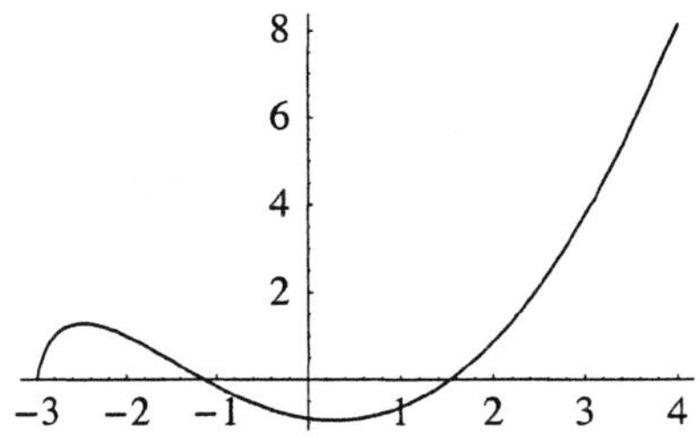

Example 1.2.2

DSolve will also solve initial value problems. The differential equation and the initial condition are given to DSolve as a list. For example, in order to solve

$$y' + 3y = \frac{t}{y}, \quad y(0) = 1,$$

we'll enter

```
Clear[y];
soln = DSolve[{y'[t] + 3 y[t] == t / y[t], y[0] == 1}, y[t], t]
```

$$\left\{\left\{y[t] \to \frac{\sqrt{e^{-6t}\,(19 - e^{6t} + 6\,e^{6t}\,t)}}{3\sqrt{2}}\right\}\right\}$$

```
y[t_] = y[t] /. soln[[1]]
```

$$\frac{\sqrt{e^{-6t}\,(19 - e^{6t} + 6\,e^{6t}\,t)}}{3\sqrt{2}}$$

(A couple of warning messages have been deleted.) Here's the graph of the solution:

```
Plot[y[t], {t, -1, 8}];
```

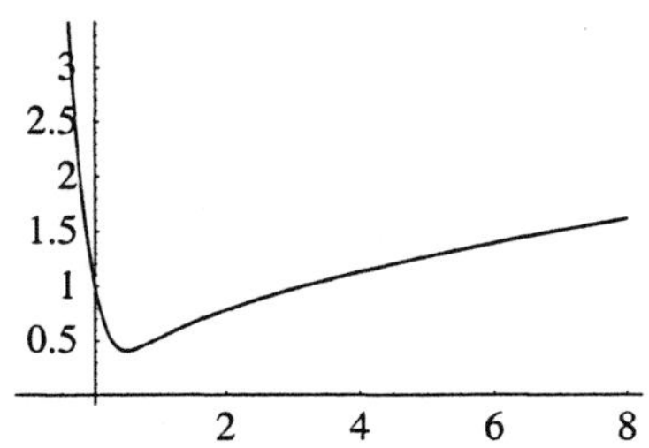

We can also use DSolve to obtain a description of the general solution in which an initial value is used as the arbitrary parameter.

```
Clear[y, y0];
DSolve[{y'[t] + 3 y[t] == t / y[t], y[0] == y0}, y[t], t] // Simplify
```

$$\left\{\left\{y[t] \to -\frac{\sqrt{-1 + 6\,t + e^{-6t}\,(1 + 18\,y0^2)}}{3\sqrt{2}}\right\}, \left\{y[t] \to \frac{\sqrt{-1 + 6\,t + e^{-6t}\,(1 + 18\,y0^2)}}{3\sqrt{2}}\right\}\right\}$$

Note that here DSolve has returned two solutions, one negative and the other positive. The correct solution is the one whose sign is consistent with the sign of the initial value y_0.

Exercises

1. a–g) Use `DSolve` to find the general solution of each of the equations in Problem 2 of Section 1.2. Compare the result to the form of the solution given there.

2. a–g) Use `DSolve` to find the solution of each of the initial value problems in Problem 3 of Section 1.2. Compare the result to the form of the solution given there.

1.3 Plotting Solution Families

Since solutions of a differential equation typically comprise a family of functions describable in terms of one or more parameters, it is worth looking at how we can use *Mathematica* to plot representative members of such a family of functions. Our purpose is to plot several members of a general solution in order to get a rough graphical view of the general solution family.

Example 1.3.1

The initial value problem

$$y' + y = 2\,e^{-t}\cos 2t, \;\; y(0) = y_0$$

can be solved by use of an integrating factor, yielding the solution

$$y = e^{-t}(\sin 2t + y_0).$$

We can also find this solution with `DSolve` by entering

```
Clear[y, y0];
soln = DSolve[
    {y'[t] + y[t] == 2 e^-t Cos[2 t], y[0] == y0}, y[t], t] // Simplify
{{y[t] → e^-t (y0 + Sin[2 t])}}
```

```
y[t_] = y[t] /. soln[[1]]
e^-t (y0 + Sin[2 t])
```

What we would like to do now is plot this as a function of t for several specific values of y_0, say $-2, -1, 0, 1, 2, 3$. To do this, first create a list of initial values by entering

```
y0 = Range[-2, 3]
{-2, -1, 0, 1, 2, 3}
```

Notice that `y[t]` is now a *list* of solutions:

```
y[t]
```

```
{e^-t (-2 + Sin[2 t]), e^-t (-1 + Sin[2 t]), e^-t Sin[2 t],
 e^-t (1 + Sin[2 t]), e^-t (2 + Sin[2 t]), e^-t (3 + Sin[2 t])}
```

Now the solution curves can be plotted on the interval $[-2, 4]$ by entering

```
Plot[Evaluate[y[t]], {t, -2, 4}];
```

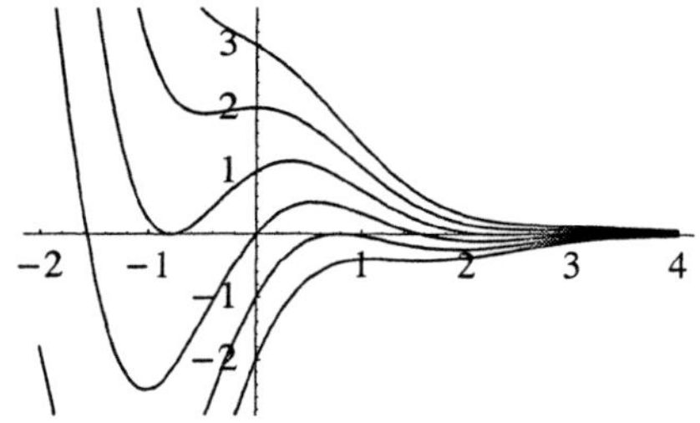

(The use of `Evaluate` is necessary here. Without it, `Plot` interprets `y[t]` as a single list-valued expression rather than a list of numeric-valued expressions.)

■ Example 1.3.2

The initial value problem

$$y' - \frac{1}{3+t}\, y = t, \;\; y(0) = y_0$$

can be solved with `DSolve` by entering

```
Clear[y, y0];
soln =
 DSolve[{y'[t] - y[t] / (3 + t) == t, y[0] == y0}, y[t], t] // Simplify
```

$$\left\{\left\{y[t] \to \frac{1}{3}\ (3+t)\ (3\,t + y0 + 3\,\mathrm{Log}[27] - 9\,\mathrm{Log}[3+t])\right\}\right\}$$

```
y[t_] = y[t] /. soln[[1]]
```

$$\frac{1}{3}\ (3+t)\ (3\,t + y0 + 3\,\mathrm{Log}[27] - 9\,\mathrm{Log}[3+t])$$

The solutions with initial values $y_0 = -2, -1, 0, 1, 2, 3$ can be plotted on the interval $[-2, 4]$ by entering

```
y0 := Range[-2, 3];
Plot[Evaluate[y[t]], {t, -3, 4}];
```

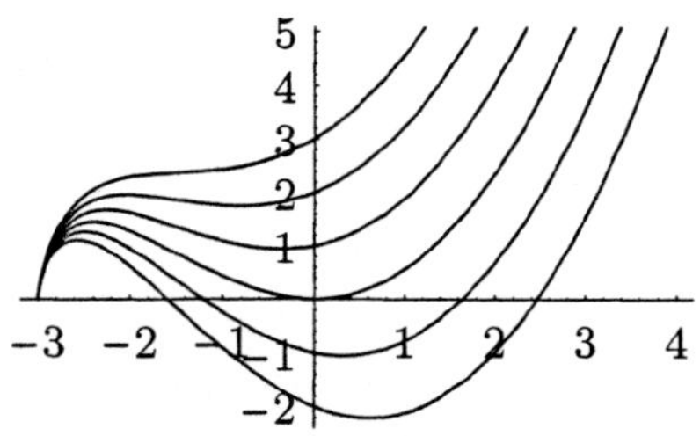

■ Exercises

1. Plot the solution on the indicated interval for each indicated value of y_0.

a) $y' + 4\,t\,y/(t^2+1) = 6\,t,\ y(0) = y_0$ $[-3, 3]$ $y_0 = -10, -9, \ldots, 9, 10$

b) $y' + y = 2\cos t,\ y(0) = y_0$ $[-1, 2\,\pi]$ $y_0 = -5, -4, \ldots, 4, 5$

c) $y'\,y = \sin t,\ y(0) = y_0$ with $y_0 > 0$ $[-3\,\pi, 3\,\pi]$ $y_0 = 0.25, 0.5, \ldots, 2.5$

Tip for (c): Enter the values of y_0 as `Range[1,10]/4`.

1.4 Solutions by Integration

The solution of

$$y' + k\,y = f,\ y(0) = y_0,$$

can be expressed in terms of a definite integral as

$$y(t) = e^{-kt}\left(y_0 + \int_0^t e^{ks} f(s)\,ds\right).$$

This is a form of the *variation of constants formula*. In the following we will use *Mathematica* to help us evaluate the solution and compare it graphically to the input function f in some instances where evaluation of the integral in terms of elementary functions is either extremely tedious or literally impossible to do by hand.

Mathematica's `Integrate` command can be used to perform indefinite and definite integration symbolically—whenever that is possible in terms of elementary functions and a large collection of known "special functions." The solution formula described above can be defined by entering

```
Clear[y0, k, f];
Soln[f_][t_] := e^(-k t) (y0 + ∫_0^t e^(k s) (f /. t -> s) ds)
```

This particular definition emphasizes the role of f as an "input function." We can already verify that the result satisfies the differential equation as follows:

```
y[t_] = Soln[f[t]][t]
e^(-k t) (y0 + ∫_0^t e^(k s) f[s] ds)

y'[t] + k y[t] == f[t] // Simplify
True
```

■ Example 1.4.1

To find the solution of

$$y' + 2y = \cos \pi t, \ y(0) = 1,$$

we first observe that $k = 2$, $y_0 = 1$, and $f(t) = \cos \pi t$, and then enter

```
k := 2; y0 := 1;
f[t_] := Cos[π t];
y[t_] = Soln[f[t]][t] // Simplify
```

$$\frac{2\,\text{Cos}[\pi\,t] + e^{-2t}\,(2 + \pi^2 + e^{2t}\,\pi\,\text{Sin}[\pi\,t])}{4 + \pi^2}$$

The solution can then be plotted along with the input function f by entering

```
Plot[{f[t], y[t]}, {t, 0, 4}, PlotStyle → {Hue[0], Thickness[.008]}];
```

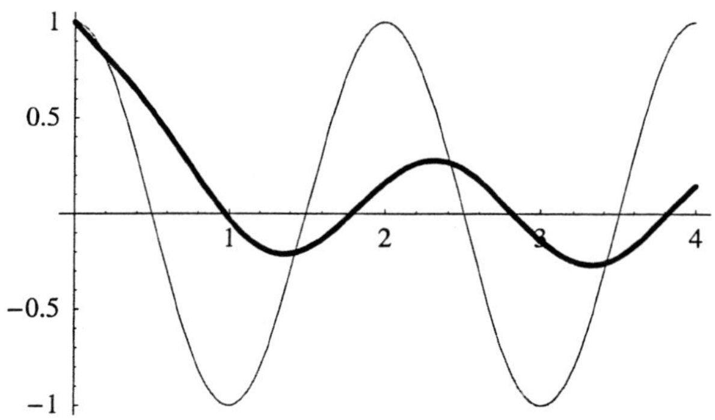

■ Example 1.4.2

To find the solution of

$$y' + 3y = \sqrt{t}, \ y(0) = 2,$$

we first observe that $k = 3$, $y_0 = 2$, and $f(t) = \sqrt{t}$, and then enter

```
k := 3; y0 := 2; f[t_] := √t;
y[t_] = Soln[f[t]][t] // Simplify
```

$$\frac{1}{18}\,e^{-3t}\left(36 + 6\,e^{3t}\,\sqrt{t} - \sqrt{3\,\pi}\,\text{Erfi}\left[\sqrt{3}\,\sqrt{t}\right]\right)$$

Here, the solution could not be expressed in terms of elementary functions and is given in terms of the *special function* `Erfi`, the *imaginary error function*. The solution is plotted along with the input function f by entering

```
Plot[{f[t], y[t]}, {t, 0, 4},
  PlotStyle → {Hue[0], Thickness[.008]}];
```

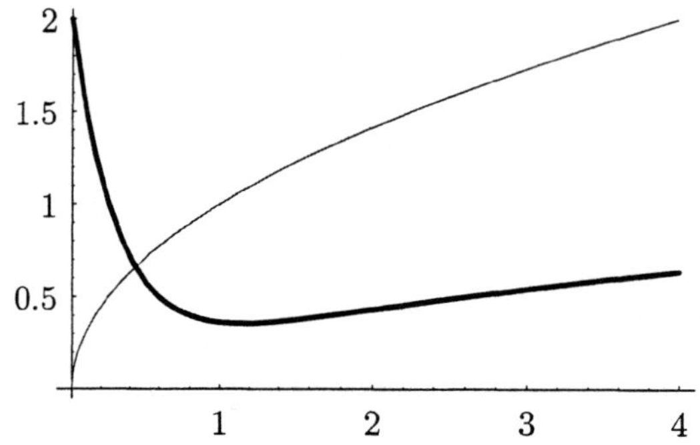

■ Example 1.4.3

Consider now the initial-value problem

$$y' + y = \cos(2\pi\sqrt{t}), \; y(0) = 1.$$

First observe that $k = 1$, $y_0 = 1$, and $f(t) = \sqrt{t}$, and then enter

```
k := 1; y0 := 1;
f[t_] := Cos[2 π √t];
y[t_] = Soln[f[t]][t] // Simplify
```

$$e^{-t}\left(1 + \int_0^t e^s \operatorname{Cos}\left[2\pi\sqrt{s}\right] ds\right)$$

Here, the solution could be expressed neither in terms of elementary functions nor any of the standard special functions. In order to compute values of the solution (which is necessary for plotting its graph), we must use some type of numerical approximation procedure. One approach to this is the subject of the next section.

■ Exercises

1. Repeat the process above for the following problems.

   a) $y' + y = \cos 4t \sin 2t$, $y(0) = 0$ (Plot on $[0, 2\pi]$.)

   b) $y' + y = \sin^4 t$, $y(0) = 1$ (Plot on $[0, 3\pi]$.)

   c) $y' + 2y = \sqrt{t} - t/4$, $y(0) = 1$ (Plot on $[0, 10]$.)

   d) Attempt to do the same for $y' + y = \cos t^3$, $y(0) = 1$. What is going on here?

2. For the differential equation in Problem 1b, create a single plot that shows the input function $\sin^4 t$ along with solutions with initial values $y_0 = -1, 0$, and 1. What do you observe about these solutions?

1.5 Using NDSolve

Mathematica has a built-in command named `NDSolve`, that computes a numerical approximation to the solution of an initial-value problem. Basic use of `NDSolve` is illustrated in the following example.

Example 1.5.1

Consider the initial-value problem from Example 1.4.3:

$$y' + y = \cos(2\pi\sqrt{t}),\ \ y(0) = 1.$$

To compute a numerical approximation of the solution on the interval [0, 5], for instance, enter

```
Clear[y];
numsoln =
 NDSolve[{y'[t] + y[t] == Cos[2 π √t], y[0] == 1}, y[t], {t, 0, 5}]
{{y[t] → InterpolatingFunction[{{0., 5.}}, <>][t]}}
```

As was the case with `DSolve`, the result is a rule embedded within a nested list. So to get a function we can plot, we apply the rule as follows:

```
y[t_] = y[t] /. numsoln[[1]]
InterpolatingFunction[{{0., 5.}}, <>][t]
```

Now we can plot the (numerical) solution.

```
Plot[y[t], {t, 0, 5}];
```

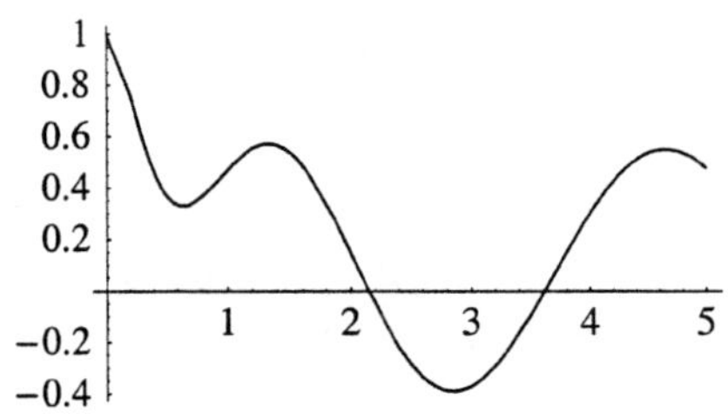

■ Exercises

1. Repeat the process of Example 1.5.1 on the following problems.

a) $y' + y\cos t = 1 - y^2,\ \ y(0) = 0,\ \ 0 \le t \le 20$

b) $y' = y\,(\sin^2 t)\,(1 - \frac{y}{3+\cos t}),\ \ y(0) = 2,\ \ 0 \le t \le 20$

c) $y' + e^{-t}\cos(10\,t)\,y = e^{-y^2},\ \ y(0) = 2,\ \ 0 \le t \le 8$

1.6 Implicit Solutions

It is common to encounter differential equations that can be solved by integration to the extent of obtaining an algebraic equation satisfied by the solution. Such an algebraic equation is called an *implicit solution*.

Example 1.6.1

Suppose we want to find the general solution of

$$y' = \frac{t \sin t}{1 + 4y^3}.$$

Since the equation is *separable*, we rewrite it in the form

$$(1 + 4y^3)\, dy = t \sin t\, dt.$$

Next we integrate each side of the equation, adding an arbitrary constant to the right side.

```
Clear[y];
∫(1 + 4 y³) ⅆy == ∫t Sin[t] ⅆt + C[1]
```

```
y + y⁴ == C[1] - t Cos[t] + Sin[t]
```

This implicit solution is not an equation that we should attempt to solve. However, we can plot solution curves by first noting that solution curves are contours (or level curves) of the following function:

```
f[t_, y_] = ∫(1 + 4 y³) ⅆy - ∫t Sin[t] ⅆt
```

```
y + y⁴ + t Cos[t] - Sin[t]
```

Mathematica's `ContourPlot` function will create a plot of several solution curves.

```
ContourPlot[f[t, y], {t, 0, 12}, {y, -2, 2},
  ContourShading → False, AspectRatio → Automatic, PlotPoints → 40];
```

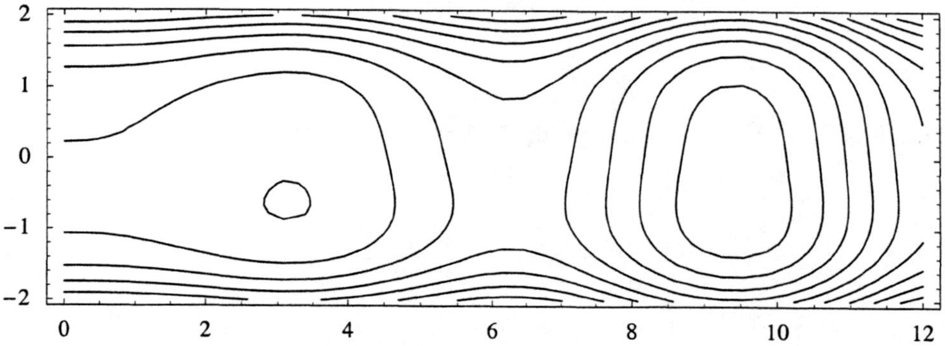

A surface plot is also interesting to see.

```
Plot3D[f[t, y], {t, 0, 12}, {y, -2, 2},
  PlotPoints → 30, BoxRatios → {2, 1, 1}];
```

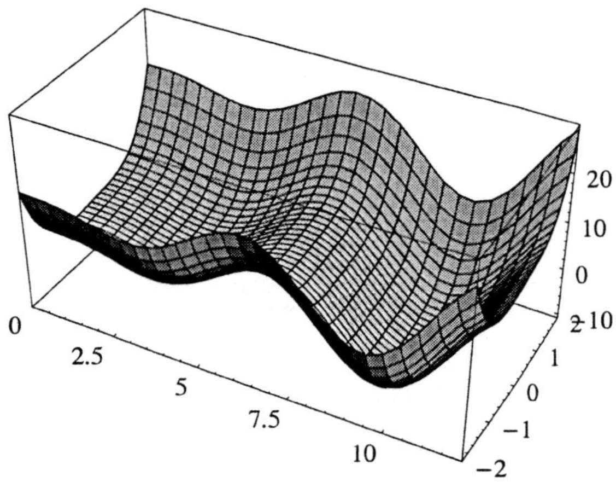

■ Exercises

1. As in Example 1.6.1, create a plot of several solution curves for each of the following equations.

   a) $y' = \dfrac{1+\sin t}{2+\cos y}$

   b) $y' = e^{-y^2} \arctan t$ (*Suggestion*: Plot on $-4 \le t \le 4$, $-1 \le y \le 2$.)

   c) $y' = (2+\sin y)(\sin t + \cos 2t)$ (*Suggestion*: Plot on $0 \le t \le 2\pi$, $-\pi \le y \le \pi$.)

Chapter 2

Linear First-Order Equations and Applications

2.1 Linear First-Order Equations

Recall that an initial-value problem involving a linear, first-order differential equation has the form

$$y' + p\,y = f, \quad y(t_0) = y_0,$$

where p and f are continuous functions of the independent variable. Moreover, recall that the solution is given by the variation of constants formula:

$$y(t) = e^{-P(t)}\left(y_0 + \int_{t_0}^{t} e^{P(s)}\, f(s)\, ds\right), \text{ where } P(t) = \int_{t_0}^{t} p(\tau)\, d\tau\ .$$

The `DETools` package contains a command called `FOLDSolve` (as in *First-Order-Linear-DSolve*) that computes this solution. To access the package, we enter

```
<< DiffEqs`DETools`
```

Information about `FOLDSolve` is obtained as follows:

```
? FOLDSolve
```

```
FOLDSolve[equation, y[t], t, options] solves a first-order
  linear (FOL) differential equation for the expression y[t].
```

If no initial point is specified, `FOLDSolve` returns a general solution; for instance:

```
FOLDSolve[ y'[t] + y[t] == t Sin[t], y[t], t] // Simplify
```

$$\frac{1}{2}\left(2\, e^{-t}\, C[1] + Cos[t] - t\, Cos[t] + t\, Sin[t]\right)$$

(Note that `FOLDSolve` differs from `DSolve` in that it returns an expression rather than a list of rules.) To specify an initial point (t_0, y_0); we give a list containing the differential equation and the initial condition; for instance:

```
FOLDSolve[{ y'[t] + y[t] == t Sin[t], y[0] == 1}, y[t], t] // Simplify
```

$$\frac{1}{2}\left(e^{-t} + Cos[t] - t\, Cos[t] + t\, Sin[t]\right)$$

■ Example 2.1.1

Let's look at the equation

$$y' + t\,y = t^5\,.$$

The solution in terms of an arbitrary initial point (t_0, y_0) is

```
FOLDSolve[{y'[t] + t y[t] == t^5, y[t0] == y0}, y[t], t]
```

$$e^{\frac{1}{2}(-t^2+t0^2)}\left(-8+e^{\frac{t^2}{2}-\frac{t0^2}{2}}(8-4t^2+t^4)+4t0^2-t0^4+y0\right)$$

The solution in terms of an arbitrary initial value y_0 at $t_0 = 0$ is

```
FOLDSolve[{y'[t] + t y[t] == t^5, y[0] == y0}, y[t], t]
```

$$e^{-\frac{t^2}{2}}\left(-8+e^{\frac{t^2}{2}}(8-4t^2+t^4)+y0\right)$$

which can be tidied up some as follows:

```
% // Expand // Simplify
```

$$8-4t^2+t^4+e^{-\frac{t^2}{2}}(-8+y0)$$

The solution with the initial condition $y(0) = 3$, for example, is

```
FOLDSolve[{y'[t] + t y[t] == t^5, y[0] == 3}, y[t], t] // Expand //
 Simplify
```

$$8-5e^{-\frac{t^2}{2}}-4t^2+t^4$$

■ Example 2.1.2

Let's now look at the initial-value problem

$$y' + t^2 y = \cos t, \ \ y(0) = 0.$$

`FOLDSolve` can only express the solution of this problem as an integral:

```
FOLDSolve[{y'[t] + t^2 y[t] == Cos[t], y[0] == 0}, y[t], t]
```

$$e^{-\frac{t^3}{3}}\int_0^t e^{\frac{s^3}{3}}\,\mathrm{Cos}[s]\,ds$$

(`DSolve` does no better. Try it.) One (very inefficient) way around this difficulty is to compute the integral numerically with `NIntegrate`:

```
y[t_] := e^(-t^3/3) NIntegrate[e^(s^3/3) Cos[s], {s, 0, t}]

Plot[y[t], {t, 0, 9}];
```

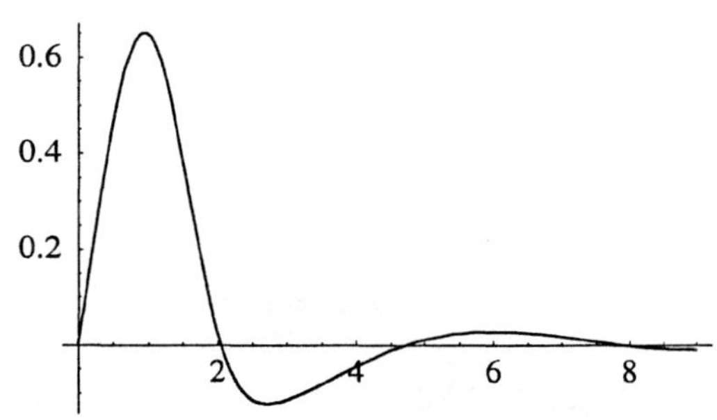

■ Exercises

1. Use `FOLDSolve` to solve each of the following problems.

   a) $y' + y\tan t = 4\cos^3 t,\ \ y(0) = y_0$

   b) $(1+t^3)y' + 3t^2 y = \sin t,\ \ y(0) = y_0$

   c) $t y' + 2y = t^4,\ \ y(1) = 1$

   d) $y' + p(t)y = f(t),\ \ y(t_0) = y_0$

2.2 Baseball Trajectories

Consider the flight of a projectile under the assumption of a drag force that is proportional to velocity. Let $V(t) = (v_1(t), v_2(t))$ be the projectile's velocity vector at time t. With V_0 denoting the initial *speed*, the initial velocity vector is $V(0) = V_0(\cos\theta,\ \sin\theta)$, where θ is the initial angle of inclination. By Newton's second law, the velocity components satisfy the following initial-value problems:

$$\begin{aligned} m v_1' &= -k v_1, & v_1(0) &= V_0 \cos\theta; \\ m v_2' &= -mg - k v_2, & v_2(0) &= V_0 \sin\theta. \end{aligned}$$

We can reduce the number of parameters by setting $\beta = k/m$. Then the equations become

$$\begin{aligned} v_1' &= -\beta v_1, & v_1(0) &= V_0 \cos\theta; \\ v_2' &= -g - \beta v_2, & v_2(0) &= V_0 \sin\theta. \end{aligned}$$

If we assume that the initial position is $(x(0), y(0)) = (x_0, y_0)$, then, once the velocity components are found, the position coordinates are obtained by simply integrating as follows:

$$x(t) = x_0 + \int_0^t v_1(s)\,ds \ \text{ and } y(t) = y_0 + \int_0^t v_2(s)\,ds.$$

These computations are executed as follows:

```
Needs["DiffEqs`DETools`"]

v1[t_] = FOLDSolve[{v'[t] == -β v[t], v[0] == V0 Cos[θ]}, v[t], t]
```

$e^{-t\beta}\,\mathrm{V0}\,\mathrm{Cos}[\theta]$

```
v2[t_] = FOLDSolve[{v'[t] == -g - β v[t], v[0] == V0 Sin[θ]}, v[t], t]
```

$e^{-t\beta}\left(-g\left(-\frac{1}{\beta}+\frac{e^{t\beta}}{\beta}\right)+\mathrm{V0}\,\mathrm{Sin}[\theta]\right)$

$\mathtt{x[t\_]} = \int_0^t \mathtt{v1[s]}\,ds$

$\mathrm{V0}\left(\frac{1}{\beta}-\frac{e^{-t\beta}}{\beta}\right)\mathrm{Cos}[\theta]$

$\mathtt{y[t\_]} = \int_0^t \mathtt{v2[s]}\,ds$ `/. g → 32 // Expand // Simplify`

$\frac{32 - 32e^{-t\beta} - 32t\beta + (1-e^{-t\beta})\,\mathrm{V0}\,\beta\,\mathrm{Sin}[\theta]}{\beta^2}$

For the flight of a baseball in air, "typical" values for β range from about 0.10 to 0.20. With $\beta = 0.15$ and with $V_0 = 147$ ft/s and $\theta = \pi/4$, the trajectory is plotted as follows:

```
path = {x[t], y[t]} /. {β → .15, V0 → 147, θ → π / 4};
ParametricPlot[Evaluate[path], {t, 0, 5.7},
 PlotRange → {{0, 400}, {0, Automatic}}, AspectRatio → Automatic];
```

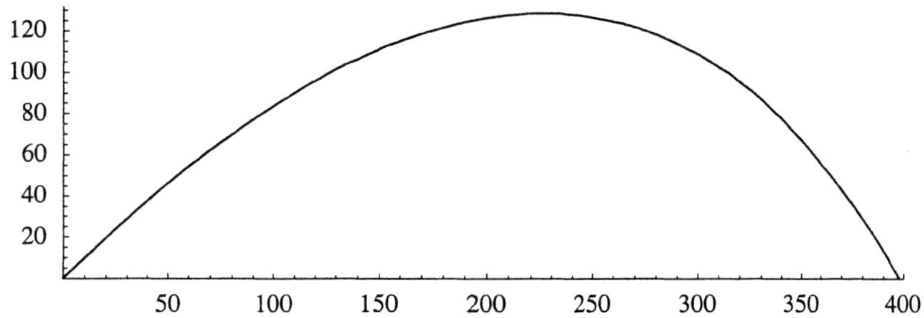

To see the effect that varying β has on the trajectory, we could enter:

```
paths = {x[t], y[t]} /.
    {β → {.10, .15, .20, .30}, V0 → 147, θ → π / 4} // Transpose;
ParametricPlot[Evaluate[paths], {t, 0, 5.7},
  PlotRange → {{0, 400}, {0, Automatic}}, AspectRatio → Automatic];
```

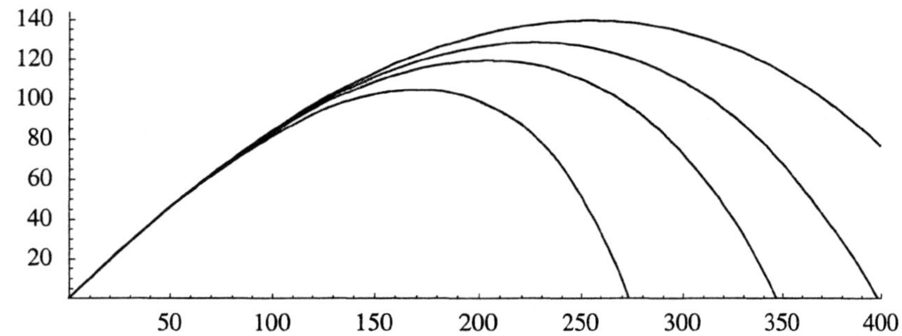

To see the effect of varying θ, we could enter:

```
paths = {x[t], y[t]} /.
    {β → .15, V0 → 147, θ → {.15 π, .2 π, .25 π, .3 π}} // Transpose;
ParametricPlot[Evaluate[paths], {t, 0, 6},
  PlotRange → {{0, 420}, {0, Automatic}}, AspectRatio → Automatic];
```

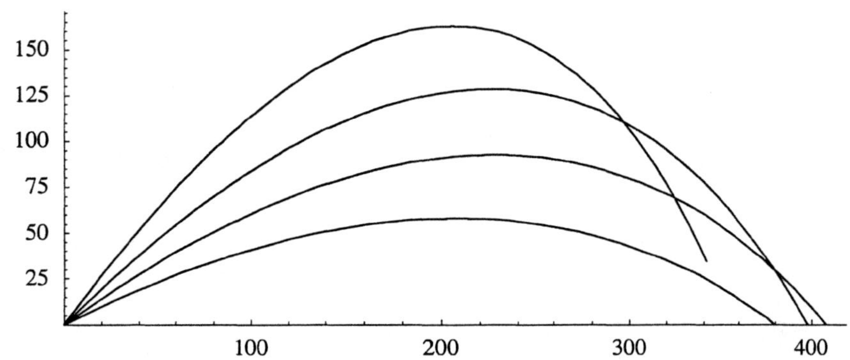

The effect of varying V_0 should be surprising:

```
paths = {x[t], y[t]} /.
    {β → .15, V0 → {130, 140, 150}, θ → π / 4} // Transpose;
ParametricPlot[Evaluate[paths], {t, 0, 6},
  PlotRange → {{0, 420}, {0, Automatic}}, AspectRatio → Automatic];
```

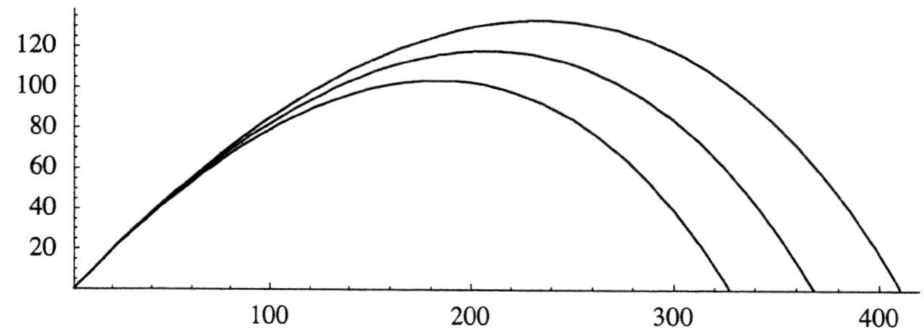

■ Example 2.2.1

Suppose that a home run must clear a 30 ft wall that is 350 ft from home plate. Given that $\beta = 0.15$, with an initial angle of $\theta = \frac{\pi}{4}$, the following indicates that the ball must have an initial speed of at least (roughly) 140 ft/s in order to clear the wall.

```
paths =
  {x[t], y[t]} /. {β → .15, V0 → {130, 140, 150}, θ → π / 4} // Transpose;
wall = {350, 30 - 5 t};
ParametricPlot[Evaluate[Append[paths, wall]], {t, 0, 6},
  PlotRange → {{0, 420}, {0, Automatic}}, AspectRatio → Automatic,
  PlotStyle → {{}, {}, {}, {Thickness[.012], RGBColor[0, .5, 0]}}];
```

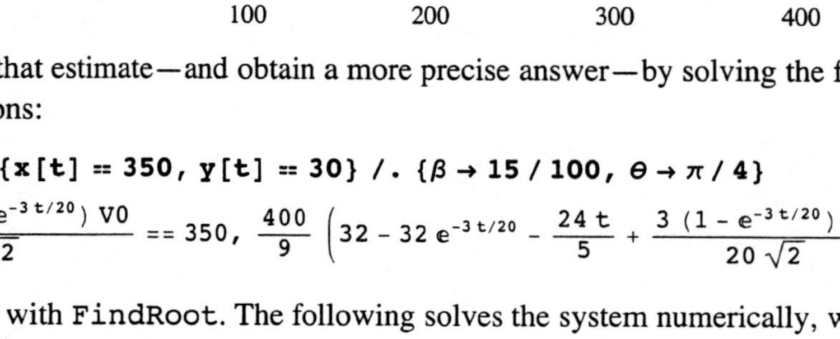

We can confirm that estimate—and obtain a more precise answer—by solving the following system of equations:

```
eqns = {x[t] == 350, y[t] == 30} /. {β → 15 / 100, θ → π / 4}
```

$$\left\{\frac{\left(\frac{20}{3} - \frac{20}{3} e^{-3t/20}\right) V0}{\sqrt{2}} == 350,\ \frac{400}{9}\left(32 - 32 e^{-3t/20} - \frac{24 t}{5} + \frac{3\left(1 - e^{-3t/20}\right) V0}{20\sqrt{2}}\right) == 30\right\}$$

This can be done with `FindRoot`. The following solves the system numerically, with initial guesses of 4 and 140 given for t and V_0, respectively:

```
FindRoot[eqns, {t, 4}, {V0, 140}]
{t → 5.0332, V0 → 140.092}
```

With the same β but an initial angle of $\theta = \frac{\pi}{6}$ (more of a "line drive"), the ball must have an initial speed of at least (roughly) 147 ft/s to clear the wall:

```
paths =
  {x[t], y[t]} /. {β → .15, V0 → {144, 147, 150}, θ → π/6} // Transpose;
wall = {350, 30 - 5 t};
ParametricPlot[Evaluate[Append[paths, wall]], {t, 0, 6},
   PlotRange → {{0, 420}, {0, 120}}, AspectRatio → Automatic,
   PlotStyle → {{}, {}, {}, {Thickness[.012], RGBColor[0, .5, 0]}}];
```

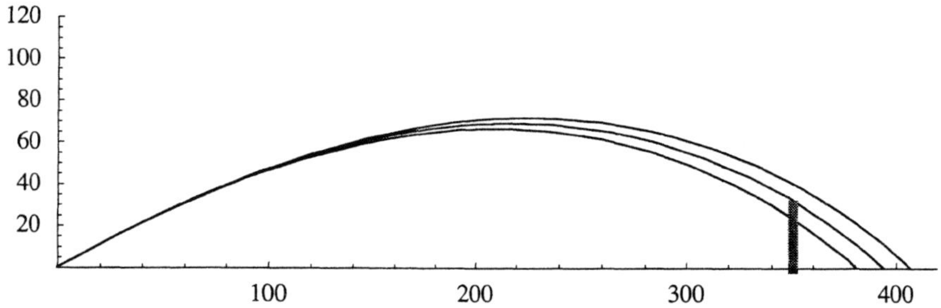

Again, we will also compute the solution by solving a system of equations:

```
eqns = {x[t] == 350, y[t] == 30} /. {β → 15 / 100, θ → π / 6}
```

$$\left\{\frac{1}{2}\sqrt{3}\left(\frac{20}{3}-\frac{20}{3}e^{-3t/20}\right)V0 == 350,\right.$$
$$\left.\frac{400}{9}\left(32-32e^{-3t/20}-\frac{24t}{5}+\frac{3}{40}\left(1-e^{-3t/20}\right)V0\right) == 30\right\}$$

```
FindRoot[eqns, {t, 5}, {V0, 147}]
{t → 3.5718, V0 → 146.154}
```

■ Exercises

1. Investigate the effect of (roughly) 10% variation in the drag coefficient k with the values $\beta = 0.15,\ 0.165$, and 0.182 (using $V_0 = 150\,\mathrm{ft}/s$ and $\theta = \pi/4$). A 10% variation in the drag coefficient can arise due to variation in temperature, humidity, altitude, and atmospheric pressure. For example, a typical value of k in Miami may be about 10% larger than a typical value of k in Denver. Estimate from the graphs the differences in how far the ball flies corresponding to the changes in β. You will need to zoom in a bit on the area where the ball hits the ground by adjusting the time interval.

2. The gravitational coefficient g varies slightly with altitude. Using $g = 32$, 32.1, and 32.2 ft/s^2, investigate the effect of varying g on the flight of a baseball. Use $\beta = 0.15$, $V_0 = 150\,\mathrm{ft}/s$ and $\theta = \pi/4$. *Suggestion*: After recomputing `y[t]` in terms of g, build the `paths` list as indicated below. (This is necessary because `x[t]` doesn't depend on g, causing the "`Transpose` trick" used above to fail. See Remark 2 below.)

```
paths = {{x[t], y[t]} /. g → 31, {x[t], y[t]} /. g → 32.1,
     {x[t], y[t]} /. g → 32.2} /. {β → .15, V0 → 147, θ → π / 4};
```

3. Suppose that a submarine is at a depth of 100 m and fires a missile with initial speed 100 m/s and angle of inclination $\theta = \pi/3$. It is known from experiments that $\beta = 0.39$ for this type of missile under water. With what velocity does the missile emerge from the surface of the water? What is the speed and angle of inclination at that instant?

4. Suppose that you're on the surface of a small planet with a thick atmosphere. You throw a stone with mass 0.5 kg with initial speed $V_0 = 30$ m/s and angle of inclination $\theta = \pi/4$, and it lands 70 meters away, 7 seconds later. Find the gravitational constant g for this planet and the stone's drag coefficient k in this atmosphere.

5. It is commonly known—and not difficult to prove—that in the absence of a drag force (i.e., in a vacuum), the horizontal distance that a projectile will fly is maximized by $\theta = \pi/4$ (for fixed initial velocity). Experimentally determine whether this fact is true under the influence of nonzero drag. If it is not true, what then is the optimal angle of inclination?

Remarks

1. `DSolve` will find the projectile path in one step from a pair of second-order differential equations as follows:

```
Clear[x, y];
{x[t_], y[t_]} = {x[t], y[t]} /.
  DSolve[{ x''[t] == -β x'[t], y''[t] == -g - β y'[t], x'[0] == V0 Cos[θ],
    y'[0] == V0 Sin[θ], x[0] == 0, y[0] == 0}, {x[t], y[t]}, t][[1]]
```

$$\left\{\frac{e^{-t\beta}\,(-1+e^{t\beta})\ \text{V0}\ \text{Cos}[\theta]}{\beta},\ -\frac{e^{-t\beta}\,(g-e^{t\beta}\,g+e^{t\beta}\,g\,t\,\beta+\text{V0}\,\beta\,\text{Sin}[\theta]-e^{t\beta}\,\text{V0}\,\beta\,\text{Sin}[\theta])}{\beta^2}\right\}$$

2. The use of `Transpose` in some of the preceding commands perhaps needs some explanation. First notice that `Transpose[`*matrix*`]` "reflects" the entries of *matrix* across its diagonal, creating a matrix whose ith column contains the entries of the ith row of *matrix*. For instance, if we define a matrix a as follows:

```
a = {{11, 12}, {21, 22}};
a // MatrixForm
```

$$\begin{pmatrix} 11 & 12 \\ 21 & 22 \end{pmatrix}$$

and then apply `Transpose`, we get

```
Transpose[a] // MatrixForm
```

$$\begin{pmatrix} 11 & 21 \\ 12 & 22 \end{pmatrix}$$

Now notice the result of the following application of a rule that replaces a symbol with a list in each member of another list:

```
{z, 2 z} /. z → {p, q}
{{p, q}, {2 p, 2 q}}
```

Transposing the result produces a list in which each member has the same form as the original list:

```
{z, 2 z} /. z → {p, q} // Transpose
{{p, 2 p}, {q, 2 q}}
```

So, for example, if we wanted to create a list containing ordered pairs $(\sin(t+c),\ \cos(t+c))$ for each of $c = 1$ and 2, then

```
{Sin[t + c], Cos[t + c]} /. c → {1, 2}
{{Sin[1 + t], Sin[2 + t]}, {Cos[1 + t], Cos[2 + t]}}
```

does not produce the desired result, but this does:

```
{Sin[t + c], Cos[t + c]} /. c → {1, 2} // Transpose
{{Sin[1 + t], Cos[1 + t]}, {Sin[2 + t], Cos[2 + t]}}
```

This technique is not fool-proof; for example, notice what happens here (cf. Problem 2):

```
{Sin[t], Cos[t] + c} /. c → {1, 2} // Transpose

Transpose::nmtx :
 The first two levels of the one-dimensional list
   {Sin[t], {1 + Cos[t], 2 + Cos[t]}} cannot be transposed.
Transpose[{Sin[t], {1 + Cos[t], 2 + Cos[t]}}]
```

2.3 Rocket Science

We consider here a model for the vertical path of a projectile with variable mass $m(t)$ being acted upon by gravitational force $-m(t)\,g$ and drag force $-kv(t)$. We assume that the mass being lost (or gained) by the projectile is moving with a velocity c *relative to the projectile*. A somewhat delicate physics argument results in the following linear, first-order differential equation for the velocity $v(t)$:

$$mv' = -mg - kv + cm'.$$

Setting $f = cm'$ and rearranging now gives

$$mv' + kv = -mg + f.$$

In the context of a fuel-burning rocket, the term $f = cm'$ is the thrust caused by the ejection of burnt fuel gases from the rocket. Note that if fuel is burned at a constant rate, and if c is constant, then the thrust is constant.

As usual, once $v(t)$ is found, the height $y(t)$ is given by

$$y(t) = y(0) + \int_0^t v(s)\,ds.$$

■ Example 2.3.1

Consider the following problem:

A small rocket has mass 1000 kg, including 500 kg of fuel. Its engine burns fuel at a constant rate of 25 kg/sec and provides an upward thrust of 10,000 N. The rocket's drag coefficient is $k = 5$ Solve for and plot the height of the rocket as a function of time. Also find the maximum height attained by the rocket and the time when it hits the ground.

Let's first notice that the mass of the rocket is given by

$$m(t) = 1000 - 25\,t \quad \text{for } 0 \le t \le 20, \quad \text{and } m(t) = 500 \quad \text{for } t \ge 20,$$

and the thrust is given by

$$f(t) = 10000 \quad \text{for } 0 \le t \le 20 \quad \text{and } f(t) = 0 \quad \text{for } t \ge 20.$$

So our approach will be to solve one initial-value problem for $0 \le t \le 20$ and then another, simpler initial-value problem for $t \ge 20$.

Let's begin by entering each of the two differential equations in a general form:

```
eq1 := m[t] v'[t] + k v[t] == -m[t] g + f[t]
eq2 := m[t1] v'[t] + k v[t] == -m[t1] g
```

In the second equation, `t1` represents the time when the mass becomes constant and the second equation comes into effect. Now we're ready to specify the parameters in the problem:

```
g := 98 / 10; k := 5; t1 := 20;
m[t_] := 1000 - 25 t; f[t_] := 12500;
```

The first differential equation is now

```
eq1
```

$$5\,v[t] + (1000 - 25\,t)\,v'[t] == 12500 + \frac{49}{5}\,(-1000 + 25\,t)$$

and the appropriate initial condition is $v(0) = 0$. Let's now use `FOLDSolve` to find the solution.

```
Needs["DiffEqs`DETools`"];
v1[t_] = FOLDSolve[{eq1, v[0] == 0}, v[t], t] // Simplify
```

$$299\,(10 + (-5)^{4/5}\,2^{2/5}\,(-40 + t)^{1/5}) - \frac{49\,t}{4}$$

Now we integrate the velocity to obtain the height, using the initial value $y(0) = 0$:

```
y1[t_] = ∫_0^t v1[s] ds // Simplify
```

$$-\frac{29900}{3}\,(10 + (-5)^{4/5}\,2^{2/5}\,(-40 + t)^{1/5}) + \frac{1495}{6}\,(12 + (-5)^{4/5}\,2^{2/5}\,(-40 + t)^{1/5})\,t - \frac{49\,t^2}{8}$$

We have run into a slight difficulty here. The solutions given by `FOLDSolve` involve complex numbers! (Can you see why?) But exact values are always real; for example,

```
y1[10] // Simplify
```

$$-\frac{422275}{6} + 37375\ 2^{3/5}\ 3^{1/5}$$

But due to round-off error, numerical values can end up with a small imaginary part.

```
y1[10.]
```

$$191.375 + 8.188 \times 10^{-12}\ i$$

We can cure this problem by taking only the real part of the solution.

```
y1[10.] // Re
```

191.375

Now we have no difficulty plotting the solution.

```
plot1 = Plot[Re[y1[t]], {t, 0, t1}];
```

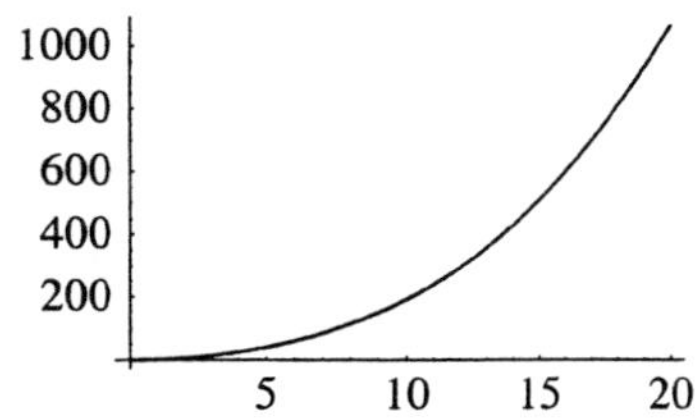

We're ready to shift to the simpler problem that takes effect when the rocket runs out of fuel. This is the equation:

```
eq2 // Simplify
```

$$5\ (980 + v[t] + 100\ v'[t]) == 0$$

The appropriate initial conditions are $v(t_1) = v_1(t_1)$ and $y(t_1) = y_1(t_1)$. Again we call on `FOLDSolve` to find the velocity.

```
v2[t_] = FOLDSolve[{eq2, v[t1] == v1[t1]}, v[t], t] // Simplify
```

$$-5\ e^{-t/100}\ ((-745 + 299\ 2^{4/5})\ e^{1/5} + 196\ e^{t/100})$$

Then we integrate to find the height, this time using $y(t_1) = y_1(t_1)$ and integrating from t_1 to t.

$$\texttt{y2[t\_] = y1[t1]} + \int_{t1}^{t} \texttt{v2[s]}\ ds\ \texttt{// Simplify}$$

$$-\frac{10}{3}\ e^{-t/100}\ (-150\ (-745 + 299\ 2^{4/5})\ e^{1/5} + e^{t/100}\ (-104935 + 37375\ 2^{4/5} + 294\ t))$$

Here's a plot of the second part of the graph:

```
plot2 = Plot[Re[y2[t]], {t, t1, 55}];
```

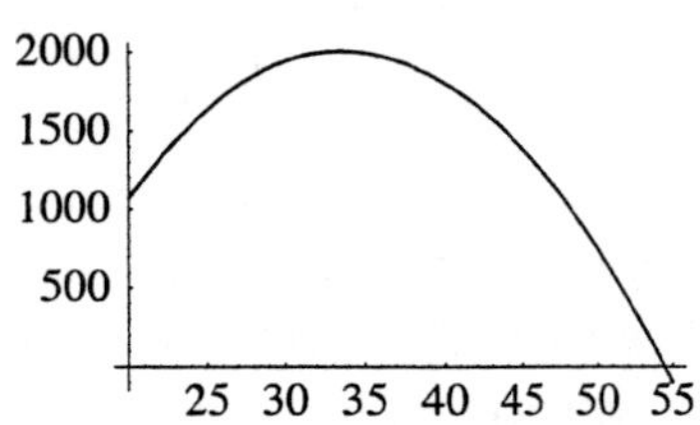

Now we use Show to piece together the entire graph.

```
Show[plot1, plot2, PlotRange → {0, 2100}];
```

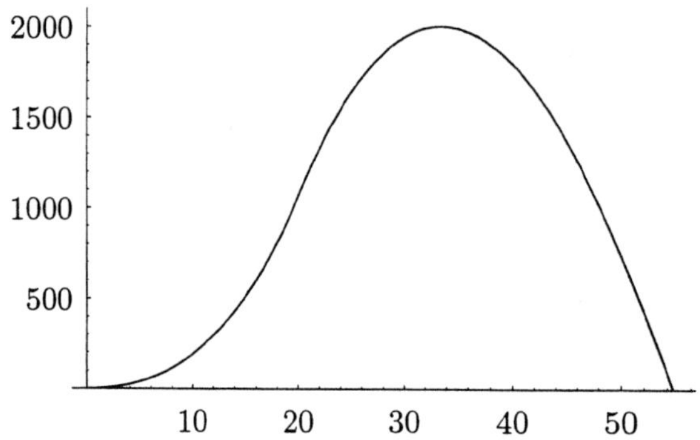

The maximum height is easy to find:

```
FindRoot[v2[t] == 0, {t, 40}]
{t → 33.5363}
```

```
y2[t] /. %
2005.53
```

So is the time when the rocket hits the ground:

```
t /. FindRoot[y2[t] == 0, {t, 68}]
54.4731
```

■ Exercises

1. Suppose that the rocket in Example 2.3.1 is equipped with a parachute that opens 20 seconds after the fuel is spent (40 seconds after lift-off). The open parachute causes the rocket's drag coefficient to increase by a factor of 20.

   a) Solve for each of the three parts of the rocket's trajectory and plot the complete trajectory.

   b) Find the time when the rocket hits the ground.

   c) With what velocity does the rocket hit the ground?

   d) Create a plot of the rocket's velocity over its entire flight.

   e) Create a plot of the rocket's acceleration over its entire flight.

2. Suppose that the engine in the rocket of Example 2.3.1 is replaced by one with the same mass that provides 16,000 N of thrust, but burns fuel at the faster rate of 40 kg/s. Will the new engine make the rocket go higher (with the same amount of fuel)? If so, by how much?

3. Consider again the rocket in Example 2.3.1. Without fuel the rocket has a mass of 500 kg. The rocket will "lift off" immediately upon ignition provided that it carries no more than $12500/9.8 - 500 \approx 776$ kg of fuel. (Why?) However, it is not obvious what amount of fuel will maximize the height attained by the rocket. Let m_0 denote the initial mass of fuel, assum-

ing that $0 < m_0 \leq 776$, so that subsequently $m(t) = 500 + m_0 - 25\,t$. Notice that the fuel is spent at time $t = m_0 / 25$ s.

a) Find the velocity v and height y in terms of m_0 for $0 \leq t \leq m_0 / 25$.

b) Evaluate each of v and y at time $t = m_0 / 25$. Plot the resulting functions of m_0 on the interval $0 < m_0 \leq 776$.

c) Observe that the greatest maximum height and velocity are attained when the rocket is loaded with the maximum of 776 kg of fuel.

d) What will the rocket actually do (initially) if it is loaded with more than 776 kg of fuel? What does the differential equation say it will do?

4. If we ignore air resistance (i.e., let $k = 0$) and set $T = f$, the initial-value problem for the velocity of a fuel-burning rocket becomes

$$v' = -g + \frac{T}{m}, \quad v(0) = 0.$$

For convenience, suppose that units are chosen so that $g = 1$ and the mass of the rocket without fuel is 1. Suppose that the initial mass of fuel is μ, so that the initial mass of the rocket, including fuel, is $1 + \mu$. Also, let r be the constant rate at which fuel is burned, so that the fuel is spent at time $t = \mu / r$ and, during $0 \leq t \leq \mu / r$, the mass is $m(t) = 1 + \mu - r\,t$.

a) Show that the "burnout" velocity of the rocket at time $t = \mu / r$ is

$$V(\mu) = \frac{1}{r}\,(T \ln(1 + \mu) - \mu),$$

provided that this quantity is positive. (A negative or zero value means that the rocket has too little thrust to lift off.)

b) Observe that $V(0) = 0$ and $V(\mu) \to -\infty$ as $\mu \to \infty$, and show that if $T > 1$, then $V(\mu)$ is increasing for $0 < \mu < T - 1$ and decreasing for $\mu > T - 1$. Conclude that $V(\mu)$ attains a positive, absolute maximum value of $\frac{1}{r}\,(1 - T + T \ln T)$ at $\mu = T - 1$. Consequently, the burnout velocity can never exceed $\frac{1}{r}\,(1 - T + T \ln T)$, regardless of how much fuel the rocket carries.

5. Suppose that a spherical raindrop falls through the atmosphere, gathering water vapor (with zero velocity) at a rate proportional to the raindrop's surface area. Suppose also that the raindrop is initially at rest with radius r_0.

a) Show that the radius of the raindrop is given by $r_0 + ct$ for some constant c.

b) Ignoring air resistance, show that the raindrop's acceleration at time t is $\frac{g}{4}\left(1 + 3\left(\frac{r_0}{r}\right)^4\right)$.

[This problem is borrowed from G. F. Simmons, *Differential Equations with Applications and Historical Notes*, Second Edition (New York: McGraw-Hill, 1991).]

2.4 Opening a Parachute

Consider the problem of an object with mass m falling under the force of gravity with linear drag due to air resistance. The velocity $v(t)$ of the object satisfies the differential equation

$$mv' + kv = -mg.$$

Suppose that the object is a parachutist who weighs 192 lbs (including parachute pack). The parachutist thus has mass $192/32 = 6$ slugs (approximately). Suppose also that the parachutist's drag coefficient is $k = 3$ with parachute closed and $k = 24$ with parachute open. If the parachute is opened completely and instantaneously at time $t = t_o$ seconds, then we have

$$k(t) = 3, \text{ if } 0 \le t < t_o; \text{ and } k(t) = 24, \text{ if } t \ge t_o.$$

Let's first solve the problem assuming that the parachute opens at time $t_o = 12$ s. We will assume that the parachutist's initial velocity (upon exiting the airplane) is 0. Let v_1 denote the velocity for the first 12 seconds. We solve as follows for v_1:

```
Needs["DiffEqs`DETools`"];

v1[t_] = FOLDSolve[{6 v'[t] + 3 v[t] == -192, v[0] == 0}, v[t], t]
```

$-32\, e^{-t/2}\, (-2 + 2\, e^{t/2})$

Letting y_1 denote the height for the first 12 seconds, we then have

```
y1[t_] = 1000 + ∫_0^t v1[s] ds // Simplify
```

$8\, (141 - 16\, e^{-t/2} - 8\, t)$

Now we'll define

```
to := 12
```

and compute the velocity v_2 after the parachute is open as follows:

```
v2[t_] = FOLDSolve[
    {6 v'[t] + 24 v[t] == -192, v[to] == v1[to]}, v[t], t] // Simplify
```

$-8\, e^{-4\,t}\, (-8\, e^{42} + 7\, e^{48} + e^{4\,t})$

Then the height after the parachute is open is

```
y2[t_] = y1[to] + ∫_to^t v2[s] ds // Simplify
```

$2\left(221 - \frac{56}{e^6} - 8\, e^{42-4\,t} + 7\, e^{48-4\,t} - 4\,t\right)$

We can now plot the height over the entire descent as follows:

```
FindRoot[y2[t] == 0, {t, 50}]
```

$\{t \to 55.2153\}$

```
Plot[If[t < to, y1[t], y2[t]], {t, 0, 55.2}, PlotRange → {0, 1000}];
```

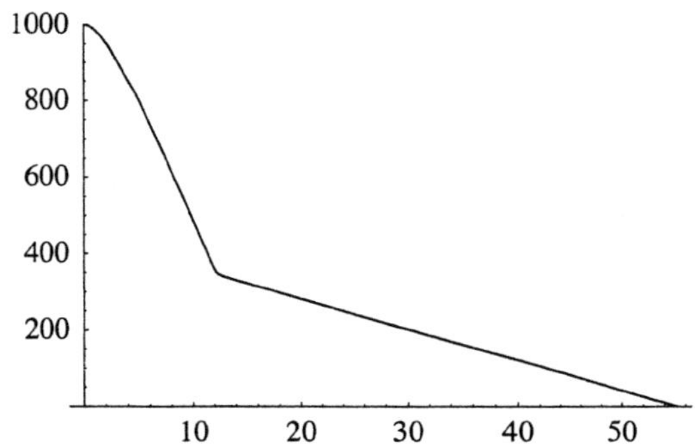

An interesting consideration here is the force that the parachutist feels when the parachute opens. So let's plot mass × acceleration as follows:

```
Plot[If[t < to, 6 v1′[t], 6 v2′[t]], {t, 0, 55.2}, PlotRange → All];
```

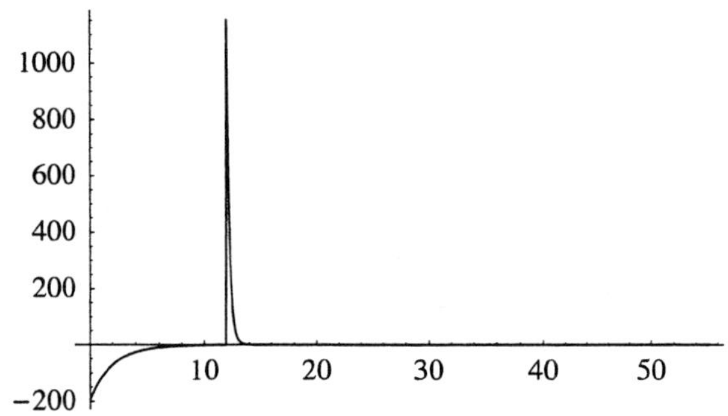

In fact, the force (in pounds) at the instant when the parachute opens is

```
6 v2′[to] // N
1340.19
```

Ouch! Such a force would cause serious injury to the parachutist, to say the least. Actually this points to a problem with our model—no parachute can open instantaneously.

So let's conjure a different model for the drag coefficient during the interval when the parachute is opening. Let's assume that it takes two seconds for the parachute to open completely. One way of modeling the drag coefficient during those two seconds is with a function of the form

$$\frac{a}{b - c(t-t_o)},$$

where we will choose the coefficients a, b, and c so that $k(t)$ will be continuous. Thus we want

$$\frac{a}{b - c\,(0)} = 3 \text{ and } \frac{a}{b - c(2)} = 24.$$

Using these equations we solve for a and b in terms of c:

```
Solve[{a/b == 3, a/(b - 2 c) == 24}, {a, b}]
{{a → 48 c/7, b → 16 c/7}}
```

A nice choice is $c = 7$, $a = 48$, and $b = 16$, which yields

$$k(t) = 3, \text{ if } 0 \le t \le t_o;\; k(t) = \frac{48}{16-7\,(t-t_o)}, \text{ if } t_o \le t \le t_o + 2;\; \text{ and } k(t) = 24, \text{ if } t \ge t_o + 2.$$

This function is defined and plotted as follows:

```
k[t_] := Which[t < to, 3, t < to + 2, 48/(16 - 7 (t - to)), t ≥ to, 24];
Plot[k[t], {t, 0, 30}];
```

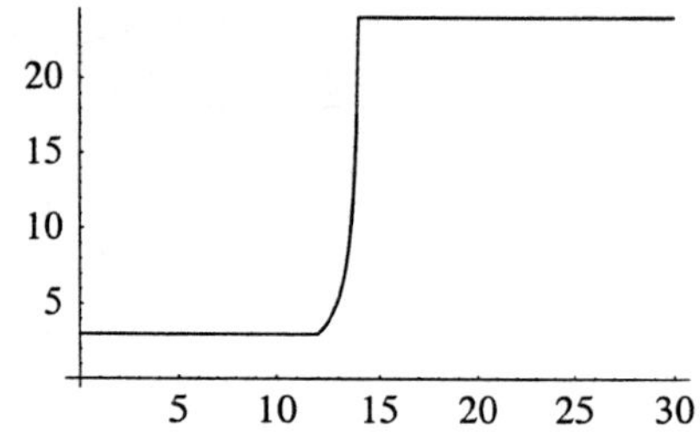

Now let's solve our new problem. First:

```
v1[t_] = FOLDSolve[{6 v'[t] + 3 v[t] == -192, v[0] == 0}, v[t], t]
```

$$-32\, e^{-t/2}\,(-2 + 2\, e^{t/2})$$

```
y1[t_] = 1000 + ∫_0^t v1[s] ds // Simplify
```

$$8\,(141 - 16\, e^{-t/2} - 8\, t)$$

Then, while the parachute is opening, we have:

```
v2[t_] = FOLDSolve[{6 v'[t] + 48/(16 - 7 (t - to)) v[t] == -192, v[to] == v1[to]},
  v[t], t] // Simplify
```

$$-\frac{1}{e^6}\,(2\,(-1)^{6/7}\,2^{3/7}\,(-100 + 7\,t)\,(-(-100 + 7\,t)^{1/7} + e^6\,(8\,(-1)^{1/7}\,2^{4/7} - 7\,(-100 + 7\,t)^{1/7})))$$

```
y2[t_] = y1[to] + ∫_to^t v2[s] ds // Simplify
```

$$\frac{1}{15\,e^6}\,(2\,(-448 + 10000\,(-1)^{6/7}\,2^{3/7}\,(-100 + 7\,t)^{1/7} - 1400\,(-1)^{6/7}\,2^{3/7}\,t\,(-100 + 7\,t)^{1/7} + 49\,(-1)^{6/7}\,2^{3/7}\,t^2\,(-100 + 7\,t)^{1/7} + e^6\,(173324 + 70000\,(-1)^{6/7}\,2^{3/7}\,(-100 + 7\,t)^{1/7} - 200\,t\,(120 + 49\,(-1)^{6/7}\,2^{3/7}\,(-100 + 7\,t)^{1/7}) + 7\,t^2\,(120 + 49\,(-1)^{6/7}\,2^{3/7}\,(-100 + 7\,t)^{1/7}))))$$

After the parachute is open:

```
v3[t_] = FOLDSolve[{6 v'[t] + 24 v[t] == -192, v[to + 2] == v2[to + 2]},
  v[t], t] // Simplify
```

$$e^{-4\,t}\,(4\,2^{4/7}\,e^{50} + 28\,(-2 + 2^{4/7})\,e^{56} - 8\,e^{4\,t})$$

```
y3[t_] = y2[to + 2] + ∫_(to+2)^t v3[s] ds // Simplify
```

$$\frac{1}{15}\,e^{-6-4\,t}\,(-15\,2^{4/7}\,e^{56} - 105\,(-2 + 2^{4/7})\,e^{62} + 7\,(-128 + 2^{4/7})\,e^{4\,t} + e^{6+4\,t}\,(5398 + 49\,2^{4/7} - 120\,t))$$

Now we find the landing time:

```
FindRoot[y3[t], {t, 50}]
{t → 45.5718}
```

and then plot the height:

```
Plot[Which[t < to, y1[t], t < to + 2, y2[t], t ≥ to + 2, y3[t]],
  {t, 0, 45.6}];
```

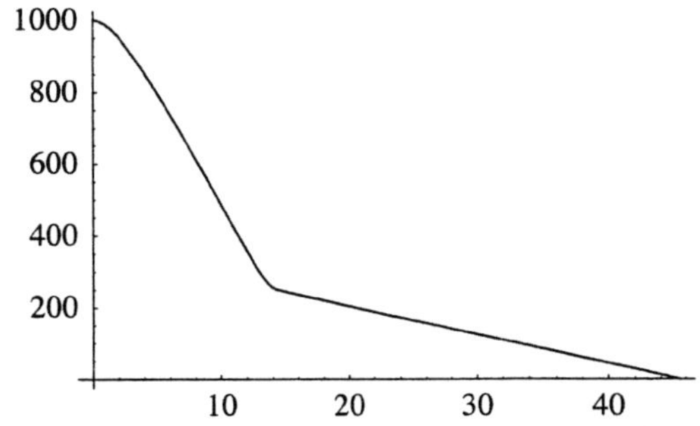

So what about the force on the parachutist? This is plotted as follows:

```
Plot[Which[t < to, 6 v1'[t], t ≤ to + 2, 6 v2'[t], t ≥ to + 2, 6 v3'[t]],
  {t, 0, 45.6}, PlotRange → All];
```

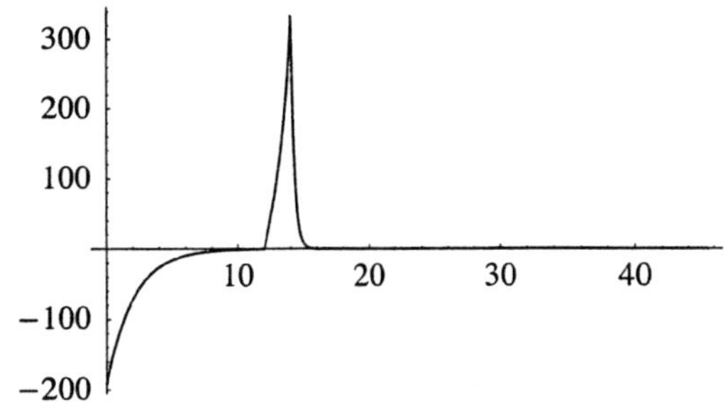

The greatest force (in pounds) is

```
6 v3'[to + 2] // N
345.058
```

which is much less painful than before, though still hardly pleasant. Also note that this model's prediction of the landing time is about 10 seconds sooner than the previous model's prediction.

■ Exercises

1. (a) Given the scenario described above, in which the parachute opens between times $t = 12$ and $t = 14$ seconds, compute the parachutist's velocity at each of $t = 8,\ 12,\ 14,\ 25$, and 45.6 seconds (the last of these being the landing time). What is the parachutist's terminal velocity?

   (b) At what time $t > 14$ does the parachutist's velocity become within 1ft/s of her open-parachute terminal velocity?

2. Using the "two-second model" described above, experimentally determine (to the nearest tenth of a second) how long the parachutist can wait to open the parachute and still have a landing velocity within 1 ft/s of her open-parachute terminal velocity. How high above the ground is she at that moment?

2.5 RL Circuits

Assuming linear device characteristics, the current $I(t)$ in an RL circuit with source (as depicted below) satisfies the differential equation

$$L\,I'(t) + R\,I = v_s(t),$$

where $v_s(t)$ denotes the source voltage. This is a linear, first-order differential equation for the current $I(t)$.

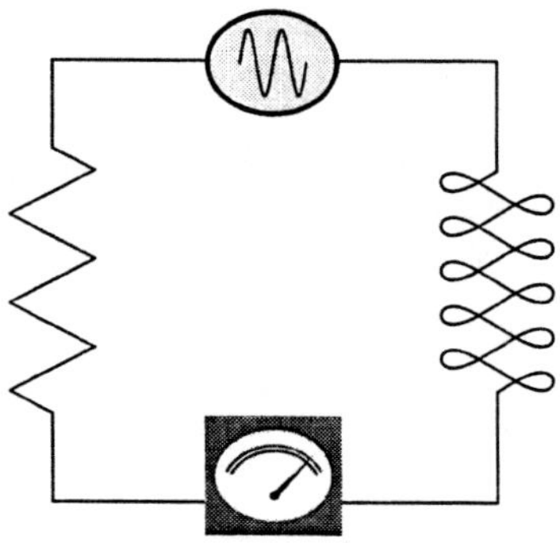

So we define the differential equation as follows:

```
RLiDE := L i'[t] + R i[t] == v[t]
```

(Note that in *Mathematica* a capital `I` denotes the imaginary number $i = \sqrt{-1}$; so we use a lower-case `i` for current.) Let us assume that the initial current is zero, i.e., $I(0) = 0$. We use `FOLDSolve` to solve for the current:

```
<< DiffEqs`DETools`

current[t_] = FOLDSolve[{RLiDE, i[0] == 0}, i[t], t] // Simplify
```

$$\frac{-e^{-\frac{R t}{L}} R + R \operatorname{Cos}[t\,\omega] + L\,\omega \operatorname{Sin}[t\,\omega]}{R^2 + L^2\,\omega^2}$$

■ Example 2.5.1

Consider a sinusoidal source of the form $v_s(t) = \sin(\omega\, t)$.

```
v[t_] := Cos[ω t];
```

Then the current becomes:

```
current[t] // Simplify // Apart
```

$$-\frac{e^{-\frac{R t}{L}} R}{R^2 + L^2 \omega^2} + \frac{R \operatorname{Cos}[t \omega] + L \omega \operatorname{Sin}[t \omega]}{R^2 + L^2 \omega^2}$$

The first term here is *transient*—it approaches 0 as $t \to \infty$ (and quickly if R/L is large). The second term is the *steady state*, which can be rewritten as

$$\texttt{SScurrent[t\_] :=} \frac{1}{\sqrt{R^2 + L^2 \omega^2}} \operatorname{Cos}\left[\omega t - \operatorname{ArcTan}\left[\frac{\omega L}{R}\right]\right]$$

Note that the amplitude of the steady-state current is $1 / \sqrt{R^2 + L^2 \omega^2}$ and the phase angle is $\tan^{-1}(\frac{\omega L}{R})$. The following is a plot of the source voltage, the current, and the steady-state current for a particular set of parameter values that results in a phase angle of approximately 1.11 radians, or about 63°.

```
paramvals = {R → .25, L → .5, ω → 1};
Plot[Evaluate[{v[t], current[t], SScurrent[t]} /. paramvals],
  {t, 0, 4 π}, PlotStyle → {Hue[0], Hue[.67], {}}];
```

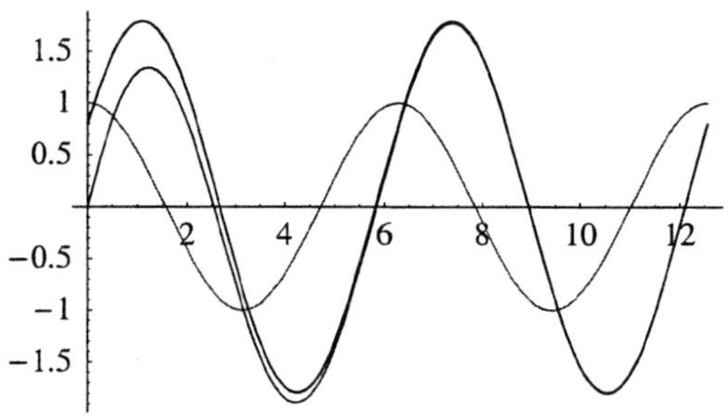

```
ArcTan[ω L / R] /. paramvals
```

```
1.10715
```

In the next plot, the phase angle is approximately 0.245 radians, or about 14°.

```
paramvals = {R → 1, L → .25, ω → 1};
Plot[Evaluate[{v[t], current[t], SScurrent[t]} /. paramvals],
  {t, 0, 4 π}, PlotStyle → {Hue[0], Hue[.67], {}}];
```

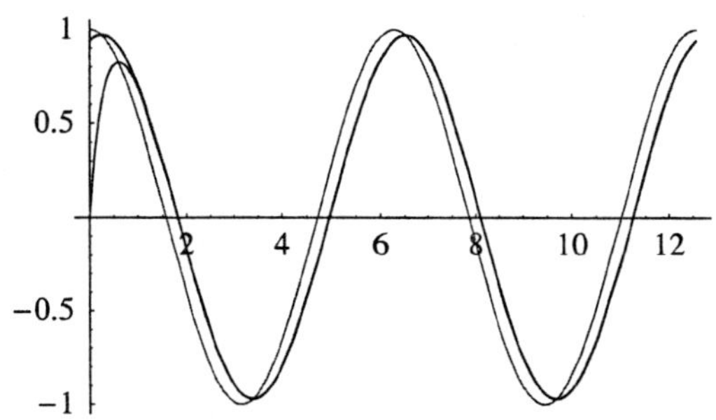

```
ArcTan[ω L / R] /. paramvals
```

```
0.244979
```

■ Example 2.5.2

Let's consider now a "ramp" source voltage of the following form, in which the voltage increases linearly to 1 while $0 \le t \le a$ and then turns off.

```
v[t_] := If[t < a, t / a, 0];
Plot[v[t] /. a → 5, {t, 0, 10}, Ticks → {{}, {1}}];
```

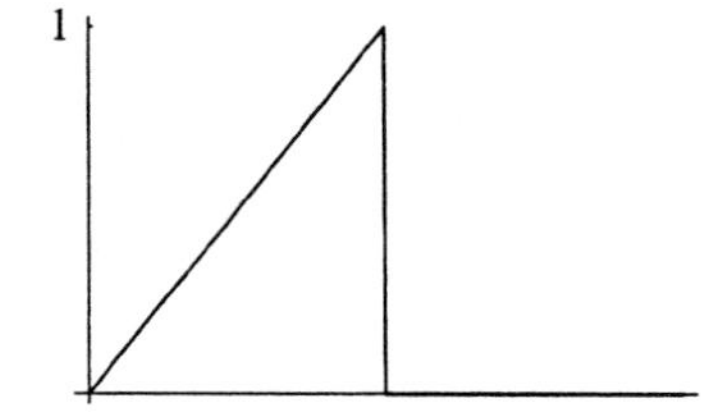

We first solve for the current $I_1(t)$ during $0 \le t \le a$:

```
current1[t_] =
 FOLDSolve[{L i'[t] + R i[t] == t / a, i[0] == 0}, i[t], t] // Simplify
```

$$\frac{\left(-1+e^{-\frac{R t}{L}}\right) L+R t}{a R^2}$$

Next we solve for the current $I_2(t)$ for $t \ge a$, using the continuity condition $I_2(a) = I_1(a)$:

```
current2[t_] = FOLDSolve[
   {L i'[t] + R i[t] == 0, i[a] == current1[a]}, i[t], t] // Simplify
```

$$\frac{e^{\frac{R(a-t)}{L}}\left(\left(-1+e^{-\frac{a R}{L}}\right) L+a R\right)}{a R^2}$$

Now we piece together the current for all $t \ge 0$ as follows:

```
current[t_] = If[t < a, Evaluate[current1[t]], Evaluate[current2[t]]]
```

$$\text{If}\left[t<a, \frac{\left(-1+e^{-\frac{R t}{L}}\right) L+R t}{a R^2}, \frac{e^{\frac{R(a-t)}{L}}\left(\left(-1+e^{-\frac{a R}{L}}\right) L+a R\right)}{a R^2}\right]$$

Here is a plot of the source voltage and the current for a specific set of parameter values:

```
paramvals = {R → .25, L → .5, a → 3};
Plot[Evaluate[{v[ t], current[t]} /. paramvals],
  {t, 0, 12}, PlotStyle → {Hue[0], {}}];
```

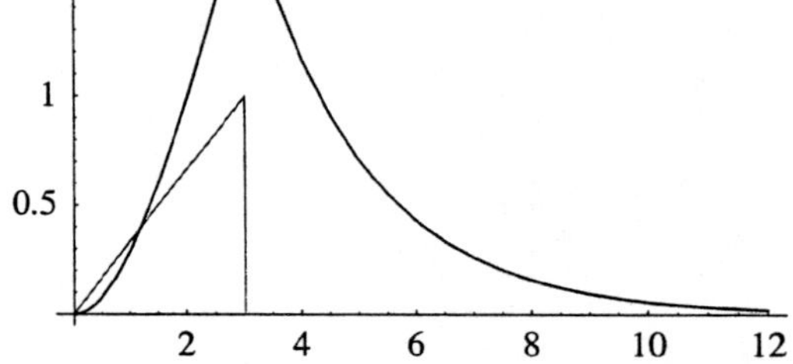

■ Exercises

Find and plot the current satisfying $I(0) = 0$ for each of the following source voltages.

1. $v_s(t) = \begin{cases} \sin t, & \text{for } 0 \le t \le 2\pi \\ 0, & \text{for } t > 2\pi \end{cases}$

2. $v_s(t) = \begin{cases} |\sin t|, & \text{for } 0 \le t \le 2\pi \\ 0, & \text{for } t > 2\pi \end{cases}$

3. $v_s(t) = e^{-t}$

4. $v_s(t) = \begin{cases} 1, & \text{for } 0 \le t \le 1 \\ -1, & \text{for } 1 < t \le 2 \\ 0, & \text{for } t > 2 \end{cases}$

Chapter 3

Nonlinear First-Order Equations and Applications

3.1 Direction Fields, Isoclines, and Solution Curves

The *Mathematica* package `Graphics`PlotField`` contains a function `PlotVector-Field` that can be used to plot the direction field for a differential equation. The function `NDSolve` returns a numerical solution to an initial value problem. In the following example we will use these to create a plot showing the direction field and a sample solution curve for a differential equation.

■ Example 3.1.1

To plot the direction field for the equation

$$y' = t^2 - y^2,$$

first load the package by entering

```
<< Graphics`PlotField`
```

and then enter

```
dirfield = PlotVectorField[{1, t² - y²}, {t, 0, 3},
   {y, 0, 2}, Axes → True, ScaleFunction → (1 &)];
```

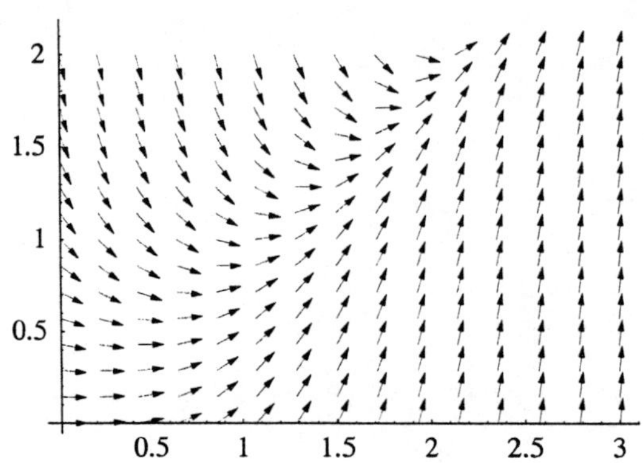

The following plot shows the direction field on the rectangle described by $0 \le t \le 2$ and $0 \le y \le 2$. The number of arrows is controlled with the `PlotPoints` option, and the lengths of the arrows can be adjusted with the `ScaleFactor` option.

```
dirfield =
  PlotVectorField[{1, t² - y²}, {t, 0, 3}, {y, 0, 2}, Axes → True,
   ScaleFunction → (1 &), PlotPoints → {9, 7}, ScaleFactor → .2];
```

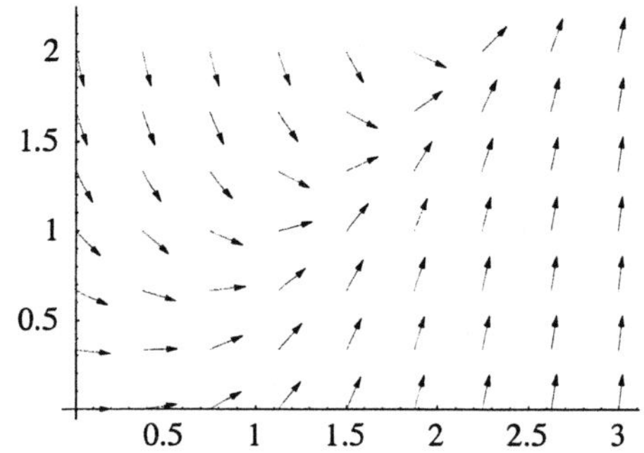

`NDSolve` returns a numerical solution to an initial-value problem in the form of an object called an `InterpolatingFunction`. For example, we'll enter the following to compute a numerical solution (for $0 \le t \le 2$) of the initial-value problem

$$y' = t^2 - y^2, \quad y(0) = 1.$$

```
soln = y[t] /. NDSolve[{y'[t] == t² - y[t]², y[0] == 1}, y[t], {t, 0, 3}]
{InterpolatingFunction[{{0., 3.}}, <>][t]}
```

Now to plot the solution we enter

```
curve1 = Plot[Evaluate[soln],
   {t, 0, 3}, PlotStyle → {{Thickness[.007], Hue[0]}}];
```

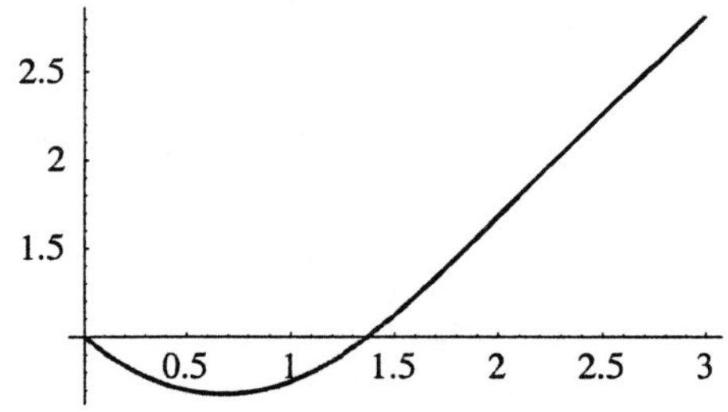

The direction field and the solution curve can be superimposed with `Show`.

```
Show[dirfield, curve1, PlotRange → {{0, 3.1}, {0, 2.2}}];
```

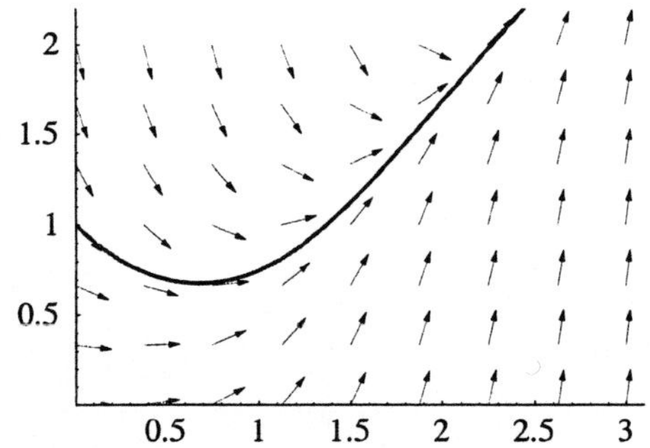

Isoclines

Recall that the **isoclines** of a differential equation

$$y' = f(t, y)$$

are curves described by $f(t, y) = m$, where m is constant. This means that isoclines are precisely the level curves—or contours—of the function f.

Example 3.1.2

Consider the differential equation

$$y' = t^3 + y^3.$$

The isoclines of this equation are the graphs of $t^3 + y^3 = m$, i.e., the contours of $f(t, y) = t^3 + y^3$. Let's first look at the surface $z = f(t, y)$.

```
f[t_, y_] := t^3 + y^3;
Plot3D[f[t, y], {t, -1.5, 1.5}, {y, -1.5, 1.5}, BoxRatios → {1, 1, 1}];
```

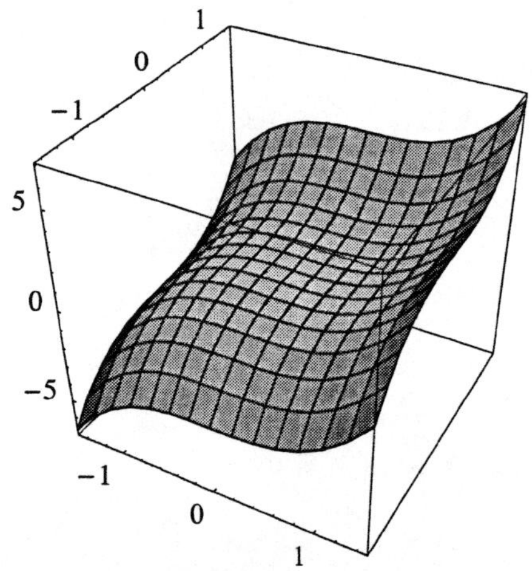

Now let's use `ContourPlot` to plot a collection of contours of f. These will be isoclines of the differential equation.

```
isoclines = ContourPlot[f[t, y], {t, -1.5, 1.5}, {y, -1, 1},
   ContourShading → False, AspectRatio → Automatic];
```

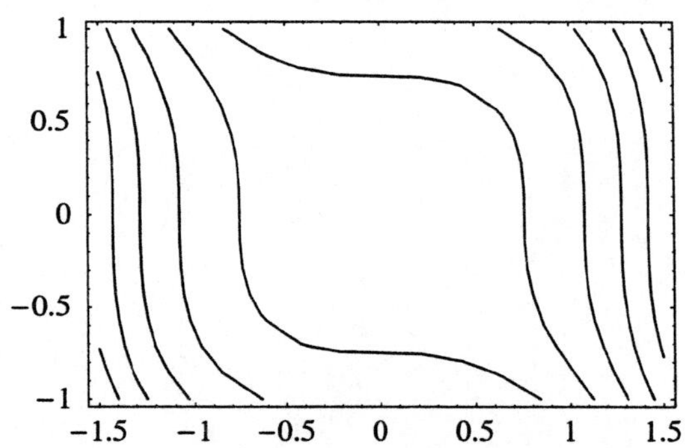

To plot the direction field and then superimpose the isoclines and the direction field, enter

```
<< Graphics`PlotField`;
dirfield = PlotVectorField[{1, f[t, y]}, {t, -1.5, 1.5}, {y, -1, 1},
   Axes → True, ScaleFunction → (1 &), DisplayFunction → Identity];
Show[dirfield, isoclines, DisplayFunction → $DisplayFunction];
```

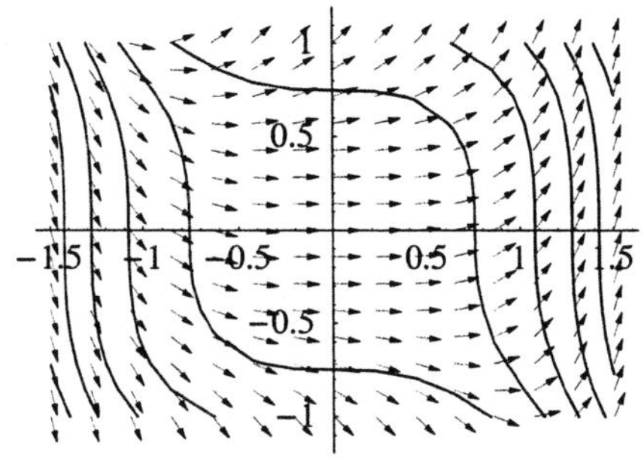

Notice in this plot the defining property of isoclines; that is, along any particular isocline, the direction field has constant slope. Now, finally, let's plot a typical solution curve and include it in the plot as well.

```
soln = y[t] /.
  NDSolve[{ y'[t] == f[t, y[t]], y[0] == -.6}, y[t], {t, -1.5, 1.5}]
{InterpolatingFunction[{{-1.5, 1.5}}, <>][t]}

curve = Plot[Evaluate[soln], {t, -1.5, 1.5},
   PlotStyle → {{Hue[0], Thickness[.01]}}];
```

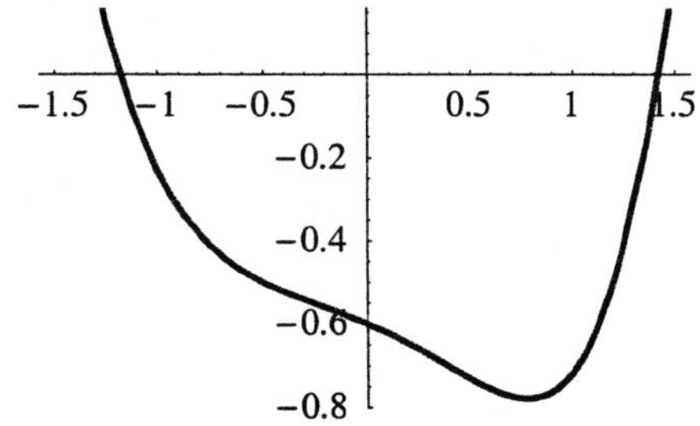


```
Show[dirfield, isoclines, curve, DisplayFunction → $DisplayFunction ];
```

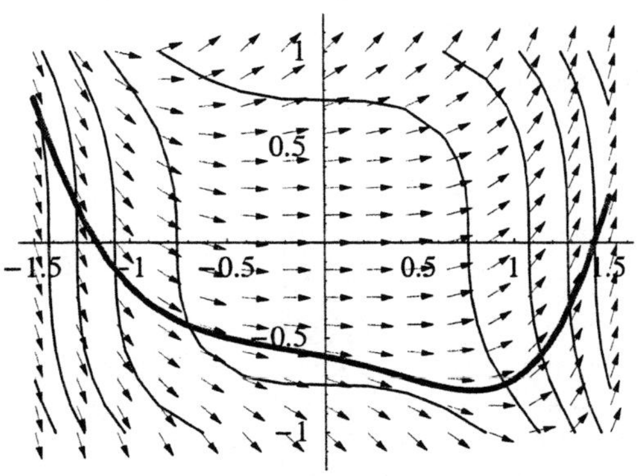

■ Exercises

1. Reproduce Example 3.1.1, and then repeat for initial values $y(0) = 0.5$ and $y(0) = 2$, generating `curve2` and `curve3`. Then enter

```
Show[dirfield,curve1,curve2,curve3];
```

to display these three solution curves at once on top of the direction field.

2. (a) Plot the direction field for the equation

$$y' = y\ (2 - y)$$

on the rectangle $0 \le t \le 3,\ -1 \le y \le 3$.

(b) For each of $y_0 = 3,\ 2,\ 0.1$, and -0.004, plot the solution of the initial value problem

$$y' = y\,(2 - y),\quad y(0) = y_0\,.$$

(c) Use `Show` to display the four curves from part (b) on top of the direction field from (a) on the interval $0 \le t \le 3$.

3. Create a plot like the last one in Example 3.1.2 for each of the following equations. In each case, create the plot over the suggested rectangle and include the solution curve satisfying $y(0) = 0$.

(a) $y' = t^2 + y^2$ $\qquad -2 \le t \le 2,\ -2 \le y \le 2$

(b) $y' = t^2 - y^2$ $\qquad -2 \le t \le 2,\ -2 \le y \le 2$

(c) $y' = t^2 / (1 + y^2)$ $\qquad -2 \le t \le 2,\ -1 \le y \le 1$

(d) $y' = \cos(t\,y)$ $\qquad -4 \le t \le 4,\ -2 \le y \le 2$

(e) $y' = \cos t + \sin y$ $\qquad -4 \le t \le 4,\ -\pi \le y \le \pi$

3.2 DEPlot

The `DEGraphics` package contains a function called `DEPlot` that plots direction fields, solution curves, and isoclines automatically.

```
<< DiffEqs`DEGraphics`
```

```
? DEPlot
```

```
DEPlot[f, {t,tmin,tmax}, {y,ymin,ymax}, options]. By
  default, four solution curves are drawn. The InitialPoints
  option can be used to specify a list of starting points for
  curves drawn by DEPlot. Each initial point may give rise
  to multiple curves as specified by the MaxCurves option.
```

The following creates a basic plot for the equation $y' = t^3 + y^3$.

```
DEPlot[t³ + y³, {t, -2, 2}, {y, -2, 2}];
```

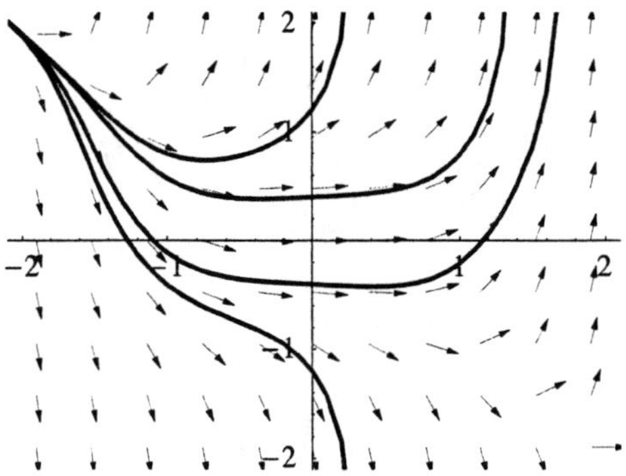

To show specific solution curves through $(0, -1)$, $(0, 0)$, and $(0, 1)$, we use the `InitialPoints` option.

```
DEPlot[t³ + y³, {t, -2, 2}, {y, -2, 2},
  InitialPoints → {{0, -1}, {0, 0}, {0, 1}}];
```

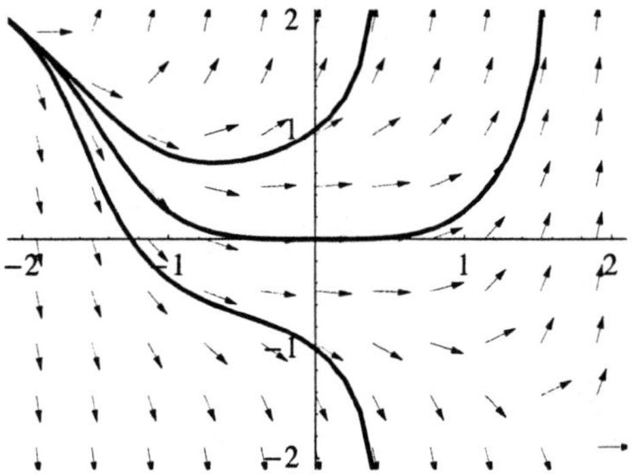

With the `MaxCurves` option, we can make `DEPlot` automatically generate starting points for additional solution curves. Here we get five curves from a single initial point.

```
DEPlot[t³ + y³, {t, -2, 2}, {y, -2, 2},
  InitialPoints → {{0, 0}}, MaxCurves → 5];
```

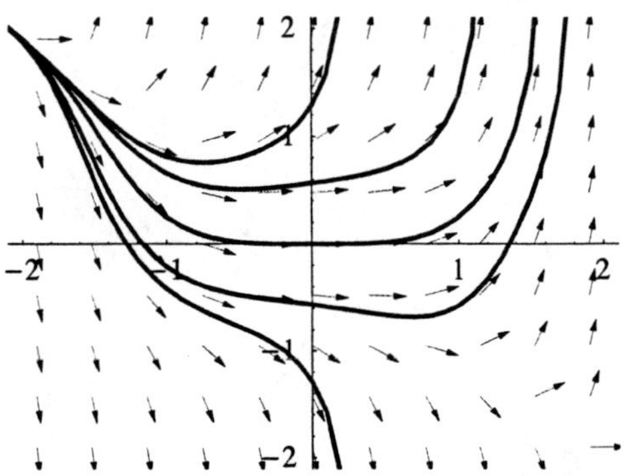

The Gap option controls the spacing between the curves. (The default is Gap → 0.1.)

```
DEPlot[t^3 + y^3, {t, -2, 2}, {y, -2, 2},
  InitialPoints → {{0, 0}}, MaxCurves → 9, Gap → .05];
```

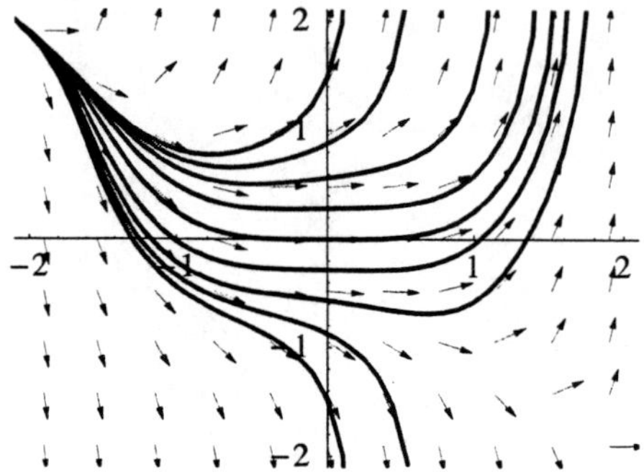

We can also include isoclines by setting Isoclines → True.

```
DEPlot[t^3 + y^3, {t, -2, 2}, {y, -2, 2}, Isoclines → True];
```

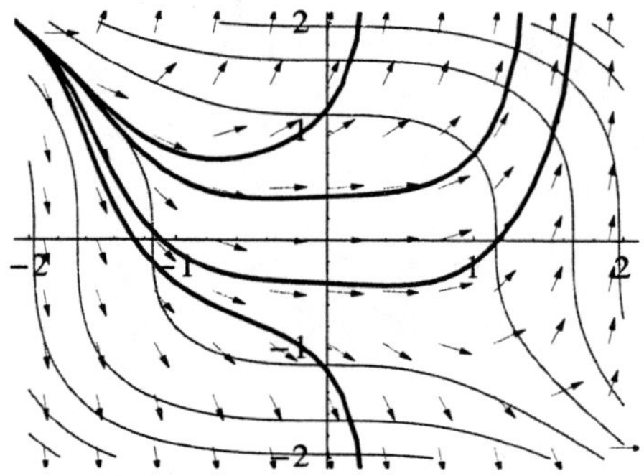

Other options can be used to customize the plot. Some of these are used in the following plot for the equation $y' = t^2 - y^2$:

```
DEPlot[t^2 - y^2, {t, -2, 2}, {y, -2, 2}, Arrows → {9, 7},
  ArrowColor → GrayLevel[0], ScaleArrows → 1.25,
  PlotStyle → {AbsoluteThickness[2], Hue[0]},
  Isoclines → {-1, 0, 1}, AspectRatio → Automatic, Frame → True];
```

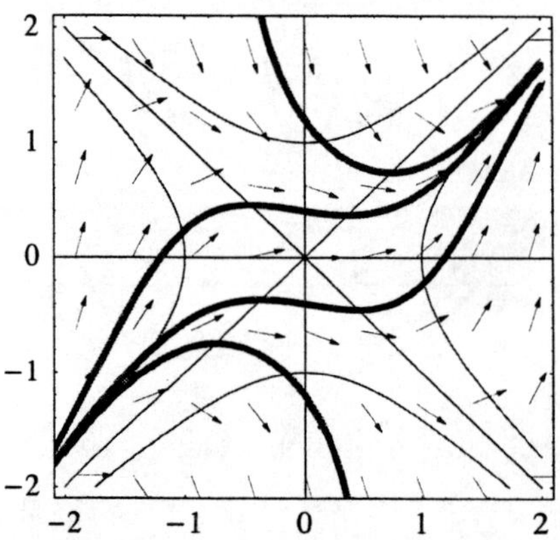

The following is a list of all options (with defaults) for `DEPlot`.

```
Options[DEPlot]
{ArrowColor → RGBColor[0, 0.6, 0.6], Arrows → {11, 9},
 ScaleArrows → 1, AxesLabel → None, AspectRatio → 0.75, Axes → True,
 Frame → False, FrameTicks → Automatic, InitialPoints → Automatic,
 PlotStyle → Thickness[0.006], Ticks → Automatic, UseVectorField → False,
 ShowField → True, MaxCurves → {1}, Gap → 0.1, ShowEquilibriumPoints → False,
 ShowStartPoints → False, Isoclines → False, Method → RungeKutta,
 StepSize → 0.1, DisplayFunction → (Display[$Display, #1] &)}
```

■ Exercises

1. For each of the following equations, use `DEPlot` to create a picture of the direction field on the rectangle $-3 \le t \le 3$, $-2 \le y \le 2$, with isoclines and a few solution curves. Experiment with `InitialPoints` (and other options) to get a picture to your liking.

   (a) $y' = t^2 + y^2 - 1$
   (b) $y' = 1 - t\,y$
   (c) $y' = (t - y)/(1 + t^2\,y^2)$
   (d) $y' = t/(0.1 + t^2 + y^2)$
   (e) $y' = y\cos(t - y)$

3.3 The Logistic Population Model

An important population model is the *logistic equation*

$$P' = k\,P\,(1 - M^{-1}\,P).$$

Here M is the *environmental carrying capacity*. Since the equation reduces to the exponential growth model $P' = kP$ if $M = \infty$, the constant k is called the *intrinsic growth rate*.

Because of the context here, we are only concerned with nonnegative solutions. The character of the solution curves can be observed in the simple case where $k = M = 1$. For this we'll use `DEPlot` as follows:

```
<< DiffEqs`DEGraphics`;
DEPlot[p (1 - p), {t, 0, 6}, {p, 0, 2}, InitialPoints → {{0, 1}},
  MaxCurves → 7, Gap → .05, Arrows → 6, Ticks → None];
```

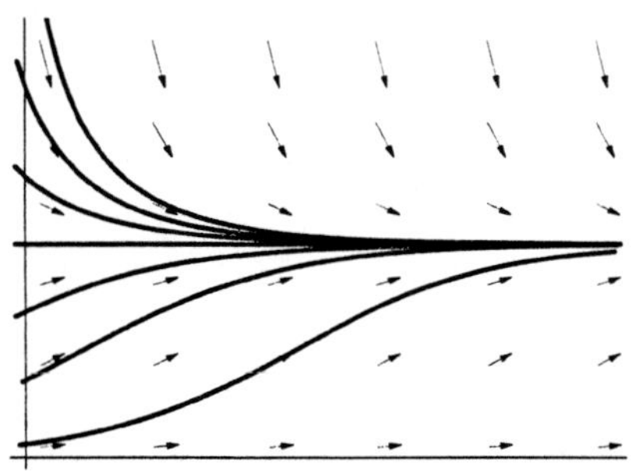

Notice that each curve approaches the horizontal asymptote/equilibrium solution $P = 1$.

An explicit, "analytic" solution can be obtained in either of two ways. First, the equation is *separable* (provided that k and M are constant). The separated form is

$$\frac{dP}{P(1-M^{-1}P)}=k\,dt.$$

To solve this we can use `SeparableDSolve` in the `DETools` package.

```
<< DiffEqs`DETools`
```

```
? SeparableDSolve
```

```
SeparableDSolve[ LHS dx == RHS dy, options].   Returns
  an implicit general solution in the form F[x,y]== C.
```

In order to get the solution satisfying a given initial value $P(0)=P_0$, we enter

```
SeparableDSolve[(p (1 - p / m))^-1 dp == k dt, InitialPoint → {p0, 0}]
Solve[%, p] // Simplify
```

```
-k t + Log[p] - Log[-m + p] == Log[p0] - Log[-m + p0]
```

$$\left\{\left\{p\to\frac{e^{k\,t}\,m\,p0}{m+(-1+e^{k\,t})\,p0}\right\}\right\}$$

(Note that the `InitialPoint` option is specified as `{p0,0}`—rather than `{0,p0}`—to be consistent with the left and right sides of the separated equation.)

The logistic equation is also a Bernoulli equation, since it can be written in the form

$$P'-k\,P=-k\,M^{-1}\,P^2.$$

So we can also solve it with the `BernoulliDSolve` function in the `DETools` package.

```
? BernoulliDSolve
```

```
BernoulliDSolve[equation, y[t], t, options]  solves a Bernoulli-
  type differential equation for the expression y[t].
```

```
BernoulliDSolve[
  {p'[t] == k p[t] (1 - p[t] / m), p[0] == p0}, p[t], t] // Simplify
```

$$\frac{e^{k\,t}\,m\,p0}{m+(-1+e^{k\,t})\,p0}$$

■ Harvesting

Consider the logistic model with constant harvesting at a rate of R individuals per unit time:

$$P'=k\,P\,(1-M^{-1}P)-R.$$

By introducing the *dimensionless* variable $y(t)=M^{-1}\,P(k^{-1}\,t)$, we obtain a simpler equation with only one parameter:

$$y'=y(1-y)-r,\ \text{where}\ r=k^{-1}\,M^{-1}\,R.$$

This is verified as follows:

```
y[t_] = m^-1 p[k^-1 t];
y'[t] == y[t] (1 - y[t]) - k^-1 m^-1 R
```

$$\frac{p'[\frac{t}{k}]}{k\,m} == -\frac{R}{k\,m} + \frac{p[\frac{t}{k}]\left(1 - \frac{p[\frac{t}{k}]}{m}\right)}{m}$$

The equilibrium solutions are

```
Clear[y];
eqsolns = y /. Solve[y (1 - y) - r == 0, y]
```

$$\left\{\frac{1}{2}\left(1-\sqrt{1-4\,r}\right),\ \frac{1}{2}\left(1+\sqrt{1-4\,r}\right)\right\}$$

provided that $r \leq 1/4$. If $r > 1/4$, then no equilibrium solution exists.

The following command produces an animation that shows how the solution curves change as r is varied from 0 to 0.35.

```
Needs["DiffEqs`DEGraphics`"];
startpts = {{0, 0}, Re[eqsolns]} // Transpose;
Do[DEPlot[p (1 - p) - r, {t, 0, 6}, {p, 0, 1.5},
   InitialPoints → startpts, MaxCurves → 5, Gap → .025,
   Ticks → None, ShowEquilibriumPoints → True, ArrowColor →
    If[r < .25, RGBColor[0, .6, .6], RGBColor[.5, .3, .2]], PlotStyle →
    {Thickness[.008], If[r < .25, RGBColor[0, .6, .4], Hue[0]]}],
  {r, 0, .35, .025}];
```

Eight of the frames produced are shown below.

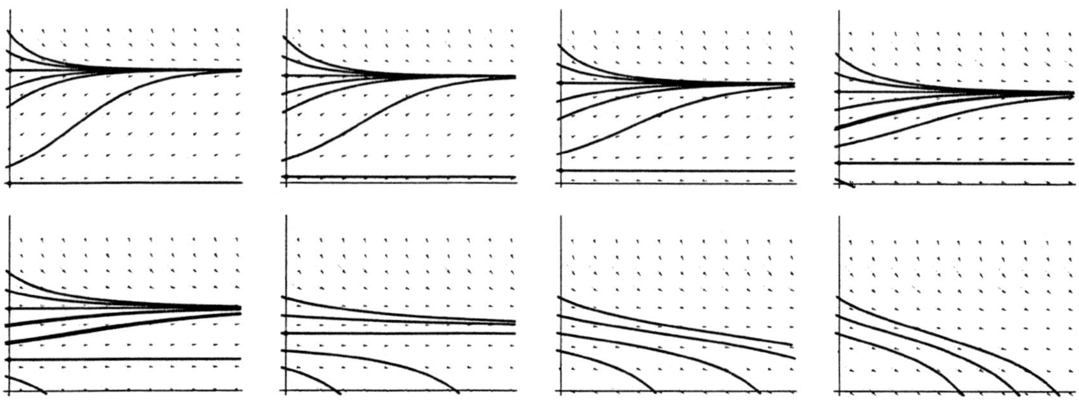

You should notice that the two equilibrium solutions merge when $r = 1/4$ and then disappear. The population is harvested into extinction if $r > 1/4$. The harvesting rate corresponding to $r = 1/4$ is called the *maximum sustainable yield*.

■ Seasonal Coefficients

Here we'll consider a population governed by the logistic model where the constants k and M are allowed to vary with time. In particular, we will examine a situation in which time is measured in years and both k and M vary seasonally.

Let's suppose first that the population reproduces most rapidly in the spring and that the food supply is greatest in the autumn. Reasonable models of k and M might then be

```
k[t_] := k0 + α Cos[2 π (t - .25)]
m[t_] := m0 + β Cos[2 π (t - .75)]
```

where k0, α, M0, and β are positive constants with $\alpha \leq$ k0 and $\beta \leq$ M0 (so that $k(t)$ and $M(t)$ are never negative). Let's use the following values and see what solutions of the equation look like.

```
k0 = .5; α = .5; m0 = 10; β = 5;
Plot[{k[t], m[t]}, {t, 0, 2}];
```

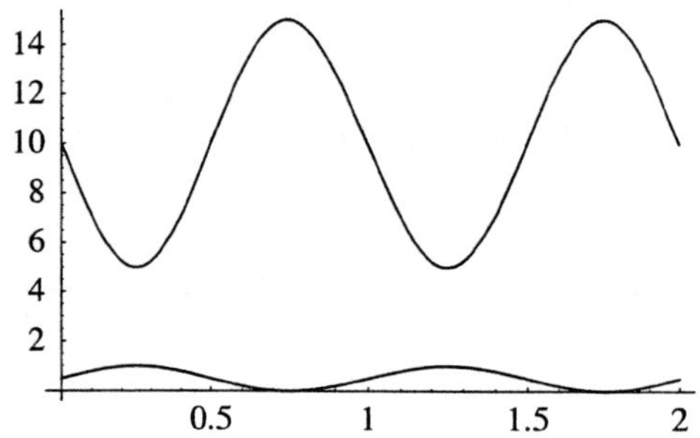


```
<< DiffEqs`DEGraphics`

DEPlot[k[t] p (1 - p/m[t]), {t, 0, 12},
  {p, 0, 12}, InitialPoints → {{0, 3}, {0, 7}, {0, 11}},
  MaxCurves → 3, Gap → .1, ShowField → False];
```

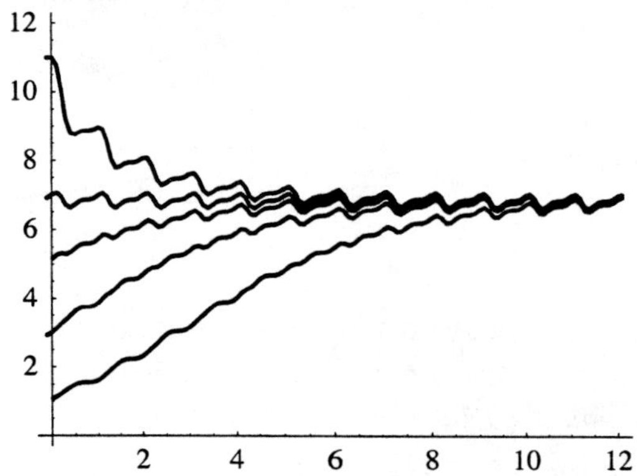

So it appears that all positive solutions approach the same periodic solution as $t \to \infty$. To get a better look at this periodic solution, let's specify an initial point at (0, 6.9):

```
DEPlot[k[t] p (1 - p / m[t]), {t, 0, 2},
  {p, 6, 7.2}, InitialPoints → {{0, 6.9}},
  MaxCurves → 1, ShowField → False, StepSize → .05];
```

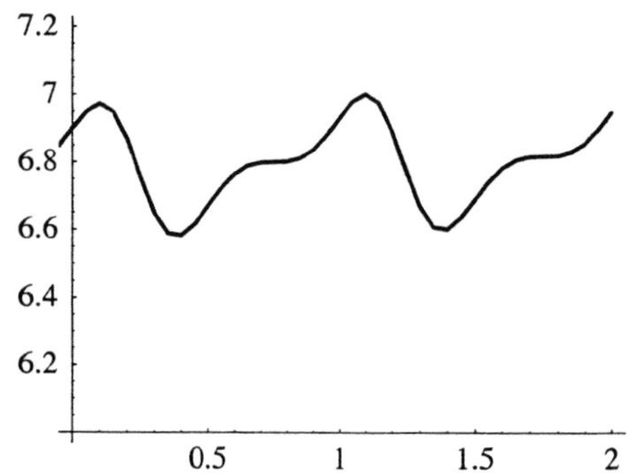

From this we observe that the population peaks in the winter of each year, then declines rapidly and bottoms out in the spring.

■ Exercises

1. Consider a population whose reproductive rate peaks in the spring, while the food supply peaks in both the spring and autumn:

```
k0 = 1; α = .75; m0 = 10; β = 5;
k[t_] := k0 + α Cos[2 π (t - .25)];
m[t_] := m0 + β Cos[4 π (t - .75)];
Plot[{k[t], m[t]}, {t, 0, 2}];
```

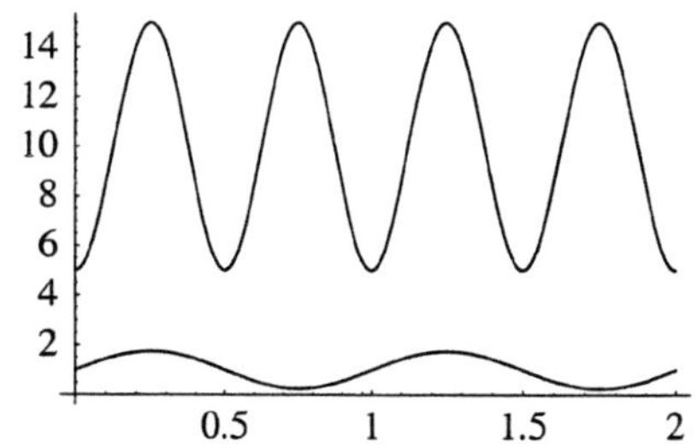

Analyze this model graphically. What do you observe?

2. Experiment with other possible scenarios of seasonality and (hopefully) discover some other interesting or surprising behavior. Can you conjecture any general principles from your observations?

3. Propose and analyze a hypothetical population model of your own that involves *seasonal harvesting*.

Curve Fitting Issues

4. The purpose of this exercise is to look at the way the behavior of the solution of the logistic equation depends upon the parameters P_0, k, and M. We will do this by varying each of these individually while holding the other two fixed. Begin by entering

```
p[p0_, k_, m_][t_] := m p0 / (p0 + (m - p0) e^(-k t))
```

(a) Create and plot a family of functions depending upon p0 with fixed k = .25 and m = 10 as follows:

```
p0Curves = Table[p[p0, .25, 10][t], {p0, 2, 20, 2}];
Plot[Evaluate[p0Curves], {t, 0, 20}, PlotRange → {0, 20}];
```

(b) Create and plot a family of functions depending upon k with fixed p0 = 1 and m = 10 as follows:

```
kCurves = Table[p[1, k, 10][t], {k, 0, 1, .1}];
Plot[Evaluate[kCurves], {t, 0, 20}, PlotRange → {0, 10}];
```

(c) Create and plot a family of functions depending upon m with fixed k = 0.5 and p0 = 1 as follows:

```
mCurves = Table[p[1, .5, m][t], {m, 1, 10, 1}];
Plot[Evaluate[mCurves], {t, 0, 20}, PlotRange → {0, 10}];
```

5 The purpose if this exercise is to demonstrate the sensitivity in the calculation of M from three observations of the population at equally spaced times. Suppose that a population is precisely determined by the logistic function $P(t)$ represented by

```
p[t_] := 40 / (1 + 39 e^(-.5 t))
```

Note that the true limiting value of $P(t)$ is $M = 40$. Now let's suppose that we wish to calculate M from observations P_0, P_1, and P_2 made at equally spaced times t_0, $t_1 = t_0 + h$, and $t_2 = t_0 + 2h$, where $h > 0$. What we would like to do is investigate how M depends upon t_0 and upon h if there is some small random error present in our observations.

First plot the graph of $P(t)$:

```
Plot[p[t], {t, 0, 20}, PlotRange → {0, 45}];
```

Now create a "noisy" version of $P(t)$ by entering

```
noisyP[t_] := p[t] * Random[Real, {1 - error, 1 + error}]
```

In effect, this introduces a random, uniformly distributed, relative error of ± `error`. It is this noisy function from which we will make our "observations." Look at its graph (with `error=` 0.05) by entering

```
error := .05;
Plot[noisyP[t], { t, 0, 20}, PlotRange -> { 0, 45}];
```

Now let's solve for M (and k) as follows:

```
Solve[{ m p0 / (p0 + (m - p0) e^(-k h)) == p1, m p0 / (p0 + (m - p0) e^(-k 2 h)) == p2}, {k, m}]
```

$$\left\{\{m \to 0,\ k \to 0\},\ \left\{m \to -\frac{p1\,(p0\,p1 - 2\,p0\,p2 + p1\,p2)}{-p1^2 + p0\,p2},\ k \to \frac{\text{Log}\left[\frac{p0\,p2 - p1\,p2}{p0\,(p1 - p2)}\right]}{h}\right\}\right\}$$

So we see that M is given in terms of the observations by

$$M = \frac{P_1\,(P_0\,P_1 - 2\,P_0\,P_2 + P_1\,P_2)}{P_1^2 - P_0\,P_2}.$$

Now let's define M as a function of t_0 (i.e., the time when the first observation is made) based on noisy observations:

```
m[t0_] :=
 Module[{p0 = noisyP[t0], p1 = noisyP[t0 + h], p2 = noisyP[t0 + 2 h]},
   p1 (p0 p1 - 2 p0 p2 + p1 p2) / (p1^2 - p0 p2)]
```

For a fixed h, this function depends only upon the time at which the first observation is made (and of course the error in the observations). To make sure everything is working properly, first plot the graph of `m[t]` with $h = 3$ and *exact* observations by entering

```
error := 0; h := 3;
Plot[m[t1], {t1, 0, 20}, PlotRange → {0, 80}];
```

You should see the graph of the constant function $M(t) = 40$, which reflects the fact that *if the observations are exact then it does not matter when they are made*. Now set `error`:= 0.02 and plot again. Then repeat with $h = 2$ and $h = 1$. Interpret and comment on the results.

6. In this exercise we'll explore the problem of determining the value of M from a large number of observations. We will do this by minimizing the *least-squares error* between the logistic model and the observed data.

(a) Begin by defining

```
p[p0_, k_, m_][t_] := m p0 / (p0 + (m - p0) e^(-k t))
```

Now define the particular logistic function

```
exactP[t_] := p[1, .5, 40][t]
```

and a corresponding noisy observation function:

```
noisyP[t_] := exactP[t] * Random[Real, { .95, 1.05}]
```

Obtain plots of these functions by entering

```
curve = Plot[exactP[t], {t, 0, 20}];

noisyCurve = Plot[noisyP[t], { t, 0, 20}, PlotRange → All];
```

(b) Set

```
nobs := 10
```

and then define a list of nobs observations, made at times $t = 0,\ 1,\ \ldots,$ nobs−1 by entering

```
obs = Table[{t, noisyP[t]}, {t, 0, nobs - 1, 1}]
```

Create a plot of these data points by entering

```
obsPlot = ListPlot[obs, PlotStyle → PointSize[.015],
   PlotRange → {{0, nobs + 1}, {0, 40}}];
```

No logistic curve will interpolate the data points; so we will try to find the logistic curve that gives the *best fit* in the sense that the sum of the squares of the errors at the data points is minimized. Assuming that the first observation is correct, this ***least squares error*** is a function of k and M, which we define by

$$\texttt{LSErr[k\_, m\_] :=} \sum_{n=2}^{nobs} (\texttt{p[obs[[1, 2]], k, m][n - 1] - obs[[n, 2]]})^2$$

To minimize LSErr with respect to k and M, we can use FindMinimum:

```
Clear[k, m];
best = FindMinimum[LSErr[k, m], {k, 1}, {m, obs[[nobs, 2]]}]
```

Here we are specifying 1 as an initial guess for k and the last observation as an initial guess for m. How good is the resulting estimate of M?

Now enter

```
k = k /. best[[2, 1]]; m = m /. best[[2, 2]];
```

and create a plot of the predicted logistic curve:

```
prediction = Plot[p[obs[[1, 2]], k, m][t], {t, 0, 20}];
```

Finally Show the predicted curve, the true curve, and the observation points on the same graph:

```
Show[prediction, curve, obsPlot];
```

For the sake of comparison, use the formula in Exercise 2 for M to define

```
m3[n_] :=
 Module[{p0 = obs[[n, 2]], p1 = obs[[n + 1, 2]], p2 = obs[[n + 2, 2]]},
       p1 (2 p0 p2 - p1 p0 - p1 p2) / (p0 p2 - p1^2)]
```

and create a list of predictions of M based on three consecutive observations by entering

```
mlist = Table[m3[t], {t, 1, nobs - 2}]
```

How many of the numbers in mlist give a reasonably good prediction of $M = 40$?

(c) Repeat 2b a few times. Since the observations contain random error, the results should vary somewhat, but hopefully not too much.

(d) Repeat the process once each with nobs:= 8, 6, and 4. What do you observe?

3.4 The Spruce Budworm Problem

An equation that has become quite popular in recent years is the following model of spruce budworm populations in Canadian forests. [*Nature*, Vol. 269, p. 471 (1977)]

$$\frac{dw}{dt} = k\,w\left(1 - \frac{w}{M}\right) - \rho\,\frac{w^2}{\eta^2 + w^2}\,.$$

Note that the first term on the right side is just the usual logistic growth term, where k is the intrinsic growth rate and M is the environmental carrying capacity. The second term models predation by a population of birds, which is assumed constant. The idea is that if the budworm population is small, the birds won't bother with the budworms and will seek out other food that provides more nourishment for their effort. Thus the birds will feed heavily on the budworms only if the budworm population is large enough; however, the birds can eat only so many budworms, regardless of how many there are. The following plot indicates how the predation term takes these factors into account:

```
Plot[w^2/(η^2 + w^2) /. η → .3, {w, 0, 1}, PlotRange → All, Ticks → None,
  AxesLabel → {"budworm density", "predation rate" }];
```

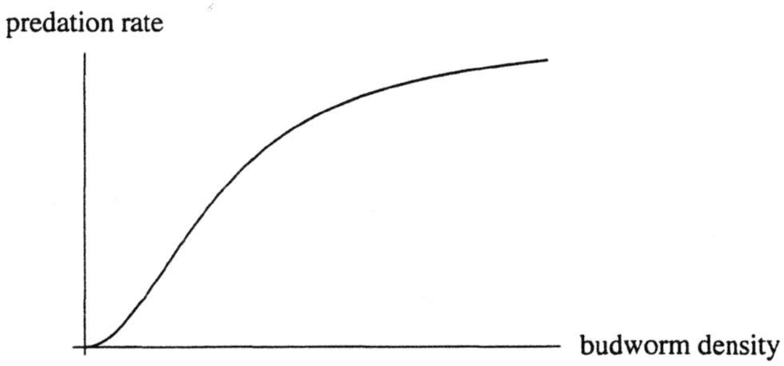

The interesting effect of this type of predation is illustrated as follows. For some parameter values, the equation has three equilibrium solutions, including one that is small and stable.

```
Plot[w (1 - w) - .2 w^2/(.01 + w^2), {w, 0, .8}, Ticks → None];
```

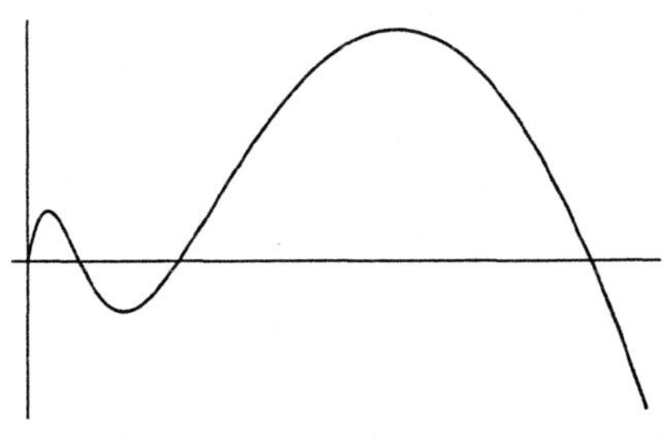

For other parameter values, there is only one equilibrium solution, which is large and stable.

```
Plot[1 - w - .2 w / (.02 + w^2), {w, 0, 1}, Ticks → None, PlotRange → All];
```

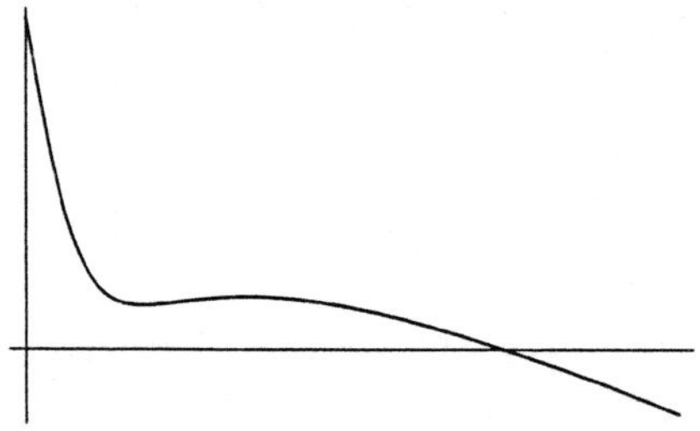

As parameter values change, the small, stable equilibrium solution can vanish and be suddenly replaced by a much larger equilibrium solution. This is the phenomenon known as the "outbreak threshold."

The carrying capacity M and the parameter η in the predation term each depend upon the foliage density of the spruce trees. This is because greater foliage density provides more food for the budworms *and* makes it more difficult for the birds to find the budworms. So it is reasonable to assume that M and η each depend linearly on the foliage density, which implies that M and η are related to each other in a linear way.

The following command creates a sequence of graphs that illustrate how the equilibrium solutions can change along with a parameter. Half of the frames produced are shown below.

```
Do[Plot[5 w ((1 - w / m) - .2 w / (η^2 + w^2)) /. {m → .25 + s, η → .1 s},
  {w, 0, 1.5}, PlotRange → {-1, 1}, Ticks → None,
  MaxBend → 1], {s, .5, 1.25, .075}]
```

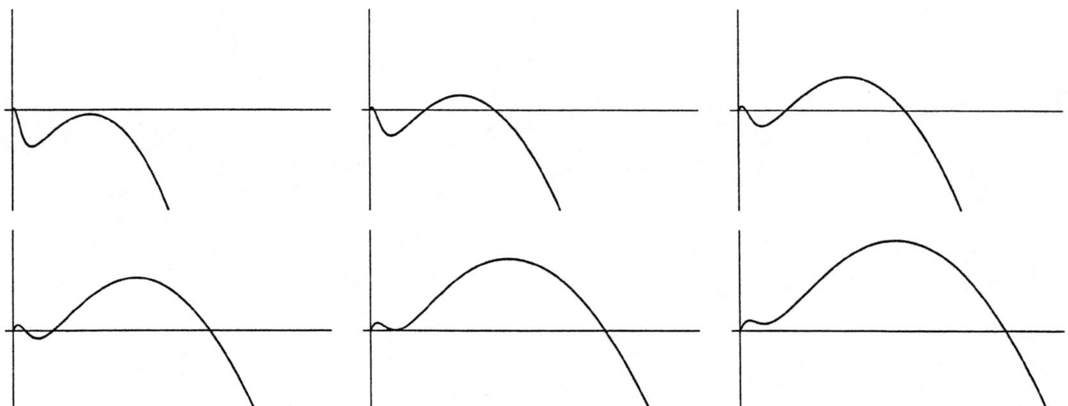

Corresponding solution curves, each with a small initial budworm population, are plotted as follows. Again, half of the resulting frames are shown.

```
<< DiffEqs`DEGraphics`
```

```
Do[DEPlot[Evaluate[10 w ((1 - w / m) - .2 w/(η^2 + w^2)) /. {m → .25 + s, η → .1 s}],
  {t, 0, 4}, {w, 0, 1.5}, InitialPoints → {{0, .1}}, StepSize → .025,
  ScaleArrows → .5, Ticks → None], {s, .5, 1.25, .075}]
```

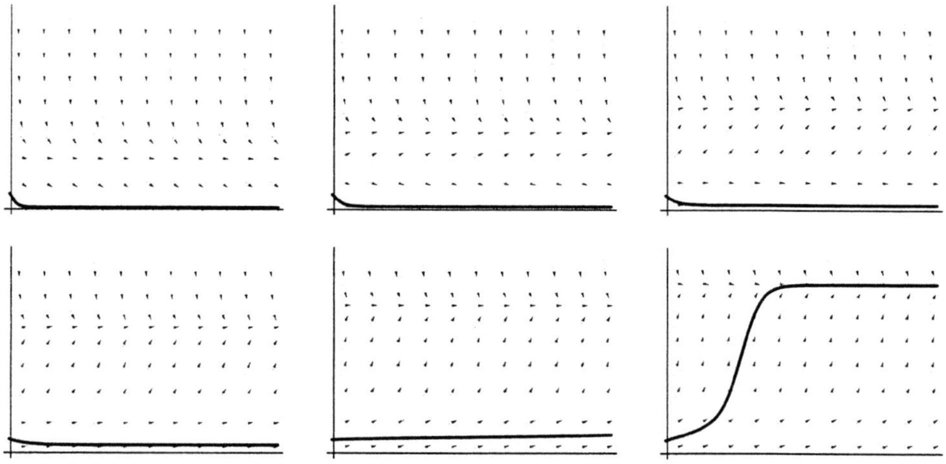

These plots illustrate the sudden "blooming" of the budworm population that can occur as a parameter crosses an outbreak threshold. Now let's create a plot in which the outbreak threshold is crossed as M and η each increase linearly in time (along with the foliage density).

```
DEPlot[Evaluate[10 w ((1 - w / m) - .2 w/(η^2 + w^2)) /. {m → .25 + .5 t, η → .01 + .05 t
  {t, 0, 4}, {w, 0, 2}, ScaleArrows → .67,
  InitialPoints → {{0, .05}}, StepSize → .01];
```

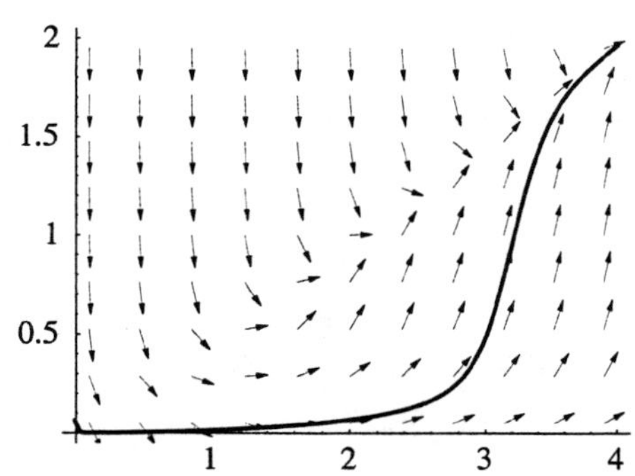

After such an outbreak occurs, trees would be quickly stripped of their leaves by the budworms, which would cause rapid decrease of M and η. So it is easy to imagine that a type of periodic behavior can occur. In fact, budworm outbreaks tend to occur in Canadian forests about every 29 years.

In Section 5.3, we will revisit the spruce budworm problem and include in it a second differential equation that governs the foliage density.

3.5 A Half-Wave Rectifier Circuit

A simple half-wave rectifier circuit consisting of a diode, a capacitor, and an alternating voltage source can be viewed as an RC circuit with a nonlinear resistor (the diode). The purpose of the circuit is to convert an alternating current source (AC) to a *direct* current source (DC).

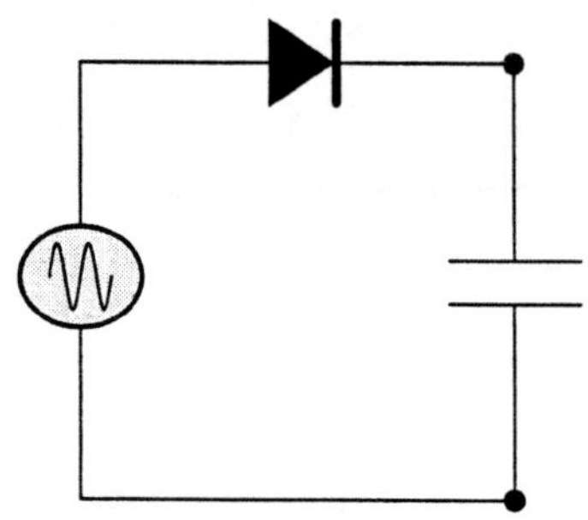

Assume that the voltage source is $v_s(t) = A \sin \omega t$. Then Kirchhoff's voltage law gives the equation

$$V_d + C^{-1} Q = A \sin\omega t,$$

where V_d is the voltage across the diode and Q is the charge on the capacitor. Since the current in the loop is $I = Q'$, differentiating with respect to t produces the differential equation

$$V_d' + C^{-1}\, \gamma(V_d) = \omega\, A\cos\omega t,$$

where $I = \gamma(V_d)$ describes the diode's *voltage-current characteristic*. For an *ideal* diode,

$$I = \gamma(V_d) = I_0(e^{\eta V_d} - 1),$$

where I_0 is the *saturation current* and η is a temperature-dependent parameter. Typical parameter values at room temperature are $I_0 = 10^{-5}$ amps and $\eta = 20$ volts^{-1}. So we'll define

```
γ[v_] := 10^-5 (e^(20 v) - 1)
```

The following graph of $\gamma(V_d)$ shows that the diode has a very low resistance when the voltage across it is around 0.7 volts or higher and a very high resistance otherwise. *Thus the diode behaves like a valve that allows current to flow in only one direction.*

```
Plot[γ[v], {v, -1, .7}, AxesLabel → {voltage, current},
  PlotRange → All, Ticks → {{-1, -.5, .5}, {10}}];
```

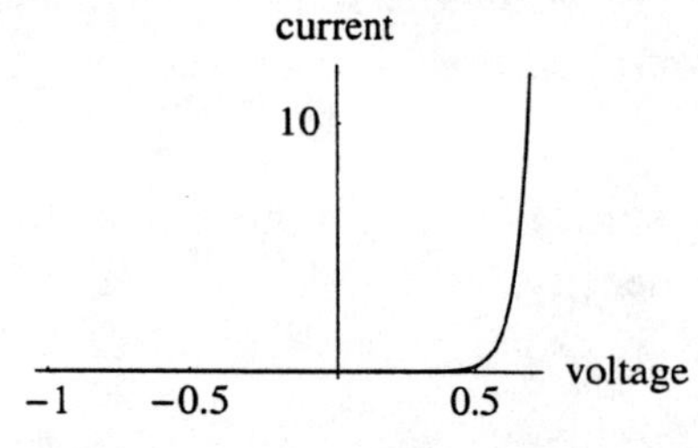

However, a magnified view reveals a small negative “leakage” current (about $-I_0$ amps) when the voltage across the diode is negative:

```
Plot[γ[v], {v, -1, .15},
  AxesLabel → {voltage, current}, PlotRange → All];
```

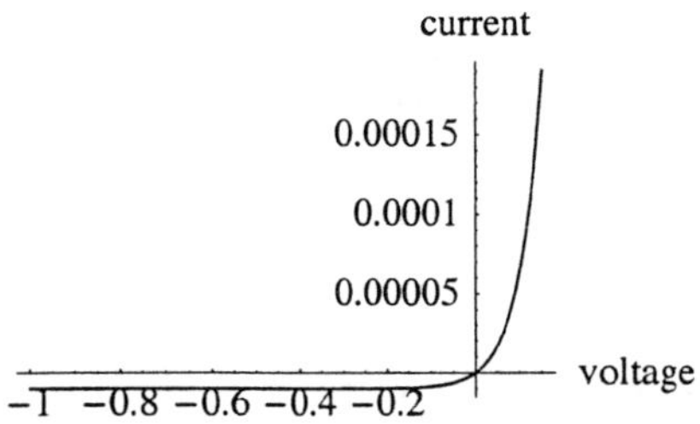

The quantities of primary interest here are the current in the loop and the voltage across the capacitor. Each of these can be recovered easily from V_d as follows:

$$I = \gamma(V_d) \text{ and } V_C = C\,(A \sin\omega t - V_d).$$

Now let’s see what V_d looks like with $C = A = \omega = 1$ and $V_d(0) = 0$. For this we’ll use `NDSolve`.

```
tmax = 10 π; c = 1; a := 1; ω = 1;
Vd = v[t] /. NDSolve[{v'[t] + c^-1 γ[v[t]] == a ω Cos[ω t], v[0] == 0},
    v[t], {t, 0, tmax}, MaxSteps → ∞][[1]]
InterpolatingFunction[{{0., 31.4159}}, <>][t]

Plot[Evaluate[Vd], {t, 0, tmax}, PlotRange → All];
```

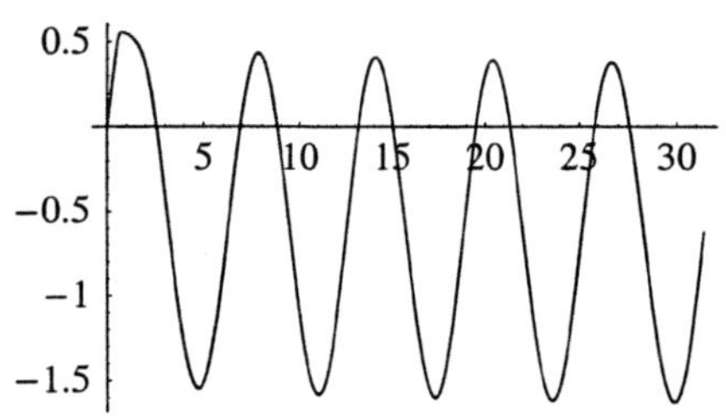

Having solved for V_d, we can now plot the current through the diode:

```
Plot[Evaluate[γ[Vd]], {t, 0, tmax}, PlotRange → All];
```

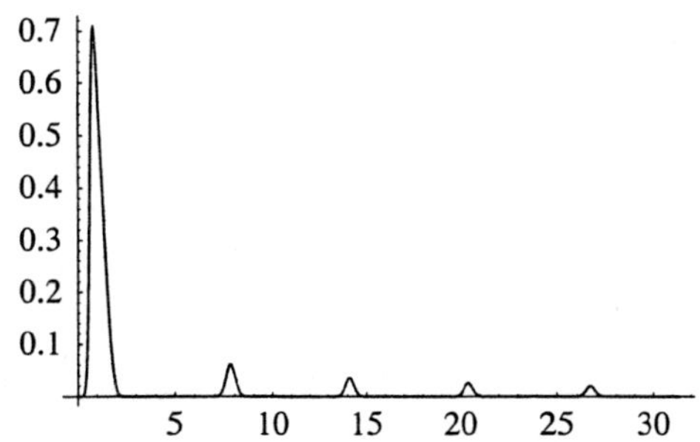

and the voltage across the capacitor:

```
Plot[Evaluate[c (a Sin[ω t] - Vd)], {t, 0, tmax}, PlotRange → All];
```

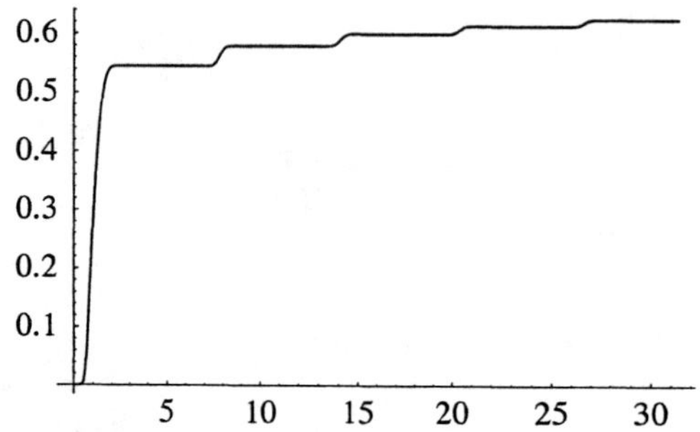

So we see that after a short time the circuit provides a near-constant positive charge to the capacitor, which could serve as a source of direct current. (We will revisit this in Section 5.4.)

Chapter 4

Approximation of Solutions

4.1 Taylor Polynomials

Consider the initial-value problem

$$y' = f(t, y), \quad y(t_0) = y_0,$$

where f has continuous partial derivatives of all orders. Local approximations to the solution can be obtained in the form of Taylor polynomials by successive differentiation of f. On its interval of convergence, the corresponding Taylor series is in fact a solution of the initial-value problem. In the following example, we illustrate the process with the construction of a fourth-degree Taylor polynomial.

Example 4.1.1

Consider the initial-value problem

$$y' = \sin(t) - \cos(y), \quad y(0) = 0.$$

We begin by defining

```
f[t_, y_] := Sin[t] - Cos[y];
y0 := 0;
```

Then we build a list of the first four derivatives of the solution.

```
derivs = Join[{y[t]}, Table[D[f[t, y[t]], {t, k - 1}], {k, 1, 4}]];
derivs // TableForm
```

```
y[t]
-Cos[y[t]] + Sin[t]
Cos[t] + Sin[y[t]] y'[t]
-Sin[t] + Cos[y[t]] y'[t]^2 + Sin[y[t]] y''[t]
-Cos[t] - Sin[y[t]] y'[t]^3 + 3 Cos[y[t]] y'[t] y''[t] + Sin[y[t]] y^(3)[t]
```

Next we substitute $t = 0$ and $y = y_0$ into the list of derivatives.

```
derivs = derivs //. {t → 0, y[0] → y0}
```

```
{0, -1, 1, y'[0]^2, -1 + 3 y'[0] y''[0]}
```

The first three entries of the list are now the values of $y(0)$, $y'(0)$, and $y''(0)$, as desired. Now we substitute those values—as well as higher-order derivative values as they become available—through the rest of the list:

```
Do[yk = D[y[t], {t, k}] /. t → 0;
  derivs = derivs //. yk → derivs[[k + 1]],
  {k, 1, 4}];
derivs
```

{0, -1, 1, 1, -4}

Finally we use the derivative values to build the degree-4 Taylor polynomial.

$$\sum_{j=0}^{4} \frac{\text{derivs}[[j+1]]}{j!} t^j$$

$$-t + \frac{t^2}{2} + \frac{t^3}{6} - \frac{t^4}{6}$$

- The `DETools` package provides a function named `SeriesDSolve` that encapsulates the process illustrated in Example 4.1.1. The computations detailed above can be done simply as follows:

```
f[t_, y_] := Sin[t] - Cos[y];
y0 := 0;

<< DiffEqs`DETools`
SeriesDSolve[{y'[t] == f[t, y[t]], y[0] == y0}, y[t], {t, 0, 4}]
```

$$-t + \frac{t^2}{2} + \frac{t^3}{6} - \frac{t^4}{6}$$

The degree-7 Taylor polynomial, for instance, is

```
SeriesDSolve[{y'[t] == f[t, y[t]], y[0] == y0}, y[t], {t, 0, 7}]
```

$$-t + \frac{t^2}{2} + \frac{t^3}{6} - \frac{t^4}{6} - \frac{t^5}{60} + \frac{41 t^6}{720} - \frac{31 t^7}{2520}$$

A list containing the Taylor polynomials of degree 3 through 7 is computed and plotted as follows:

```
taylorpolys = Table[SeriesDSolve[
    {y'[t] == f[t, y[t]], y[0] == y0}, y[t], {t, 0, n}], {n, 3, 7, 1}];
taylorpolys // TableForm
```

$$-t + \frac{t^2}{2} + \frac{t^3}{6}$$
$$-t + \frac{t^2}{2} + \frac{t^3}{6} - \frac{t^4}{6}$$
$$-t + \frac{t^2}{2} + \frac{t^3}{6} - \frac{t^4}{6} - \frac{t^5}{60}$$
$$-t + \frac{t^2}{2} + \frac{t^3}{6} - \frac{t^4}{6} - \frac{t^5}{60} + \frac{41 t^6}{720}$$
$$-t + \frac{t^2}{2} + \frac{t^3}{6} - \frac{t^4}{6} - \frac{t^5}{60} + \frac{41 t^6}{720} - \frac{31 t^7}{2520}$$

```
Plot[Evaluate[taylorpolys], {t, -2, 2}];
```

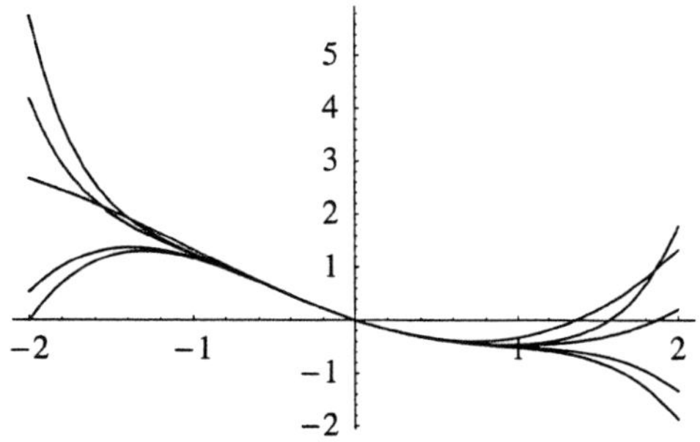

■ Exercises

For each of the following initial-value problems, use `SeriesDSolve` to create a list of the degree-3 through degree-7 Taylor polynomials of the solution. Plot the results on the interval $[-2, 2]$.

1. $y' = \cos(t\,y), \quad y(0) = 1$
2. $y' = \cos t - \sin y, \quad y(0) = 0$
3. $y' = \cos(t\sin(y)), \quad y(0) = 0$
4. $y' = \cos y - y, \quad y(0) = 0$
5. $y' = t - y^3 + y, \quad y(0) = 1$
6. $y' = e^{-ty}, \quad y(0) = 1$

4.2 Picard Iteration

Consider the initial-value problem

$$y' = f(t, y), \quad y(t_0) = y_0,$$

where f is assumed only to be continuous near the point (t_0, y_0). This problem is equivalent to the integral equation

$$y(t) = y_0 + \int_{t_0}^{t} f(s, y(s))\,ds.$$

Picard iteration generates a sequence of approximate local solutions by means of the iterative formula

$$y_{k+1}(t) = y_0 + \int_{t_0}^{t} f(s, y_k(s))\,ds, \quad k = 1, 2, 3, \ldots .$$

If the sequence $y_k(t)$ converges to a number $y(t)$ for each t in some interval containing t_0, then the function y will satisfy the integral equation and the initial-value problem on that interval.

Let's assume that $t_0 = 0$. A step of Picard iteration is then defined by

```
Clear[f];

Picardstep[y_] := y0 + ∫_0^t f[s, (y /. t → s)] ds
```

Example 4.2.1

For the initial-value problem

$$y' = \sin t - y^2, \quad y(0) = 1,$$

the associated integral equation is

$$y(t) = 1 + \int_0^t (\sin(s) - y(s)^2)\, ds.$$

To generate the approximate solutions given by Picard iteration, we'll first define

```
y0 = 1; f[t_, y_] := Sin[t] - y^2
```

A good choice for the initial approximation y_1 is the linear approximation at $t = 0$.

```
y[1][t_] = y0 + t f[0, y0]
```

$$1 - t$$

The first three iterations are computed as follows. (The last step may take several seconds.)

```
y[2][t_] = Picardstep[y[1][t]] // Simplify
```

$$2 - t + t^2 - \frac{t^3}{3} - \text{Cos}[t]$$

```
y[3][t_] = Picardstep[y[2][t]] // Simplify
```

$$-\frac{9\,t}{2} + 2\,t^2 - \frac{5\,t^3}{3} + \frac{5\,t^4}{6} - \frac{t^5}{3} + \frac{t^6}{9} - \frac{t^7}{63} + (1 + 4\,t - 2\,t^2)\,\text{Cos}[t] - \frac{2}{3}\,t\,(-3 - 3\,t + t^2)\,\text{Sin}[t] - \frac{1}{4}\,\text{Sin}[2\,t]$$

```
y[4][t_] = Picardstep[y[3][t]] // Simplify
```

$$\frac{19340269}{324} - \frac{17\,t}{32} - 2\,t^2 - \frac{113\,t^3}{12} + \frac{11\,t^4}{2} - \frac{13\,t^5}{3} + \frac{31\,t^6}{12} - \frac{4\,t^7}{3} + \frac{23\,t^8}{36} - \frac{67\,t^9}{252} + \frac{187\,t^{10}}{1890} - \frac{2\,t^{11}}{63} + \frac{19\,t^{12}}{2268} - \frac{t^{13}}{567} + \frac{t^{14}}{3969} - \frac{t^{15}}{59535} + \frac{1}{1008}\,((-15372 + 6867\,t + 28728\,t^2 - 4494\,t^3 - 5040\,t^4 + 1050\,t^5 - 28\,t^6 + 4\,t^7)\,\text{Cos}[2\,t]) - \frac{61}{324}\,\text{Cos}[3\,t] - \frac{4}{9}\,t\,\text{Cos}[3\,t] + \frac{2}{9}\,t^2\,\text{Cos}[3\,t] + 86668\,\text{Sin}[t] - \frac{119349}{2}\,t\,\text{Sin}[t] - \frac{86667}{2}\,t^2\,\text{Sin}[t] + \frac{59719}{6}\,t^3\,\text{Sin}[t] + \frac{10849}{3}\,t^4\,\text{Sin}[t] - \frac{1502}{3}\,t^5\,\text{Sin}[t] - \frac{1070}{9}\,t^6\,\text{Sin}[t] + \frac{694}{63}\,t^7\,\text{Sin}[t] + \frac{52}{21}\,t^8\,\text{Sin}[t] - \frac{52}{189}\,t^9\,\text{Sin}[t] + \text{Cos}[t]\left(-\frac{238703}{4} - 86668\,t + 29842\,t^2 + \frac{43366\,t^3}{3} - 2496\,t^4 - \frac{2158\,t^5}{3} + \frac{244\,t^6}{3} + \frac{374\,t^7}{21} - \frac{104\,t^8}{63} - \frac{40\,t^9}{189} + \frac{4\,t^{10}}{189} + \left(-61\,t + \frac{75\,t^2}{8} - \frac{7\,t^5}{6}\right)\text{Sin}[t]\right) - \frac{117}{32}\,\text{Sin}[2\,t] + 16\,t^3\,\text{Sin}[2\,t] - \frac{157}{48}\,t^4\,\text{Sin}[2\,t] + \frac{7}{72}\,t^6\,\text{Sin}[2\,t] + \frac{4}{27}\,\text{Sin}[3\,t] - \frac{17}{54}\,t\,\text{Sin}[3\,t] - \frac{1}{6}\,t^2\,\text{Sin}[3\,t] + \frac{1}{18}\,t^3\,\text{Sin}[3\,t] + \frac{1}{128}\,\text{Sin}[4\,t]$$

A plot of these (without y_1) is created as follows:

```
approxsolns = Table[y[n][t], {n, 2, 4}];
Plot[Evaluate[approxsolns], {t, -1, 1.5}, PlotRange → {-1, 5}];
```

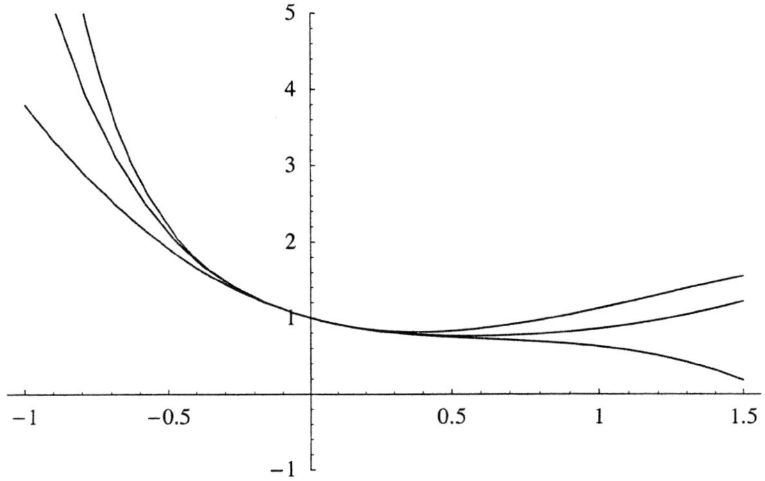

■ Picard-Taylor Iteration

For any but relatively simple equations (even the one in Example 4.2.1) Picard iteration quickly becomes impractical because of the complexity of the integrals involved. However, if we approximate the integrands with Taylor polynomials of increasing degree, we arrive at a rather efficient method for generating Taylor-polynomial approximations to the solution of the initial-value problem. (This assumes, of course, that f has continuous partial derivatives of all orders.) A step of this technique, which we'll call *Picard-Taylor iteration*, is defined as follows:

```
PicardTaylorstep[n_, y_] :=
  y0 + ∫_0^t Normal[Series[f[s, (y /. t → s)], {s, 0, n}]] ds
```

Example 4.2.2

Consider the same initial-value problem as in Example 4.2.1.

```
y0 = 1; f[t_, y_] := Sin[t] - y^2
```

We'll use the same initial approximation:

```
p[1][t_] = y0 + t f[0, y0]
1 - t
```

To generate polynomial approximations up to degree 15, we enter

```
Do[p[n][t_] = PicardTaylorstep[n - 1, p[n - 1][t]], {n, 2, 15}]
```

For instance, the first 8 of these are

```
Table[p[n][t], {n, 1, 8}] // TableForm
```

$1 - t$

$1 - t + \frac{3t^2}{2}$

$1 - t + \frac{3t^2}{2} - \frac{4t^3}{3}$

$1 - t + \frac{3t^2}{2} - \frac{4t^3}{3} + \frac{11t^4}{8}$

$1 - t + \frac{3t^2}{2} - \frac{4t^3}{3} + \frac{11t^4}{8} - \frac{23t^5}{15}$

$1 - t + \frac{3t^2}{2} - \frac{4t^3}{3} + \frac{11t^4}{8} - \frac{23t^5}{15} + \frac{131t^6}{80}$

$1 - t + \frac{3t^2}{2} - \frac{4t^3}{3} + \frac{11t^4}{8} - \frac{23t^5}{15} + \frac{131t^6}{80} - \frac{551t^7}{315}$

$1 - t + \frac{3t^2}{2} - \frac{4t^3}{3} + \frac{11t^4}{8} - \frac{23t^5}{15} + \frac{131t^6}{80} - \frac{551t^7}{315} + \frac{25267t^8}{13440}$

The following plot shows the approximations of degrees 5, 10, and 15.

```
Plot[{p[5][t], p[10][t], p[15][t]}, {t, -1, 1}];
```

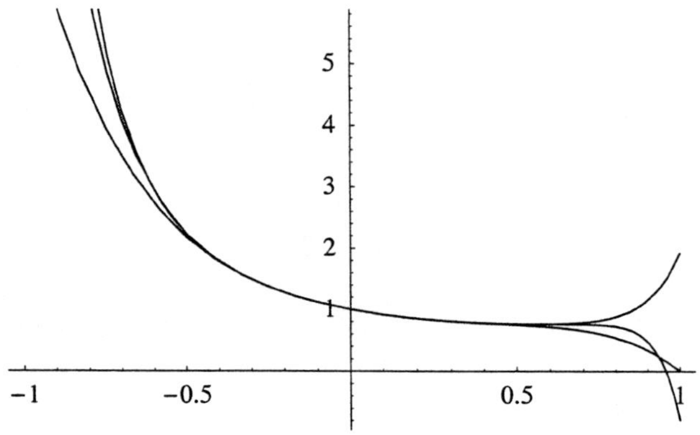

■ Exercises

1. Consider the problem

$$y' = -y^2 \sin t, \quad y(0) = 1 .$$

(a) Find the exact solution.

(b) Do 4 steps of Picard iteration, starting with $y_1(t) = 1$. Compare the last iterate y_5 with the exact solution by graphing them together on $[-2, 2]$.

(c) Use Picard-Taylor iteration to generate a degree-20 polynomial approximation. (Start with $p_1(t) = 1$.) Graph the result along with the exact solution and y_5 from part (b) (again on $[-2, 2]$). Why do you think y_5 does a better job of approximating the exact solution?

2. Consider the problem

$$y' = 1 - t\,y^2, \quad y(0) = 0 .$$

(a) Find the exact solution with `DSolve`. (Note that the solution is given in terms of "special functions.")

(b) Do 3 steps of Picard iteration, starting with $y_1(t) = t$. Compare the last iterate y_4 with the exact solution by graphing them together on $[-1.5, 2]$.

(c) Use Picard-Taylor iteration to generate a polynomial approximation of the same degree as y_4 from part (a). (Start with $p_1(t) = t$.) Graph the result on $[-1.5, 3]$ along with the exact solution and y_4. Which of the two polynomials does a better job of approximating the solution?

(d) Do 3 steps of Picard iteration, starting with $z_1(t) = \sin t$. Compare the last iterate z_4 with the exact solution by graphing them together on $[-1.5, 2]$. Add the graph of y_4 from part (b) to the plot? Which does the better job, y_4 or z_4?

4.3 Numerical Methods: Euler, Heun, and RK4

In this section we will look at three basic numerical methods for approximating the solution of

$$y' = f(t, y), \quad y(t_0) = y_0,$$

where f is continuous near the point (t_0, y_0). These three methods are Euler's method, Heun's method, and the fourth-order Runge-Kutta method.

The *Mathematica* command `NestList` provides a convenient mechanism for implementing these iterative numerical procedures. This will be illustrated in the examples that follow.

```
? NestList
```

```
NestList[f, expr, n] gives a list of the
   results of applying f to expr 0 through n times. More...
```

■ Euler's Method

A step of Euler's method can be defined as

```
Eulerstep[{t_, y_}] := {t + h, y + h f[t, y]}
```

Two steps (for example) of Euler's method can be done with the help of `NestList` as follows:

```
NestList[Eulerstep, {t0, y0}, 2] // TableForm
```

```
t0          y0
h + t0      y0 + h f[t0, y0]
2 h + t0    y0 + h f[t0, y0] + h f[h + t0, y0 + h f[t0, y0]]
```

Example 4.3.1

To apply Euler's method to the initial value problem

$$y' = -t\,y + \cos(t)\,y^2, \quad y(0) = 1,$$

we would first define the Euler step as above and then define a stepsize h and the function f.

```
h = 0.1; f[t_, y_] := -t y + Cos[t] y^2
```

Now we use `NestList` to apply the function `Eulerstep` iteratively, beginning with the starting point (0, 1). The following generates a list of thirty points that approximate points on the graph of the solution $y(t)$.

```
Eulerpoints = NestList[Eulerstep, {0, 1}, 30]

{{0, 1}, {0.1, 1.1}, {0.2, 1.2094}, {0.3, 1.32856}, {0.4, 1.45732}, {0.5, 1.59464},
 {0.6, 1.73807}, {0.7, 1.88311}, {0.8, 2.02251}, {0.9, 2.1457}, {1., 2.23878},
 {1.1, 2.28571}, {1.2, 2.27126}, {1.3, 2.18564}, {1.4, 2.02929}, {1.5, 1.81518},
 {1.6, 1.56621}, {1.7, 1.30846}, {1.8, 1.06396}, {1.9, 0.846728}, {2., 0.662671},
 {2.1, 0.511863}, {2.2, 0.391144}, {2.3, 0.296089}, {2.4, 0.222147}, {2.5, 0.165193},
 {2.6, 0.121708}, {2.7, 0.088795}, {2.8, 0.0641075}, {2.9, 0.0457702}, {3., 0.0322934}}
```

The resulting approximate points can be plotted with `ListPlot`.

```
ListPlot[Eulerpoints, PlotStyle → PointSize[.01]];
```

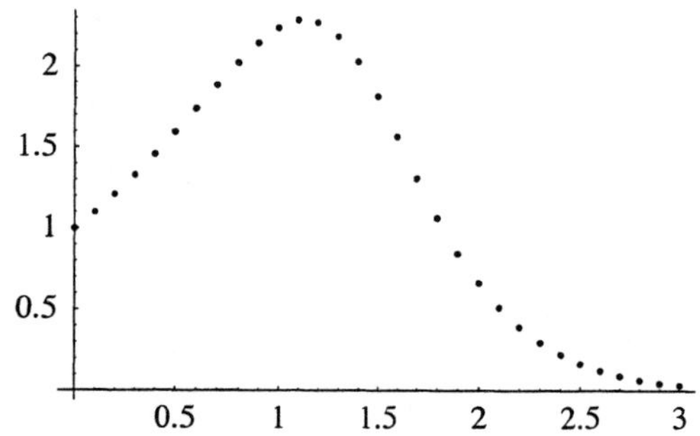

By setting `PlotJoined → True` we get a piecewise-linear interpolant of the computed points.

```
ListPlot[Eulerpoints, PlotJoined → True];
```

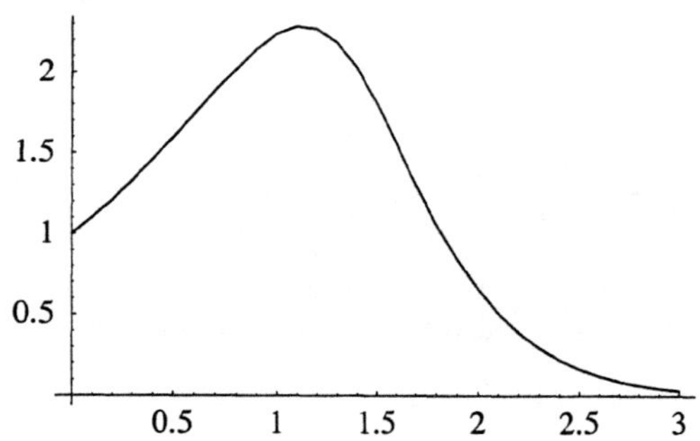

■ Heun's Method

A step of Heun's method (a.k.a. *improved Euler*) is defined by

```
Heunstep[{t_, y_}] := Module[{k1, k2},
  k1 = f[t, y];
  k2 = f[t + h, y + h k1];
  {t + h, y + h / 2 (k1 + k2)}]
```

With this defined, we can then use `NestList` and `ListPlot` in the same way as we did with Euler's method.

Example 4.3.2

Consider the same initial-value problem as in Example 4.3.1. Heun's method gives the following approximate points, which produce the graph shown below.

```
h = 0.1; f[t_, y_] := -t y + Cos[t] y^2
```

```
Heunpoints = NestList[Heunstep, {0, 1}, 30]
```

```
{{0, 1}, {0.1, 1.1047}, {0.2, 1.22009}, {0.3, 1.34668}, {0.4, 1.48433},
 {0.5, 1.63169}, {0.6, 1.78551}, {0.7, 1.9397}, {0.8, 2.08444}, {0.9, 2.20573},
 {1., 2.28632}, {1.1, 2.30871}, {1.2, 2.26038}, {1.3, 2.13923}, {1.4, 1.95598},
 {1.5, 1.73133}, {1.6, 1.4894}, {1.7, 1.25124}, {1.8, 1.03136},
 {1.9, 0.8375}, {2., 0.672084}, {2.1, 0.5342}, {2.2, 0.421179},
 {2.3, 0.32968}, {2.4, 0.25631}, {2.5, 0.197942}, {2.6, 0.151833},
 {2.7, 0.115655}, {2.8, 0.0874601}, {2.9, 0.0656409}, {3., 0.0488802}}
```

```
ListPlot[Heunpoints, PlotJoined → True];
```

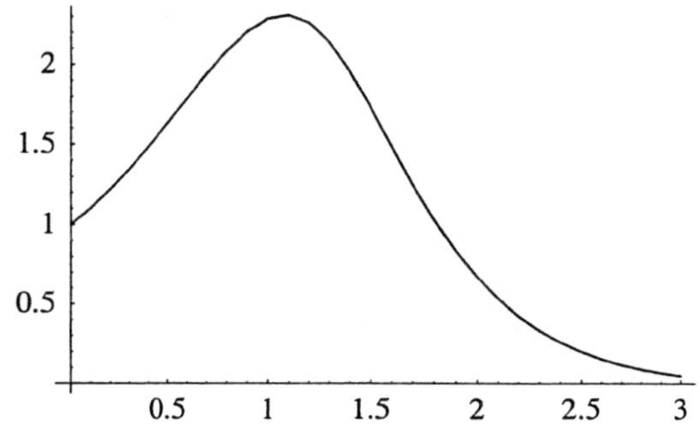

■ Fourth-Order Runge-Kutta (RK4)

A step of the fourth-order Runge-Kutta method is defined by

```
RK4step[{t_, y_}] := Module[{k1, k2, k3, k4},
  k1 = f[t, y];
  k2 = f[t + h / 2, y + h / 2 k1];
  k3 = f[t + h / 2, y + h / 2 k2];
  k4 = f[t + h, y + h k3];
  {t + h, y + h / 6 (k1 + 2 k2 + 2 k3 + k4)}]
```

Example 4.3.3

Again, let's use the same initial-value problem as in Example 4.3.1. RK4 gives the following approximate points, which produce the graph that comes after.

```
h = 0.1; f[t_, y_] := -t y + Cos[t] y^2
```

```
RK4points = NestList[RK4step, {0, 1}, 30]
{{0, 1}, {0.1, 1.10516}, {0.2, 1.22122}, {0.3, 1.34875},
 {0.4, 1.48768}, {0.5, 1.63674}, {0.6, 1.79272}, {0.7, 1.94949},
 {0.8, 2.09711}, {0.9, 2.22133}, {1., 2.30451}, {1.1, 2.3288},
 {1.2, 2.28135}, {1.3, 2.15986}, {1.4, 1.97497}, {1.5, 1.74751},
 {1.6, 1.50196}, {1.7, 1.25994}, {1.8, 1.03648}, {1.9, 0.839638},
 {2., 0.671981}, {2.1, 0.532556}, {2.2, 0.418579}, {2.3, 0.326571},
 {2.4, 0.253015}, {2.5, 0.194678}, {2.6, 0.148742}, {2.7, 0.112819},
 {2.8, 0.0849222}, {2.9, 0.0634155}, {3., 0.046963}}

ListPlot[RK4points, PlotJoined → True];
```

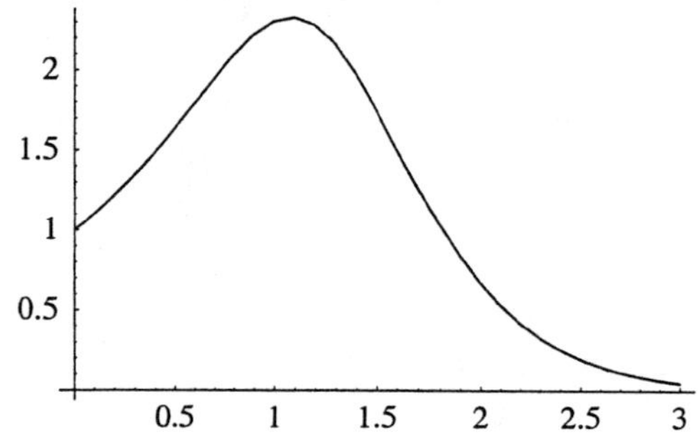

■ Building a Table

In order to compare our numerical results with an exact solution, let's look at an initial-value problem for which the exact solution is relatively easy to find:

$$y' = y(1-y), \quad y(0) = 1/10.$$

We can solve this problem with `BernoulliDSolve` from the `DETools` package.

```
Needs["DiffEqs`DETools`"];
exacty[t_] = BernoulliDSolve[
   {y'[t] == y[t] (1 - y[t]), y[0] == 1 / 10}, y[t], t] // Simplify
```

$$\frac{e^t}{9 + e^t}$$

We want to build a table with columns showing the exact values of the solution and the Euler, Heun, and RK4 approximations. For this we will combine the functions previously defined into a single step that we can apply iteratively to compute the rows of our table.

```
tablestep[{t_, v1_, v2_, v3_, v4_}] :=
 {t + h, exacty[t + h], Eulerstep[{t, v2}][[2]],
  Heunstep[{t, v3}][[2]], RK4step[{t, v4}][[2]]}
```

Now we create a table of values as follows:

```
h = 0.1; f[t_, y_] := y (1 - y)
```

```
tablevals = NestList[tablestep, {0, .1, .1, .1, .1}, 20]
```

```
{{0, 0.1, 0.1, 0.1, 0.1}, {0.1, 0.109367, 0.109, 0.109356, 0.109367},
 {0.2, 0.119495, 0.118712, 0.119471, 0.119495},
 {0.3, 0.130423, 0.129174, 0.130386, 0.130423}, {0.4, 0.142189, 0.140423, 0.142137, 0.142189},
 {0.5, 0.154828, 0.152493, 0.15476, 0.154828}, {0.6, 0.16837, 0.165417, 0.168283, 0.16837},
 {0.7, 0.18284, 0.179222, 0.182734, 0.18284}, {0.8, 0.198257, 0.193933, 0.198131, 0.198257},
 {0.9, 0.214632, 0.209565, 0.214486, 0.214632}, {1., 0.231969, 0.22613, 0.231801, 0.231969},
 {1.1, 0.25026, 0.243629, 0.250069, 0.25026}, {1.2, 0.269487, 0.262056, 0.269274, 0.269487},
 {1.3, 0.289621, 0.281395, 0.289385, 0.289621}, {1.4, 0.31062, 0.301616, 0.310361, 0.310619},
 {1.5, 0.332428, 0.32268, 0.332148, 0.332428}, {1.6, 0.354979, 0.344536, 0.354678, 0.354979},
 {1.7, 0.378193, 0.367119, 0.377873, 0.378193}, {1.8, 0.401979, 0.390353, 0.401641, 0.401979},
 {1.9, 0.426236, 0.414151, 0.425881, 0.426236}, {2., 0.450853, 0.438414, 0.450483, 0.450853}}
```

Using `TableForm`, we now display several rows of the table. Notice the use of the `TableHeadings` option.

```
TableForm[tablevals[[{1, 6, 11, 16, 21}]], TableHeadings →
  {{"h=" <> ToString[h]}, {"t", "Exact", "Euler", "Heun", "RK4"}}]
```

	t	Exact	Euler	Heun	RK4
h=0.1	0	0.1	0.1	0.1	0.1
	0.5	0.154828	0.152493	0.15476	0.154828
	1.	0.231969	0.22613	0.231801	0.231969
	1.5	0.332428	0.32268	0.332148	0.332428
	2.	0.450853	0.438414	0.450483	0.450853

■ Exercises

1. Use Heun's method with step size $h = 0.1$ to approximate the solution of

$$y' = y\cos(t\,y), \quad y(0) = 1$$

on the interval $0 \le t \le 5$. Use `ListPlot` to plot the approximate solution, both with and without `PlotJoined→True`.

2. (a) Make a table to compare the results of the Euler, Heun, and fourth-order Runge-Kutta methods, with stepsize $h = 0.1$, on the problem

$$y' = \sin(t\cos y), \quad y(0) = 4.39.$$

Your table should have five rows showing the approximations at each of $t = 1, 2, 3, 4, 5$.

(b) Repeat part (a) with $h = 0.01$.

3. One step of a numerical method with stepsize h on the problem

$$y' = -y, \quad y(0) = 1,$$

gives an approximation to e^{-h}.

(a) Use one step of Euler's method to approximate e^{-h}, and show that the leading term in the Taylor series expansion of the error is $O(h^2)$.

(b) Use one step of Heun's method to approximate e^{-h}, and show that the leading term in the Taylor series expansion of the error is $O(h^3)$.

(c) Use one step of the fourth-order Runge-Kutta method to approximate e^{-h}, and show that the leading term in the Taylor series expansion of the error is $O(h^5)$.

4. One step of a numerical method with stepsize h on the problem

$$y' = y^2, \quad y(0) = 1,$$

gives an approximation to $\frac{1}{1-h}$.

(a) Use one step of Euler's method to approximate $\frac{1}{1-h}$, and show that the leading term in the Taylor series expansion of the error is $O(h^2)$.

(b) Use one step of Heun's method to approximate $\frac{1}{1-h}$, and show that the leading term in the Taylor series expansion of the error is $O(h^3)$.

(c) Use one step of the fourth-order Runge-Kutta method to approximate $\frac{1}{1-h}$, and show that the leading term in the Taylor series expansion of the error is $O(h^5)$.

5. (a) Show that $y = 4 \arctan t$ is the solution of

$$y' = \frac{4+t+t^3}{1+t^2} - \tan(\tfrac{1}{4}y), \quad y(0) = 0.$$

(b) Show that four steps of the fourth-order Runge-Kutta method on this problem with $h = 0.25$ yields π (rounded) to five significant digits.

(c) Show that 100 steps of Heun's method on this problem with $h = 0.01$ yields π (rounded) to five significant digits, while 50 steps with $h = 0.02$ yields only four.

(d) How small must h be in order for Euler's method on this problem to yield π (rounded) to five significant digits.?

6. Consider the initial-value problem

$$y' = y\left(1 - \frac{y}{1 - a\cos 2\pi t}\right), \quad y(0) = 0.1,$$

where a is an unknown parameter in (0, 1). Suppose that the value of the solution at $t = 4$ is observed to be 0.782. Use this information to find a (to three decimal places).

Suggestion: Enter `RK4step` and use the following function.

```
solnAt4[a_] := Module[{}, h = .1; f[t_, y_] := y (1 - y/(1 - a Cos[2 π t]));
  Nest[RK4step, {0., 0.1}, 40][[2]]]
```

7. **(Chaotic behavior of Euler's method.)** Consider the initial-value problem

$$y' = -(10 + \beta(y-1))\,y, \quad y(0) = \frac{1}{10}.$$

Do the following for each of (a) $\beta = 20$, (b) $\beta = 25$, (c) $\beta = 30$, (d) $\beta = 35$, (e) $\beta = 39$.

(i) Find the exact solution, and plot its graph over [0, 3].

(ii) Compute 30 steps of Euler's method with stepsize $h = 0.1$ and plot the result.

(iii) Show both graphs together in the same plot.

4.4 Derivation of Runge-Kutta Methods

In this section we will use *Mathematica*'s symbolic computation capabilities to derive the fourth-order Runge-Kutta method. We will begin, however, with the derivation of a family of second-order Runge-Kutta methods, one of which is the familiar Heun's method. Third-order methods will be explored in the exercises. (Recall that a method is said to be an *nth-order method* if its local truncation error is $O(h^{n+1})$.)

An "n-stage" Runge-Kutta method (with stepsize h) has the form

$$\begin{aligned} k_1 &= f(t, y(t)), \\ k_2 &= f(t+\theta_1 h, y(t)+\theta_1 hk_1), \\ k_3 &= f(t+\theta_2 h, y(t)+\theta_2 hk_2), \\ &\vdots \\ k_n &= f(t+\theta_{n-1} h, y(t)+\theta_{n-1} hk_{n-1}), \\ y(t+h) \approx \hat{y}(t+h) &:= y(t)+h\,(w_1\,k_1+w_2\,k_2+\cdots+w_n\,k_n). \end{aligned}$$

The goal is to choose the $2n-1$ parameters $\theta_1, \theta_2, \ldots, \theta_{n-1}$ and $w_1, w_2, \ldots, w_n$ so that

$$\hat{y}(t+h)-y(t+h) = O(h^{n+1}).$$

To accomplish this, $\theta_1, \theta_2, \ldots, \theta_{n-1}$ and $w_1, w_2, \ldots, w_n$ should be chosen so that each coefficient in the degree-n Taylor polynomial of $\hat{y}(t+h)-y(t+h)$ (about $h=0$) is zero (under the assumption, of course, that y has $n+1$ continuous derivatives in an interval containing t).

In the following we will use the `SeriesDSolve` function described in Section 4.1 to compute Taylor polynomials.

```
Clear[f, y0, h];
<< DiffEqs`DETools`;
p[n_][h_] := SeriesDSolve[
   {η'[τ] == f[τ, η[τ]], η[t] == y}, η[τ], {τ, t, n}] /. {τ - t → h}
```

Note that the function p returns an expansion of the solution at $t+h$ as illustrated by:

```
p[2][h]
```

$$y + h\, f[t, y] + \frac{1}{2} h^2 \left(f[t, y]\, f^{(0,1)}[t, y] + f^{(1,0)}[t, y]\right)$$

■ Second-Order Methods

A two-stage Runge-Kutta method has the form

$$k_1 = f(t, y(t)),\quad k_2 = f(t+\theta h, y(t)+\theta hk_1),\quad y(t+h) \approx y(t)+h\,(w_1\,k_1+w_2\,k_2).$$

Our goal is to choose the parameters θ, w_1, w_2 so that the local truncation error is $O(h^3)$, i.e.,

$$y(t+h)-(y(t)+h\,(w_1\,k_1+w_2\,k_2)) = O(h^3).$$

So we define the approximation at $t+h$, based on the value $y(t) = y$, as

```
RK2y[h_] := y + h (k[1] w[1] + k[2] w[2]) //.
  {k[1] → f[t, y], k[2] → f[t + h θ, y + h k[1] θ]}
```

The first- and second-order terms in the error are

```
error2 = p[2][h] - Normal[Series[RK2y[h], {h, 0, 2}]] // Simplify
```

$$-\frac{1}{2} h (f[t, y] (2 (-1 + w[1] + w[2]) + h (-1 + 2 \theta w[2]) f^{(0,1)}[t, y]) + h (-1 + 2 \theta w[2]) f^{(1,0)}[t, y])$$

We require this expression to equal zero; thus we have two equations to solve for θ, w_1, and w_2:

```
Solve[{-1 + w[1] + w[2] == 0, -1 + 2 θ w[2] == 0}, {w[1], w[2]}] //
 Simplify
```

$$\left\{\left\{w[1] \to 1 - \frac{1}{2\theta}, w[2] \to \frac{1}{2\theta}\right\}\right\}$$

So any choice of θ, along with $w_1 = 1 - \frac{1}{2\theta}$ and $w_2 = \frac{1}{2\theta}$, will result in a second-order method. The choice of $\theta = 1$ gives Heun's method:

$$k_1 = f(t, y(t)),\ k_2 = f(t+h, y(t) + hk_1),\ \ y(t+h) \approx y(t) + \tfrac{1}{2} h (k_1 + k_2).$$

To investigate further the choice of θ, let's look at the third-order term in the error:

```
error3 = p[3][h] - Normal[Series[RK2y[h], {h, 0, 3}]] /.
  {w[1] → 1 - 1/(2 θ), w[2] → 1/(2 θ)} // Simplify
```

$$\frac{1}{12} h^3 \left((2 - 3\theta) f[t, y]^2 f^{(0,2)}[t, y] + 2 f^{(0,1)}[t, y] f^{(1,0)}[t, y] + 2 f[t, y] \left(f^{(0,1)}[t, y]^2 + (2 - 3\theta) f^{(1,1)}[t, y]\right) + (2 - 3\theta) f^{(2,0)}[t, y]\right)$$

The factor $2 - 3\theta$ that appears in this expression suggests the choice of $\theta = 2/3$. With $\theta = 2/3$, the third-order term in the error is

```
error3 /. θ → 2 / 3 // Simplify
```

$$\frac{1}{6} h^3 f^{(0,1)}[t, y] (f[t, y] f^{(0,1)}[t, y] + f^{(1,0)}[t, y])$$

With $\theta = 1/2$ (Heun), the third-order term in the error is

```
error3 /. θ → 1 / 2 // Simplify
```

$$\frac{1}{24} h^3 \left(f[t, y]^2 f^{(0,2)}[t, y] + 4 f^{(0,1)}[t, y] f^{(1,0)}[t, y] + 2 f[t, y] \left(2 f^{(0,1)}[t, y]^2 + f^{(1,1)}[t, y]\right) + f^{(2,0)}[t, y]\right)$$

At first glance, it appears as though the "2/3" method might be superior in general to Heun's method. The third-order term in the error *is* simpler, but not necessarily *smaller*. Which of these methods gives better results depends on the problem.

■ The Fourth-Order Method

A four-stage Runge-Kutta method has the form

$$k_1 = f(t, y(t)),\ k_2 = f(t + \theta_1 h, y(t) + \theta_1 h k_1),$$

$$k_3 = f(t + \theta_2 h, y(t) + \theta_2 h k_2),\ k_4 = f(t + \theta_3 h, y(t) + \theta_3 h k_3),$$

$$y(t + h) \approx y(t) + h (w_1 k_1 + w_2 k_2 + w_3 k_3 + w_4 k_4).$$

The goal here is to choose θ_1, θ_2, θ_3, w_1, w_2, w_3, and w_4 so that the local truncation error is $O(h^5)$, that is,

$$y(t + h) - (y(t) + h (w_1 k_1 + w_2 k_2 + w_3 k_3 + w_4 k_4)) = O(h^5).$$

We define the approximation at $t + h$, based on the value $y(t) = y$, as

```
RK4y[y0_, h_] := y + h (k[1] w[1] + k[2] w[2] + k[3] w[3] + k[4] w[4]) //.
  {k[1] → f[t, y],
   k[2] → f[t + h θ[1], y + h k[1] θ[1]],
   k[3] → f[t + h θ[2], y + h k[2] θ[2]],
   k[4] → f[t + h θ[3], y + h k[3] θ[3]]}
```

The first-order term in the error is

```
error1 = Simplify[p[1][h] - Normal[Series[RK4y[y0, h], {h, 0, 1}]]]
```

```
-h f[t, y] (-1 + w[1] + w[2] + w[3] + w[4])
```

So we begin accumulating a list of rules by defining

```
rulelist = {-1 + w[1] + w[2] + w[3] + w[4] → 0};
```

which we then apply to the second-order term in the error:

```
error2 = Simplify[p[2][h] - Normal[Series[RK4y[y0, h], {h, 0, 2}]]] //.
   rulelist // Simplify
```

$$-\frac{1}{2} h^2 (-1 + 2 w[2] \theta[1] + 2 w[3] \theta[2] + 2 w[4] \theta[3]) (f[t, y] f^{(0,1)}[t, y] + f^{(1,0)}[t, y])$$

We next expand our list of rules accordingly,

```
rulelist =
 Union[rulelist, {-1 + 2 w[2] θ[1] + 2 w[3] θ[2] + 2 w[4] θ[3] → 0}]
```

```
{-1 + w[1] + w[2] + w[3] + w[4] → 0, -1 + 2 w[2] θ[1] + 2 w[3] θ[2] + 2 w[4] θ[3] → 0}
```

and apply it to the third-order term in the error:

```
error3 = Simplify[
   p[3][h] - Normal[Series[RK4y[y0, h], {h, 0, 3}]]] //. rulelist
```

$$-\frac{1}{6} h \left(h^2 f[t, y]^2 (-1 + 3 w[2] \theta[1]^2 + 3 w[3] \theta[2]^2 + 3 w[4] \theta[3]^2) f^{(0,2)}[t, y] + f[t, y] \left(h^2 (-1 + 6 w[3] \theta[1] \theta[2] + 6 w[4] \theta[2] \theta[3]) f^{(0,1)}[t, y]^2 + 2 h^2 (-1 + 3 w[2] \theta[1]^2 + 3 w[3] \theta[2]^2 + 3 w[4] \theta[3]^2) f^{(1,1)}[t, y]\right) + h ((-3 + 6 w[2] \theta[1] + 6 w[3] \theta[2] + 6 w[4] \theta[3] + h (-1 + 6 w[3] \theta[1] \theta[2] + 6 w[4] \theta[2] \theta[3]) f^{(0,1)}[t, y]) f^{(1,0)}[t, y] + h (-1 + 3 w[2] \theta[1]^2 + 3 w[3] \theta[2]^2 + 3 w[4] \theta[3]^2) f^{(2,0)}[t, y])\right)$$

This suggests that we add two more rules to the list:

```
rulelist = Union[rulelist,
  {-1 + 3 w[2] θ[1]^2 + 3 w[3] θ[2]^2 + 3 w[4] θ[3]^2 → 0,
   -1 + 6 w[3] θ[1] θ[2] + 6 w[4] θ[2] θ[3] → 0}]
```

$$\{-1 + w[1] + w[2] + w[3] + w[4] \to 0,\ -1 + 2 w[2] \theta[1] + 2 w[3] \theta[2] + 2 w[4] \theta[3] \to 0,\ -1 + 6 w[3] \theta[1] \theta[2] + 6 w[4] \theta[2] \theta[3] \to 0,\ -1 + 3 w[2] \theta[1]^2 + 3 w[3] \theta[2]^2 + 3 w[4] \theta[3]^2 \to 0\}$$

These conditions are sufficient for the local truncation error to be $O(h^4)$.

```
(error3 //. rulelist // Simplify) //. rulelist
```

0

The fourth-order term in the error is now

```
error4 = Simplify[
   p[4][h] - Normal[Series[RK4y[y0, h], {h, 0, 4}]]] //. rulelist
```

$$-\frac{1}{24} h \left(h^3 f[t, y]^3 (-1 + 4 w[2] \theta[1]^3 + 4 w[3] \theta[2]^3 + 4 w[4] \theta[3]^3) f^{(0,3)}[t, y] + h^2 f[t, y]^2 (4 h (-1 + 3 w[3] \theta[1] \theta[2] (\theta[1] + 2 \theta[2]) + 3 w[4] \theta[2] \theta[3] (\theta[2] + 2 \theta[3])) f^{(0,1)}[t, y] f^{(0,2)}[t, y] + 3 h (-1 + 4 w[2] \theta[1]^3 + 4 w[3] \theta[2]^3 + 4 w[4] \theta[3]^3) f^{(1,2)}[t, y]) + f[t, y] \left(-24 + 24 w[1] + 24 w[2] + 24 w[3] + 24 w[4] + h^3 (-1 + 24 w[4] \theta[1] \theta[2] \theta[3]) f^{(0,1)}[t, y]^3 + 3 h^3 (-1 + 8 w[3] \theta[1] \theta[2]^2 + 8 w[4] \theta[2] \theta[3]^2) f^{(0,2)}[t, y] f^{(1,0)}[t, y] - 8 h^2 f^{(1,1)}[t, y] + 24 h^2 w[2] \theta[1]^2 f^{(1,1)}[t, y] + 24 h^2 w[3] \theta[2]^2 f^{(1,1)}[t, y] + 24 h^2 w[4] \theta[3]^2 f^{(1,1)}[t, y] + h^3 (-5 + 24 w[3] \theta[1] \theta[2] (\theta[1] + \theta[2]) + 24 w[4] \theta[2] \theta[3] (\theta[2] + \theta[3])) f^{(0,1)}[t, y] f^{(1,1)}[t, y] - 3 h^3 f^{(2,1)}[t, y] + 12 h^3 w[2] \theta[1]^3 f^{(2,1)}[t, y] + 12 h^3 w[3] \theta[2]^3 f^{(2,1)}[t, y] + 12 h^3 w[4] \theta[3]^3 f^{(2,1)}[t, y]\right) + h \left(f^{(1,0)}[t, y] \left(h^2 (-1 + 24 w[4] \theta[1] \theta[2] \theta[3]) f^{(0,1)}[t, y]^2 + 3 (-4 + 8 w[2] \theta[1] + 8 w[3] \theta[2] + 8 w[4] \theta[3] + h^2 (-1 + 8 w[3] \theta[1] \theta[2]^2 + 8 w[4] \theta[2] \theta[3]^2) f^{(1,1)}[t, y])\right) + h (h (-1 + 12 w[3] \theta[1]^2 \theta[2] + 12 w[4] \theta[2]^2 \theta[3]) f^{(0,1)}[t, y] f^{(2,0)}[t, y] + h (-1 + 4 w[2] \theta[1]^3 + 4 w[3] \theta[2]^3 + 4 w[4] \theta[3]^3) f^{(3,0)}[t, y])\right)\right)$$

After closely examining this expression, we add a few more rules to our list and then recompute the fourth-order term in the error:

```
rulelist = Union[rulelist, {
    -1 + 4 w[2] θ[1]^3 + 4 w[3] θ[2]^3 + 4 w[4] θ[3]^3 → 0,
    -1 + 3 w[3] θ[1] θ[2] (θ[1] + 2 θ[2]) +
      3 w[4] θ[2] θ[3] (θ[2] + 2 θ[3]) → 0,
    -1 + 24 w[4] θ[1] θ[2] θ[3] → 0,
    -1 + 8 w[3] θ[1] θ[2]^2 + 8 w[4] θ[2] θ[3]^2 → 0}];
```

```
(error4 //. rulelist // Simplify) //. rulelist
```

$$-\frac{1}{24} h \left(h^3 f[t, y] \left(-5 + 24 w[3] \theta[1] \theta[2] (\theta[1] + \theta[2]) + 24 w[4] \theta[2] \theta[3] (\theta[2] + \theta[3])\right) f^{(0,1)}[t, y] f^{(1,1)}[t, y] + h^3 \left(-1 + 12 w[3] \theta[1]^2 \theta[2] + 12 w[4] \theta[2]^2 \theta[3]\right) f^{(0,1)}[t, y] f^{(2,0)}[t, y]\right)$$

Two more rules are enough:

```
rulelist = Union[rulelist,
   {-5 + 24 w[3] θ[1] θ[2] (θ[1] + θ[2]) + 24 w[4] θ[2] θ[3] (θ[2] + θ[3]) → 0
    -1 + 12 w[3] θ[1]^2 θ[2] + 12 w[4] θ[2]^2 θ[3] → 0}];
```

```
(error4 //. rulelist // Simplify) //. rulelist
```

0

Now we convert our rules to equations.

```
eqns = rulelist /. {(x_ → y_) → x == y}
```

```
{-1 + w[1] + w[2] + w[3] + w[4] == 0,
 -1 + 2 w[2] θ[1] + 2 w[3] θ[2] + 2 w[4] θ[3] == 0,
 -1 + 6 w[3] θ[1] θ[2] + 6 w[4] θ[2] θ[3] == 0, -1 + 24 w[4] θ[1] θ[2] θ[3] == 0,
 -1 + 12 w[3] θ[1]^2 θ[2] + 12 w[4] θ[2]^2 θ[3] == 0,
 -1 + 3 w[2] θ[1]^2 + 3 w[3] θ[2]^2 + 3 w[4] θ[3]^2 == 0,
 -1 + 8 w[3] θ[1] θ[2]^2 + 8 w[4] θ[2] θ[3]^2 == 0,
 -1 + 4 w[2] θ[1]^3 + 4 w[3] θ[2]^3 + 4 w[4] θ[3]^3 == 0,
 -5 + 24 w[3] θ[1] θ[2] (θ[1] + θ[2]) + 24 w[4] θ[2] θ[3] (θ[2] + θ[3]) == 0,
 -1 + 3 w[3] θ[1] θ[2] (θ[1] + 2 θ[2]) + 3 w[4] θ[2] θ[3] (θ[2] + 2 θ[3]) == 0}
```

Next we'll use the first and fourth equations in the list to simplify the system of equations (by eliminating w_1 and w_4).

$$\texttt{eqns = eqns //. } \left\{w[1] \to 1 - w[2] - w[3] - w[4],\ w[4] \to \frac{1}{24\,\theta[1]\,\theta[2]\,\theta[3]}\right\}$$

$$\Big\{\text{True},\ -1 + 2 w[2] \theta[1] + \frac{1}{12 \theta[1] \theta[2]} + 2 w[3] \theta[2] == 0,$$
$$-1 + \frac{1}{4 \theta[1]} + 6 w[3] \theta[1] \theta[2] == 0,\ \text{True},\ -1 + \frac{\theta[2]}{2 \theta[1]} + 12 w[3] \theta[1]^2 \theta[2] == 0,$$
$$-1 + 3 w[2] \theta[1]^2 + 3 w[3] \theta[2]^2 + \frac{\theta[3]}{8 \theta[1] \theta[2]} == 0,$$
$$-1 + 8 w[3] \theta[1] \theta[2]^2 + \frac{\theta[3]}{3 \theta[1]} == 0,\ -1 + 4 w[2] \theta[1]^3 + 4 w[3] \theta[2]^3 + \frac{\theta[3]^2}{6 \theta[1] \theta[2]} == 0,$$
$$-5 + 24 w[3] \theta[1] \theta[2] (\theta[1] + \theta[2]) + \frac{\theta[2] + \theta[3]}{\theta[1]} == 0,$$
$$-1 + 3 w[3] \theta[1] \theta[2] (\theta[1] + 2 \theta[2]) + \frac{\theta[2] + 2 \theta[3]}{8 \theta[1]} == 0\Big\}$$

At this point the system is reduced enough for `Solve` to have an easy time finding the solution.

```
soln = Solve[eqns, {θ[1], θ[2], θ[3], w[2], w[3]}][[1]]
```

$\{\theta[3] \to 1,\ w[2] \to \frac{1}{3},\ w[3] \to \frac{1}{3},\ \theta[2] \to \frac{1}{2},\ \theta[1] \to \frac{1}{2}\}$

The resulting values of w_4 and w_1 are

$\{w[4] \to \frac{1}{24\,\theta[1]\,\theta[2]\,\theta[3]}\}$ /. soln

$\{w[4] \to \frac{1}{6}\}$

```
{w[1] → 1 - w[2] - w[3] - w[4]} /. Join[soln, %]
```

$\{w[1] \to \frac{1}{6}\}$

Thus we have arrived at *the* fourth-order Runge-Kutta method,

$$k_1 = f(t, y(t)),\ k_2 = f(t + \tfrac{1}{2}\,h,\ y(t) + \tfrac{1}{2}\,h k_1),$$

$$k_3 = f(t + \tfrac{1}{2}\,h,\ y(t) + \tfrac{1}{2}\,h k_2),\ k_4 = f(t + h,\ y(t) + h k_3),$$

$$y(t + h) \approx y(t) + \tfrac{1}{6}\,h\,(k_1 + 2\,k_2 + 2\,k_3 + k_4).$$

■ Exercises

1. Derive a one-parameter family of *third-order* Runge-Kutta methods,

$$k_1 = f(t, y(t)),$$

$$k_2 = f(t + \theta_1\,h,\ y(t) + \theta_1\,h k_1),$$

$$k_3 = f(t + \theta_2\,h,\ y(t) + \theta_2\,h k_2),$$

$$y(t + h) \approx y(t) + h\,(w_1\,k_1 + w_2\,k_2 + w_3\,k_3).$$

In particular, find the third-order methods in which (i) $\theta_1 = \frac{1}{2}$ and (ii) $\theta_1 = \frac{2}{3}$.

Chapter 5

Coupled Pairs of First-Order Equations

5.1 Direction Fields and the Phase Plane

The problem under consideration consists of a pair of autonomous, first-order differential equations

$$\frac{dx}{dt} = f(x, y), \quad \frac{dy}{dt} = g(x, y).$$

We will begin by indicating how to use *Mathematica*'s `PlotVectorField` and `NDSolve` to plot the system's direction field and a sampling of parametric solution curves in the *xy phase plane*.

To use `PlotVectorField` we first need to load the `PlotField` package.

```
<< Graphics`PlotField`
```

Example 5.1.1

Consider the linear system

$$\frac{dx}{dt} = -x + y, \quad \frac{dy}{dt} = -x - y.$$

Here we have

```
f[x_, y_] := -x + y; g[x_, y_] := -x - y;
```

These functions are the components of the system's *vector* field.

```
vecfield = PlotVectorField[
   {f[x, y], g[x, y]}, {x, -2, 2}, {y, -2, 2}, Axes → True];
```

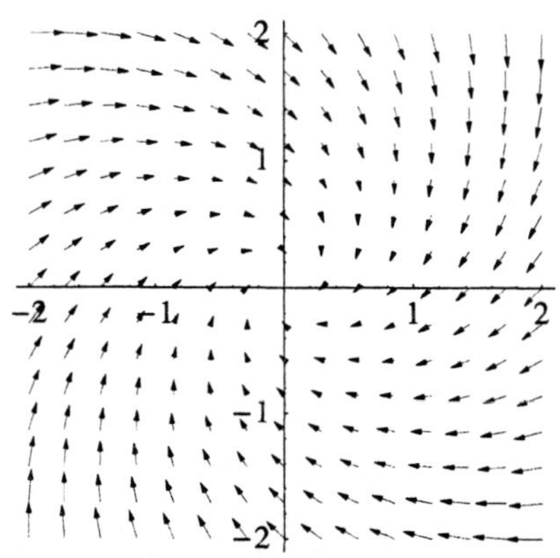

By scaling the arrows to have uniform length, we get a plot of the system's *direction* field.

```
dirfield = PlotVectorField[{f[x, y], g[x, y]},
   {x, -2, 2}, {y, -2, 2}, Axes → True, ScaleFunction → (1 &)];
```

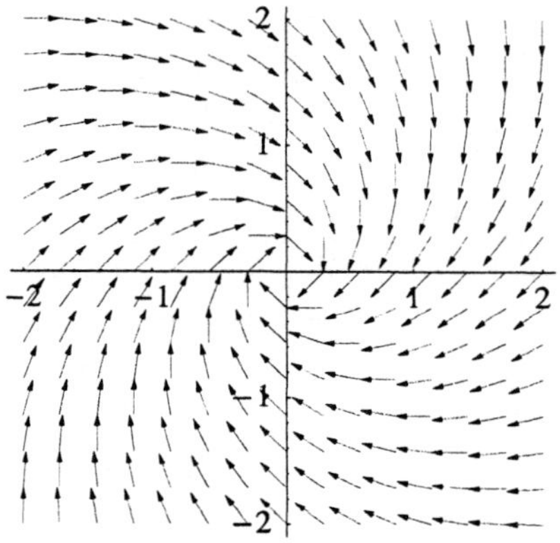

To solve the system numerically for $0 \le t \le 5$ with initial values $x(0) = -2$ and $y(0) = 2$, we can use NDSolve as follows:

```
sys := {x'[t] == f[x[t], y[t]],
  y'[t] == g[x[t], y[t]], x[0] == x0, y[0] == y0}

x0 = -2; y0 = 2;
soln = { x[t], y[t]} /. NDSolve[sys, {x, y}, {t, 0, 5}]
{{InterpolatingFunction[{{0., 5.}}, <>][t], InterpolatingFunction[{{0., 5.}}, <>][t]}}
```

This shows each of x and y plotted versus t:

```
Plot[Evaluate[soln], {t, 0, 5}, PlotStyle → {Hue[.67], Hue[0]},
  PlotRange → All, AxesLabel → {t, {x, y}}];
```

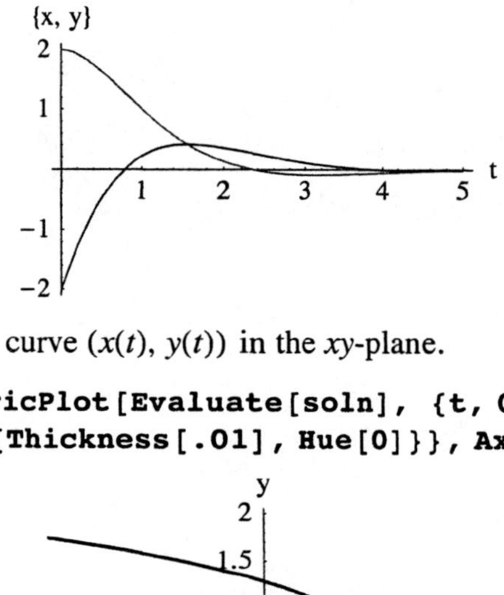

Next we'll plot the parametric curve $(x(t), y(t))$ in the xy-plane.

```
curve1 = ParametricPlot[Evaluate[soln], {t, 0, 5},
   PlotStyle → {{Thickness[.01], Hue[0]}}, AxesLabel → {x, y}];
```

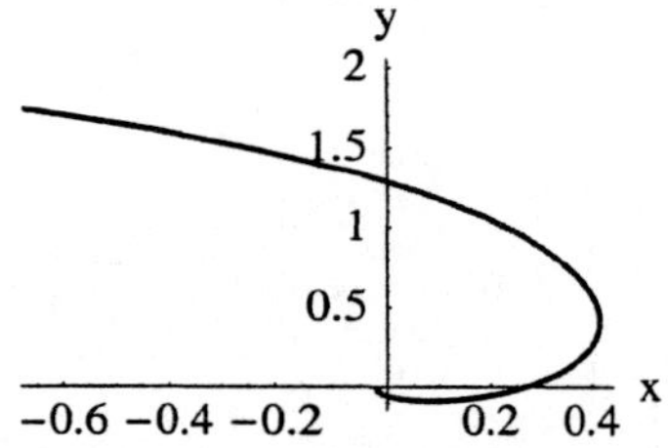

Then we superimpose this *orbit* over the direction field.

```
Show[dirfield, curve1, AxesLabel → {x, y}];
```

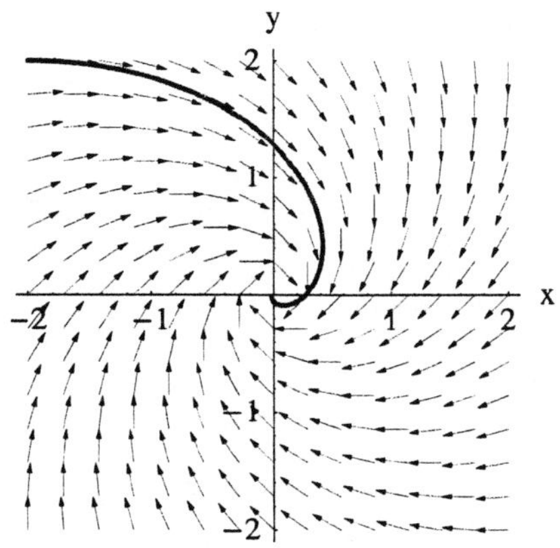

The following adds two more curves to the plot and replaces the direction field with a "cleaner" version.

```
x0 = 2; y0 = 1;
soln = {x[t], y[t]} /. NDSolve[sys, {x, y}, {t, 0, 5}];
curve2 = ParametricPlot[Evaluate[soln], {t, 0, 5}, PlotStyle →
    {{Thickness[.01], Hue[0]}}, DisplayFunction → Identity];

x0 = -1; y0 = -2;
soln = {x[t], y[t]} /. NDSolve[sys, {x, y}, {t, 0, 5}];
curve3 = ParametricPlot[Evaluate[soln], {t, 0, 5}, PlotStyle →
    {{Thickness[.01], Hue[0]}}, DisplayFunction → Identity];

dirfield =
  PlotVectorField[{f[x, y], g[x, y]}, {x, -2, 2}, {y, -2, 2},
   Axes → True, PlotPoints → 9, ScaleFunction → (1 &),
   ScaleFactor → .3, DisplayFunction → Identity];

Show[dirfield, curve1, curve2, curve3,
 AxesLabel → {x, y}, DisplayFunction → $DisplayFunction];
```

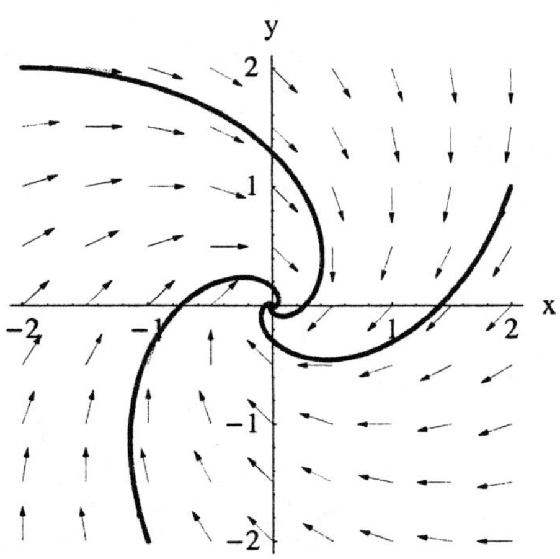

■ Using PhasePlot

The DEGraphics package provides a function named PhasePlot that draws a system's *phase portrait*, i.e., its direction field together with a collection of orbits.

```
<< DiffEqs`DEGraphics` ;
? PhasePlot
```

```
PhasePlot[{f,g}, {t,tmin,tmax}, {x,xmin,xmax}, {y,ymin,
  ymax}, options]. By default, four solution curves are drawn. The
  InitialPoints option can be used to specify a list of starting
  points for curves drawn by PhasePlot. Each initial point may
  give rise to multiple curves as specified by the MaxCurves option.
```

The following is a basic phase plot—with no options—for the system in Example 5.1.1.

```
PhasePlot[{-x + y, -x - y}, {t, -5, 5}, {x, -2, 2}, {y, -2, 2}];
```

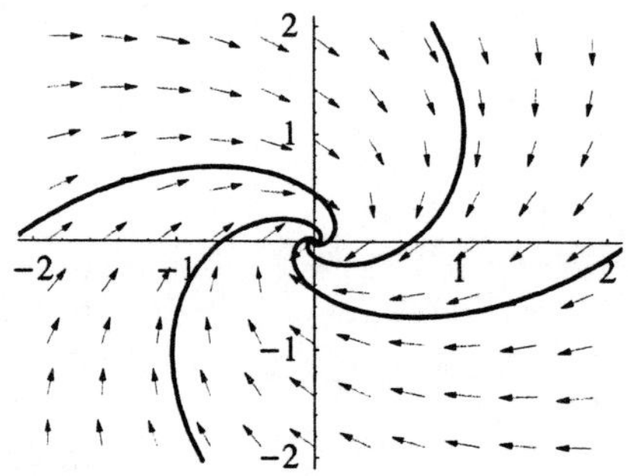

The options for PhasePlot are mainly the same as for DEPlot. In particular, one can use InitialPoints, MaxCurves, and Gap to customize the plot. We'll illustrate these in the next example.

Example 5.1.2

Consider the system

$$\frac{d\theta}{dt} = \omega, \quad \frac{d\omega}{dt} = -\sin\theta,$$

which governs the motion of a simple pendulum. For comparison, let's first use PhasePlot with no options. Notice that symmetry causes the four default curves to coincide.

```
PhasePlot[{ω, -Sin[θ]}, {t, -3, 3}, {θ, -3.5, 3.5},
  {ω, -3, 3}, AxesLabel → {θ, ω}, Ticks → {{-π, π}, {-2, 2}}];
```

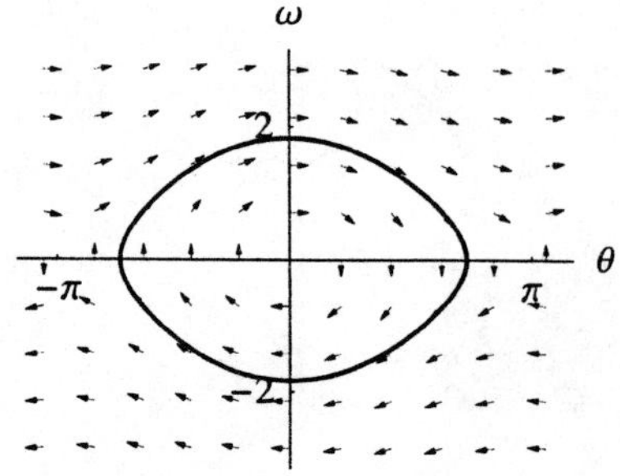

By specifying initial points and scaling the arrows, we get a much better picture.

```
PhasePlot[{ω, -Sin[θ]}, {t, -3, 3}, {θ, -3.5, 3.5}, {ω, -3, 3},
  ScaleArrows → 1.7, InitialPoints → {{0, 2}, {0, -2}},
  MaxCurves → 3, AxesLabel → {θ, ω}, Ticks → {{-π, π}, {-2, 2}}];
```

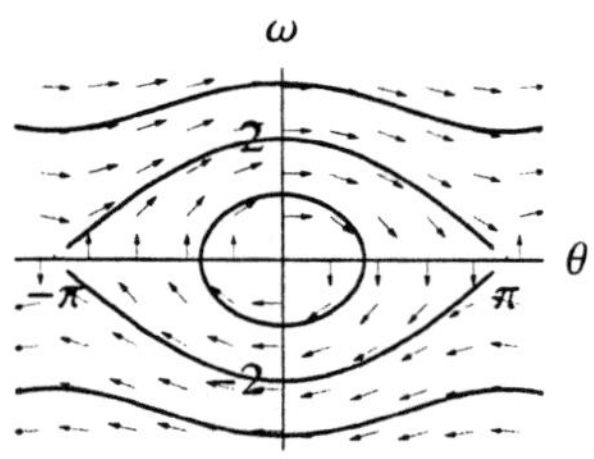

Now let's increase `MaxCurves`, decrease the `Gap` and, for the sake of variety, `Frame` the plot.

```
PhasePlot[{ω, -Sin[θ]}, {t, -3, 3}, {θ, -3.5, 3.5}, {ω, -3, 3},
  InitialPoints → {{0, 2}, {0, -2}}, MaxCurves → 9,
  Gap → .05, ScaleArrows → 1.7, AxesLabel → {θ, ω},
  FrameTicks → {{-π, π}, {-2, 2}, {}, {}}, Frame → True];
```

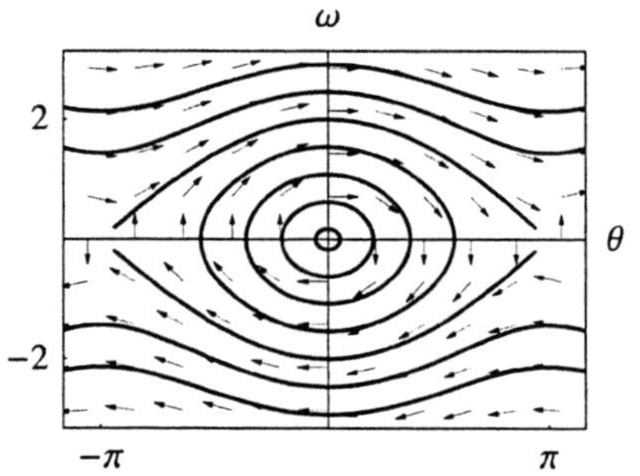

Example 5.1.3

`PhasePlot` will also plot solution curves for *nonautonomous* systems. In this case, `PhasePlot` detects the t-dependence of the vector field, and no direction field is drawn. For the system

$$\frac{dx}{dt} = -x\,y\sin t, \quad \frac{dy}{dt} = x\,y\sin t - 2\,y + 1,$$

a few solution curves are plotted as follows:

```
PhasePlot[{-.5 x y Sin[t], 1.5 x y Sin[t] - 2 y + 1},
  {t, 0, 25}, {x, 0, 1}, {y, 0, 1},
  InitialPoints → {{1, 1}, {1, 0}, {.25, 0}, {.1, 0}}];
```

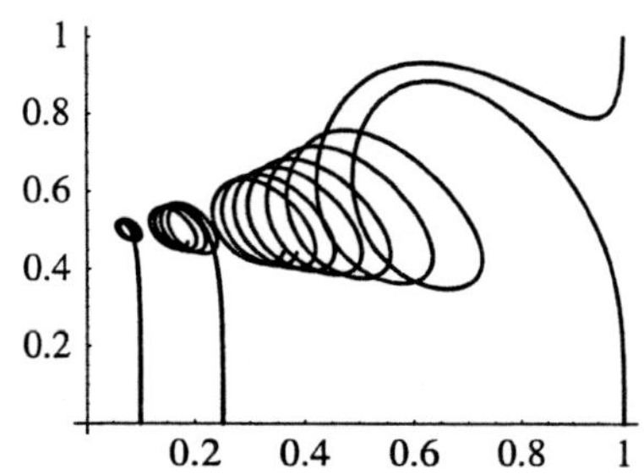

■ Nullclines and Equilibrium Points

The *nullclines* of a system $\frac{dx}{dt} = f(x, y)$, $\frac{dy}{dt} = g(x, y)$ are the curves described by

$$f(x, y) = 0 \text{ (the } x\text{-nullcline) or } g(x, y) = 0 \text{ (the } y\text{-nullcline).}$$

At any point where $f(x, y) = 0$ and $g(x, y) \neq 0$, an orbit must have a vertical tangent line, and at any point where $g(x, y) = 0$ and $f(x, y) \neq 0$, an orbit must have a horizontal tangent line. Points where $f(x, y) = g(x, y) = 0$ are *equilibrium points* (a.k.a., *critical points*).

Setting `ShowNullclines→True` causes `PhasePlot` to include a system's isoclines. With `ShowEquilibriumPoints → True`, `PhasePlot` attempts to solve for and show the equilibrium points.

Example 5.1.4

Consider the system

$$\frac{dx}{dt} = -x + y + 1, \quad \frac{dy}{dt} = -x - y.$$

Here the nullclines are the lines $-x + y + 1 = 0$ and $-x - y = 0$, and there is an equilibrium point at the intersection of these lines.

```
PhasePlot[{-x + y + 1, -x - y}, {t, -5, 5}, {x, -2, 2}, {y, -2, 2},
  ShowNullclines → True, ShowEquilibriumPoints → True];
```

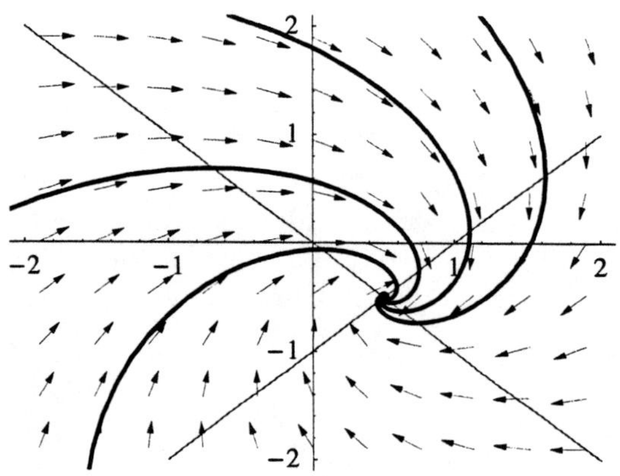

Example 5.1.5

The nullclines of the system

$$\frac{dx}{dt} = -x + y^2, \quad \frac{dy}{dt} = 1 - x^2 - y^2,$$

are the parabola $x = y^2$ and the unit circle $x^2 + y^2 = 1$. There are two equilibrium points located at the intersection of these curves.

```
PhasePlot[{-x + y^2, 1 - x^2 - y^2}, {t, -5, 5}, {x, -1.5, 2}, {y, -1.5, 1.5},
  InitialPoints → {{-1, 0}, {1, 1}, {0, -1}, {1, -.5}},
  MaxCurves → 3, ShowNullclines → True, Ticks → None,
  AspectRatio → Automatic, ShowEquilibriumPoints → True];
```

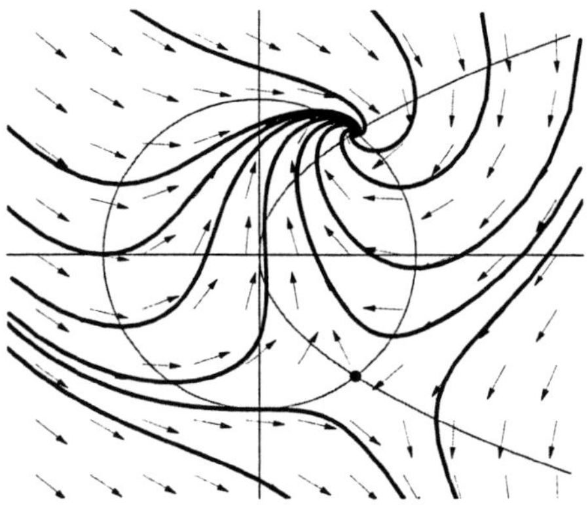

Notice that the equilibrium point in the first quadrant is *asymptotically stable*; i.e., it is the limit of all nearby solutions as $t \to \infty$. The equilibrium point in the fourth quadrant is *unstable*.

Example 5.1.6

A model of a certain hypothetical chemical reaction leads to the system

$$\frac{dx}{dt} = -x\,y^2 + 2\,y, \quad \frac{dy}{dt} = x\,y^2 - 3\,y + 1.$$

Here the x-nullclines consist of the x-axis (where $y = 0$) and the hyperbola $x\,y = 2$. The y-nullcline is the curve $x = (3\,y - 1)\,y^{-2}$. These intersect at the equilibrium point $(2, 1)$. This equilibrium point is asymptotically stable, although this is not completely obvious from the picture.

```
PhasePlot[{-x y^2 + 2 y, x y^2 - 3 y + 1}, {t, 0, 30}, {x, 0, 4},
  {y, 0, 2.5}, InitialPoints → {{1, 0}, {3, 0}, {0, 2}},
  ShowNullclines → True, ShowEquilibriumPoints → True,
  Arrows → {9, 7}, AspectRatio → Automatic];
```

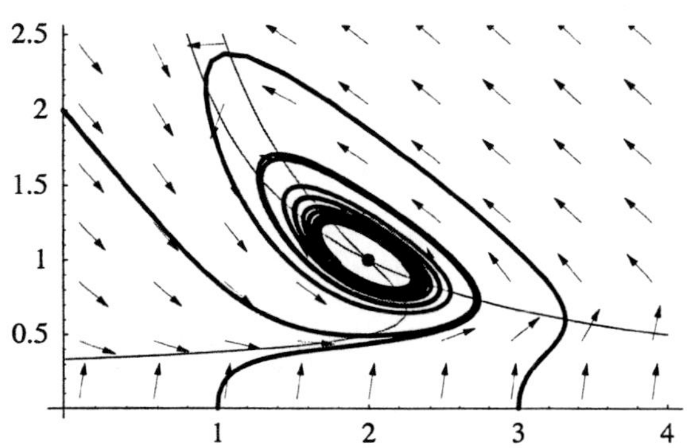

Exercises

1. Consider the following linear system in which a is a parameter:

$$\frac{dx}{dt} = x + 2y, \quad \frac{dy}{dt} = -2x + a\,y.$$

Use `PhasePlot` to create phase portraits for each of $a = -5, -4, -3, -2, -1, 0, 2, 4,$ and 8. Use `ShowNullclines → True`, and use the `PlotLabel` option to label each plot with the corresponding value of a. Then display the nine plots in a 3×3 `GraphicsArray`.

2. Consider the following system in which a is a parameter:

$$\frac{dx}{dt} = x - y^2 + a, \quad \frac{dy}{dt} = 1 - x^2 - y^2.$$

Use `PhasePlot` to create phase portraits for each of a = 1.5, 1.25, 1.125, 1, and 0.5. (Set `ShowNullclines → True` and `ShowEquilibriumPoints → True`.) In each case, use `Solve` to find the equilibrium points, and state which of the equilibrium points are asymptotically stable.

3. Enter the following to create an animation of the solution of

$$\frac{dx}{dt} = -y, \qquad x(0) = 0,$$
$$\frac{dy}{dt} = x^2 - t, \quad y(0) = y_0,$$

as y_0 varies from -4 to 1.

```
Needs["DiffEqs`DEGraphics`"];
Do[PhasePlot[{-y, x² - t}, {t, 0, 50}, {x, -3, 7},
  {y, -7, 7}, InitialPoints → {{0, y0}}, MaxCurves → 1,
  Gap → .025, Ticks → None, AxesLabel → {x, y}], {y0, -4, 1, .1}]
```

Note that this system is equivalent to the second-order problem

$$x'' = t - x^2, \; x(0) = 0, \; x'(0) = -y_0,$$

which involves a so-called *Painlevé equation*.

5.2 Predator-Prey Systems

Imagine a two-species "community" in which species P is the food source for species Q, while P feeds upon some other abundantly available food. We refer to species Q as the *predator* and species P as the *prey*. We make the following assumptions:

1. The per capita growth rate of the prey population decreases linearly with both the number of prey and the number of predators, amounting to logistic growth if there are no predators.

2. The per capita growth rate of the predator population increases linearly with the number of prey and is a negative constant if there are no prey. Thus the predator population declines exponentially if there are no prey to eat.

With p and q denoting the numbers of prey and predators, respectively, these assumptions lead to the equations

$$\frac{dp}{dt} = k_1\, p\left(1 - \frac{p}{M} - bq\right), \quad \frac{dq}{dt} = -k_2\, q\,(1 - cp).$$

We can make our lives a bit simpler here by *nondimensionalizing* the problem. To do this, we introduce dimensionless quantities $x = cp$ and $y = bq$. Then

$$\frac{dx}{dt} = c\,\frac{dp}{dt} = ck_1\,\frac{x}{c}\left(1 - \frac{x}{cM} - y\right), \text{ and } \frac{dy}{dt} = b\,\frac{dq}{dt} = -bk_2\,\frac{y}{b}\,(1 - x).$$

Thus, with $a = (cM)^{-1}$, we have the following system with 3 parameters instead of 5:

$$\frac{dx}{dt} = k_1\, x\,(1 - ax - y), \quad \frac{dy}{dt} = -k_2\, y\,(1 - x),$$

Note that we can think of $a = 0$ as corresponding to $M = \infty$. In this case the solutions are periodic, causing phase-plane orbits to be closed curves.

```
<< DiffEqs`DEGraphics`

PhasePlot[{2 x (1 - 0 x - y), -y (1 - x)}, {t, 0, 10}, {x, 0, 2.5},
  {y, 0, 2}, InitialPoints → {{1.2, 1}, {1.6, 1.25}, {2, 1.5}},
  ScaleArrows → .5, Ticks → None, ShowNullclines → True,
  ShowEquilibriumPoints → True, AxesLabel → {"Prey", "Predator"}];
```

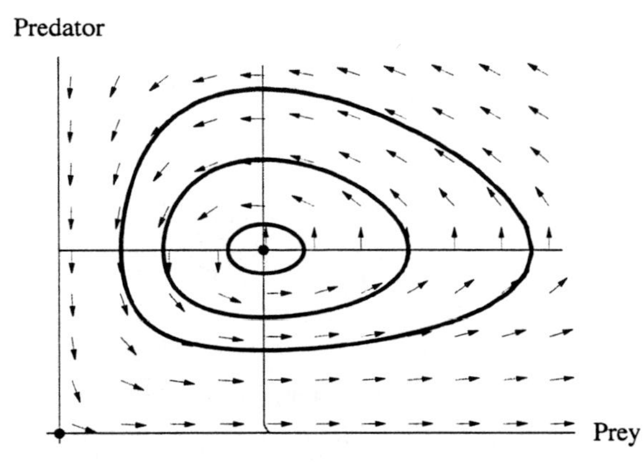

The following command creates an animation that essentially illustrates all possible behaviors of the system. Six of the frames are shown below.

```
Do[PhasePlot[{2 x (1 - a x - y), -y (1 - x)}, {t, 0, 10}, {x, 0, 2.5},
  {y, 0, 2}, InitialPoints → {{1.2, 1}, {1.6, 1.25}, {2, 1.5}},
  ScaleArrows → .5, Ticks → None,
  ShowNullclines → True, ShowEquilibriumPoints → True,
  PlotLabel → "a = " <> ToString[N[a]]], {a, 0, 2, .2}]
```

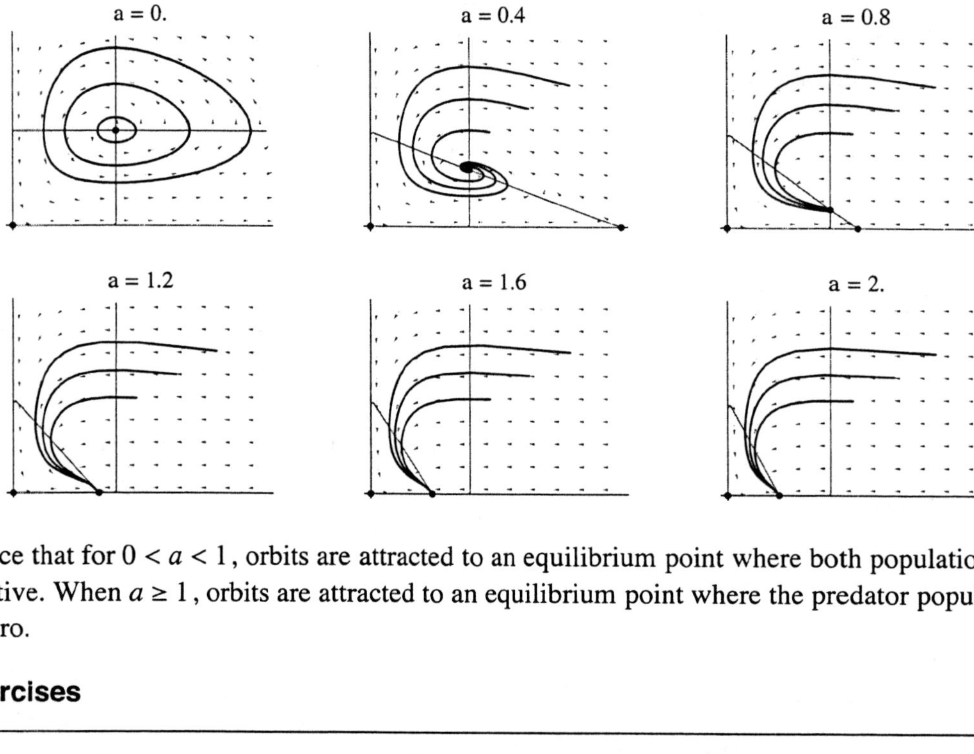

Notice that for $0 < a < 1$, orbits are attracted to an equilibrium point where both populations are positive. When $a \geq 1$, orbits are attracted to an equilibrium point where the predator population is zero.

Exercises

1. Let $k_1 = k_2 = 1$ and $a = 0$, so that the predator-prey system becomes

$$\frac{dx}{dt} = x(1-y), \quad \frac{dy}{dt} = -y(1-x).$$

Verify that $x\,y\,e^{-x-y} = c$ gives an implicit general solution to the slope equation

$$\frac{dy}{dx} = -\frac{y(1-x)}{x(1-y)}.$$

Then use `ContourPlot` to draw the system's phase portrait.

2. A model of *competition* between two species is

$$\frac{dx}{dt} = k_1\,x(1-x-a\,y), \quad \frac{dy}{dt} = k_2\,y(1-y-bx),$$

where a and b are positive parameters. Let $k_1 = k_2 = 1$ and $b = 1.5$, and create an animation showing the phase portrait for $a = 0.1, 0.2, \ldots, 2$. Plot over the rectangle given by $0 \leq x \leq 1.5$, $0 \leq y \leq 1.5$, and set `ShowNullclines →True` and `ShowEquilibriumPoints →True`. What do the phase portraits tell you about what happens to the populations in the long run? *Suggestion*: Use the default initial points, and let t range from -5 to 10 in each plot.

5.3 The Spruce Budworm Problem Revisited

Recall the spruce budworm model from Section 3.6:

$$\frac{dw}{dt} = k\,w\left(1 - \frac{w}{M}\right) - \rho\,\frac{w^2}{\eta^2 + w^2}.$$

Here we will introduce a variable y representing *foliage density* into the model. The assumptions are: (1) the carrying capacity M and the parameter η are each proportional to y, (2) budworms consume foliage at a rate that is proportional to w, and (3) y satisfies a simple logistic equation if the worm population is zero. These assumptions give rise to the system

$$\frac{dy}{dt} = k_1\,y\,(1-y) - \mu\,w, \qquad \frac{dw}{dt} = k_2\,w\left(1 - \frac{w}{\alpha y}\right) - \rho\,\frac{w^2}{\beta^2\,y^2 + w^2}.$$

The foliage density y may viewed as the average leaf size in the forest, scaled so that its maximum value is 1. Likewise, it will be useful to view w as a scaled measure of the number of budworms, where $w = 1$ might represent, say, a million budworms.

Example 5.3.1

With parameter values $k_1 = 0.25$, $k_2 = 2$, $\mu = 0.1$, $\alpha = 6$, $\rho = 1$, and $\beta^2 = 0.125$, and initial values $y(0) = 0.2$ (a sparsely foliated forest) and $w(0) = 0.01$ (not many budworms), we get the following plot.

```
<< DiffEqs`DEGraphics`;

PhasePlot[{.25 y (1 - y) - .1 w, 2 w (1 - w/(6 y)) - 1 w^2/(.125 y^2 + w^2)},
  {t, 0, 26}, {y, .00001, 1}, {w, 0, 3},
  InitialPoints → {{.2, .01}}, ScaleArrows → .6,
  ShowNullclines → True, AxesLabel → {Foliage, Budworms}];
```

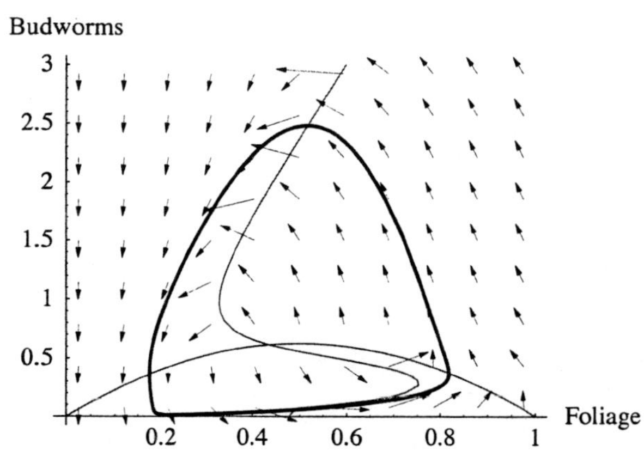

The following plots give a more complete picture of what's happening. Up to around $t = 18$ the budworm density remains small as the forest matures. Then it begins to grow *very rapidly*, because the foliage has become dense enough to provide the budworms both plenty of food and shelter from hungry birds. By $t = 22$ the budworm density has already peaked and begun to decline rapidly, since much of the foliage has been consumed. By $t = 26$, both densities are close to where they started, ready to begin the cycle anew.

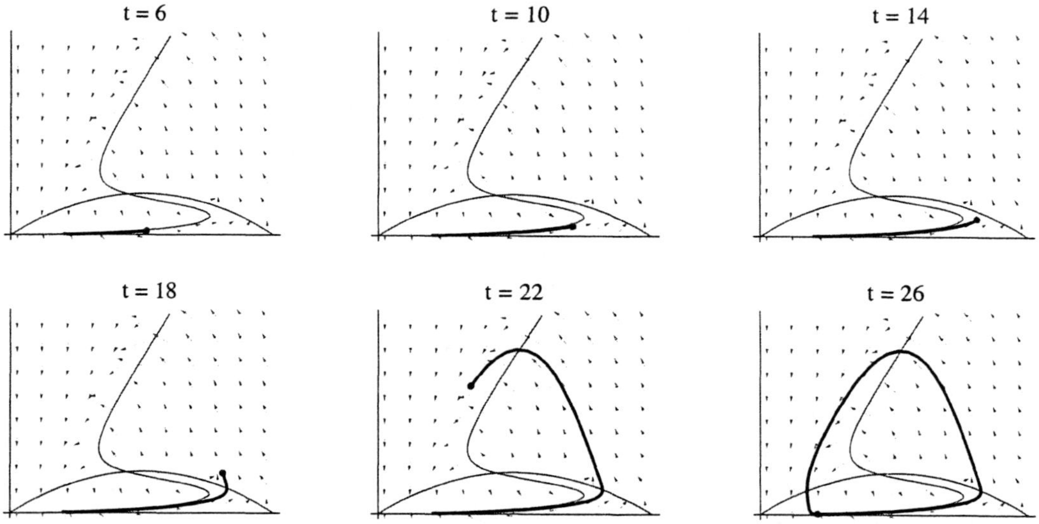

Exercises

1. Here we will investigate the effect of variations in the parameter ρ, which could be caused by changes in the size of the bird population. Let $k_1 = 0.25$, $k_2 = 2$, $\mu = 0.1$, $\alpha = 6$, $\beta^2 = 0.125$, $y(0) = 0.2$, and $w(0) = 0.01$ as above. Using the time interval $0 \le t \le 40$, create plots that show the evolution of the foliage and budworm densities for $\rho = 1.15, 1.14, 1.13, 1.12,$ and 1.11. Include equilibrium points in the plots by setting `ShowEquilibriumPoints→True`. Describe in detail what the plots imply about the effect of small changes in the bird population.
2. Repeat Problem 1 with $k_1 = 0.5$ and $\rho = 0.98,\ 0.99,\ 1,\ 1.01$.
3. Use `NDSolve` to help create plots of the foliage and budworm densities versus t for the parameter and initial values in Example 5.3.1. Describe the correspondence with the phase plane orbit.

5.4 The Half-Wave Rectifier Circuit Revisited

Recall the half-wave rectifier circuit we studied in Section 3.7. With source voltage $A\sin\omega t$, the differential equation for the voltage across the diode was

$$V_d' + C^{-1}\,\gamma(V_d) = \omega\,A\cos\omega t, \text{ where } \gamma(V_d) = I_0(e^{\eta V_d} - 1).$$

We are now going to add another loop to the circuit. Suppose that a resistance R is placed in parallel with the capacitor. This plays the role of a *load* for the rectifier circuit. Its primary effect is that it provides a path through which the capacitor can discharge.

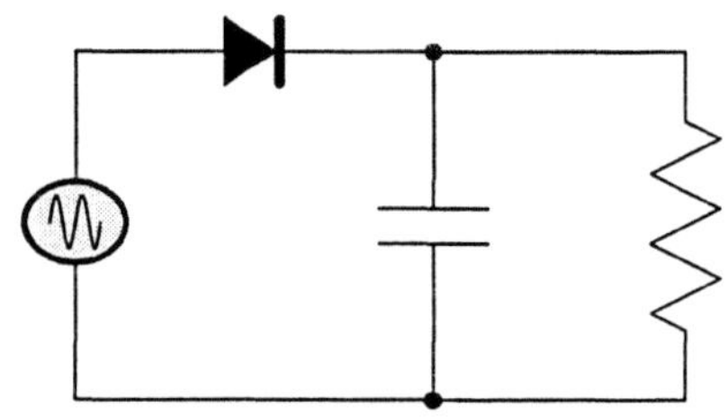

Let I_C and I_R denote the currents through the capacitor and the resistor. Kirchhoff's voltage law applied to the left and right loops in the figure results in, respectively,

$$V_d' + \tfrac{1}{C}\,I_C = \omega\,A\cos\omega t \quad\text{and}\quad R\,I_R' - \tfrac{1}{C}\,I_C = 0.$$

(Here we have assumed that positive currents through the capacitor and the resistor each flow downward in the figure, resulting in opposite orientations for I_C and I_R within the right loop.) By Kirchhoff's current law, the current through the diode must equal the sum of the currents through the capacitor and the resistor. Therefore,

$$I_C + I_R = \gamma(V_d).$$

Using this, we eliminate I_C from the differential equations, arriving at

$$V_d' + \tfrac{1}{C}\,(\gamma(V_d) - I) = \omega\,A\cos\omega t, \quad I' - \tfrac{1}{RC}\,(\gamma(V_d) - I) = 0.$$

where we've abbreviated I_R with I. After solving this system, the voltage across the resistor can be obtained through Ohm's law: $V_R = R\,I$.

Now let's define diode's current-voltage characteristic as in Section 3.7,

```
γ[v_] := 10^-5 (e^(20 v) - 1)
```

and for convenience define

```
sys := {v'[t] + c^-1 (γ[v[t]] - i[t]) == a ω Cos[ω t],
  i'[t] - (r c)^-1 (γ[v[t]] - i[t]) == 0, v[0] == 0, i[0] == 0}
```

Example 5.4.1

Suppose that the source has amplitude $A = 10$ volts and frequency 60 Hz (i.e., $\omega = 120\pi$). If $C = 2\times 10^{-4}$ farads and $R = 10^3$ ohms, the voltage across the resistor has the following graph:

```
tmax = .1; a := 10; ω = 120 π; r = 10^3; c = 2 * 10^-4;
soln = {v[t], i[t]} /.
   NDSolve[sys, {v[t], i[t]}, {t, 0, tmax}, MaxSteps → ∞][[1]];
Plot[Evaluate[r soln[[2]]], {t, 0, tmax},
  PlotRange → {0, a}, GridLines → Automatic];
```

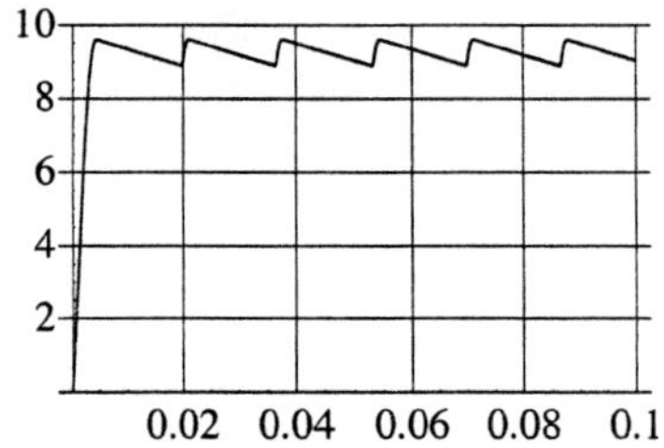

If we decrease the resistance (i.e., increase the load) to $R = 200$, then the voltage becomes more "choppy," and its average value decreases significantly.

```
tmax = .1; a := 10; ω = 120 π; r = 200; c = 2 * 10^-4;
soln = {v[t], i[t]} /.
   NDSolve[sys, {v[t], i[t]}, {t, 0, tmax}, MaxSteps → ∞][[1]];
Plot[Evaluate[r soln[[2]]], {t, 0, tmax},
  PlotRange → {0, a}, GridLines → Automatic];
```

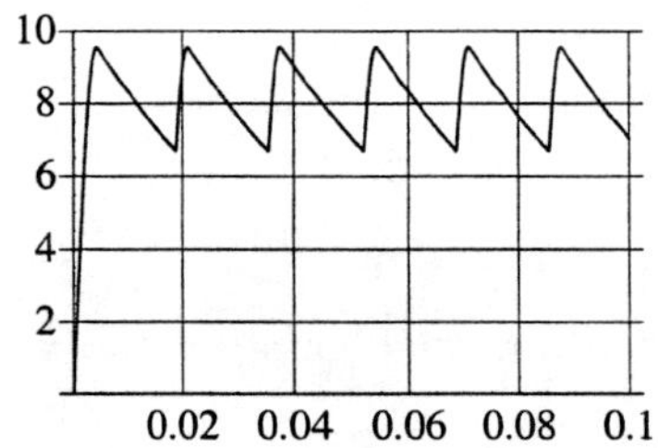

Increasing the capacitance to $C = 10^{-3}$ decreases the amplitude of the oscillations.

```
tmax = .1; a := 10; ω = 120 π; r = 200; c = 10^-3;
soln = {v[t], i[t]} /.
   NDSolve[sys, {v[t], i[t]}, {t, 0, tmax}, MaxSteps → ∞][[1]];
Plot[Evaluate[r soln[[2]]], {t, 0, tmax},
  PlotRange → {0, a}, GridLines → Automatic];
```

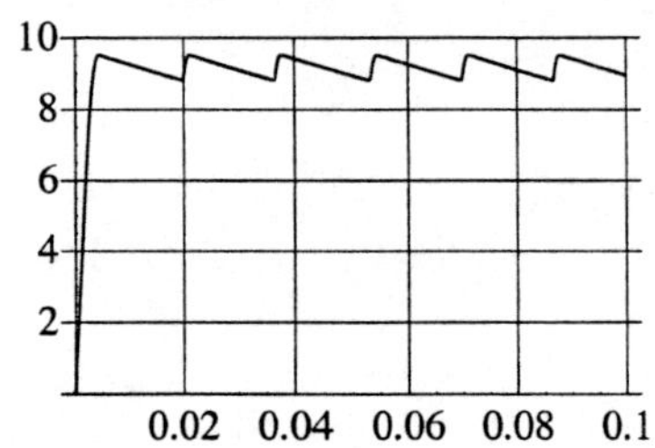

Exercises

1. Suppose that we have a 150 ohm load to which we want our half-wave rectifier circuit to supply 1.5 volts DC. Assume that the amplitude of the AC source is 2 volts and that its frequency is 60 Hz. Experimentally estimate the least capacitance C that will result in an average DC voltage of 1.5 volts and oscillations with amplitude of no more than 0.05 volts (0.1 volts "peak-to-peak.")

2. Suppose that we have a half-wave rectifier circuit with capacitance $C = 0.01$ farads and load with resistance $R = 10$ ohms. Experimentally determine the amplitude of the AC source necessary to supply an average of 3 volts DC to the load.

3. Suppose that our half-wave rectifier circuit has capacitance $C = 0.001$ farads and that the AC source voltage has amplitude 20 volts and frequency 60 Hz. For a load with resistance $R = 500$ ohms, estimate the average DC voltage across the load and the amplitude of its oscillations about the average. To what value can R be decreased before the amplitude of the oscillations becomes 1 volt? What is the average DC voltage across the load then? (For these computations, set `WorkingPrecision→20` in `NDSolve`.)

5.5 Chapter Projects

5.5.1 Two-Tank Mixing Problems

Suppose that two tanks contain well-stirred mixtures of water and some dissolved substance (salt, for instance). Suppose further that flows in and out of each tank are as indicated in the following schematic diagram.

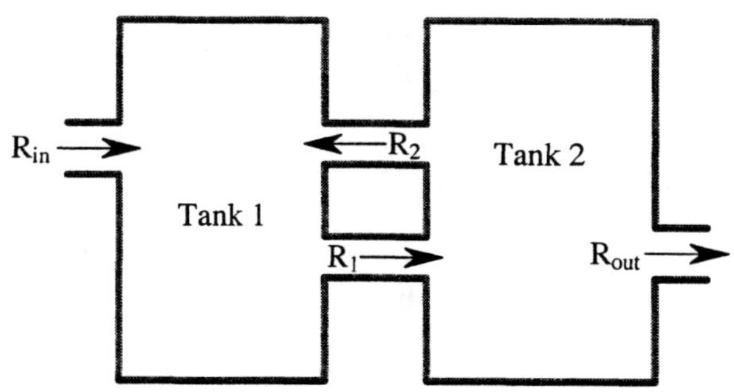

Here A_1 and A_2 are the *amounts* of the solute in the two containers, respectively, with units of mass. The rates R_{in}, R_{out}, R_1, and R_2 are understood to be *volume flow rates* (i.e., their units are volume per unit time). Let V_1 and V_2 denote the volumes of solution in the tanks. Then solute *concentrations* are given by A_i / V_i, $i = 1, 2$. We assume that the solution entering the system via the first tank has a given concentration k_{in}; thus the *mass flow rate* into the system via the first tank is $k_{in} R_{in}$. The mass flow rates corresponding to each of the other inflows and outflows have the form $R_* A_i / V_i$. These mass flow rates are indicated in the following figure.

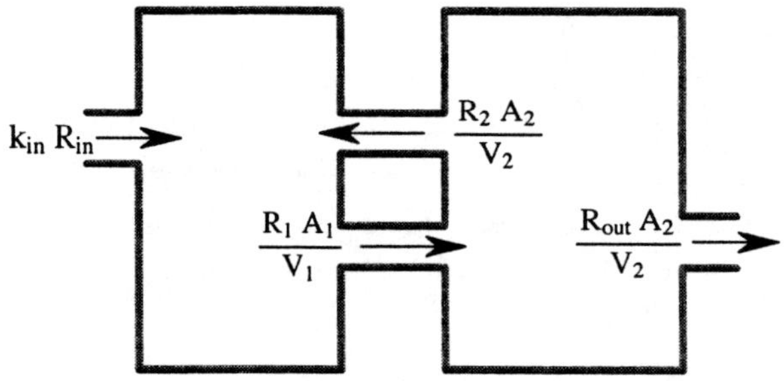

By setting A_i' = (mass flow rate in)−(mass flow rate out) for each of the two tanks, we obtain the differential equations

$$A_1' = k_{in} R_{in} - \frac{R_1}{V_1} A_1 + \frac{R_2}{V_2} A_2, \quad A_2' = \frac{R_1}{V_1} A_1 - \frac{R_2 + R_{out}}{V_2} A_2$$

We assume that the volume flow rates are constant. As a consequence the volumes are linear functions of t:

$$V_1 = V_1(0) + (R_{in} + R_2 - R_1)\, t \text{ and } V_2 = V_2(0) + (R_1 - R_2 - R_{out})\, t.$$

Thus the volumes themselves are constant if

$$R_{in} = R_{out} = R_1 - R_2\,.$$

1. For a scenario typical of the constant-volume case, let $k_{in} = 1$, $R_{in} = R_{out} = R_2 = 1$, $R_1 = 2$, and $V_1 = V_2 = 1$. The phase portrait for the system then looks like the following:

```
<< DiffEqs`DEGraphics`;
PhasePlot[{1 - 2 A1 + A2, 2 A1 - 2 A2}, {t, -1, 5}, {A1, 0, 3},
  {A2, 0, 3}, InitialPoints → {{0.3, 0}, {1, 0}, {1.75, 0}, {3, 0},
    {0, 1}, {0, 2.4}, {.5, 3}, {1.5, 3}, {2.4, 3}, {3, 1.5}},
  ShowEquilibriumPoints → True, ShowNullclines → True];
```

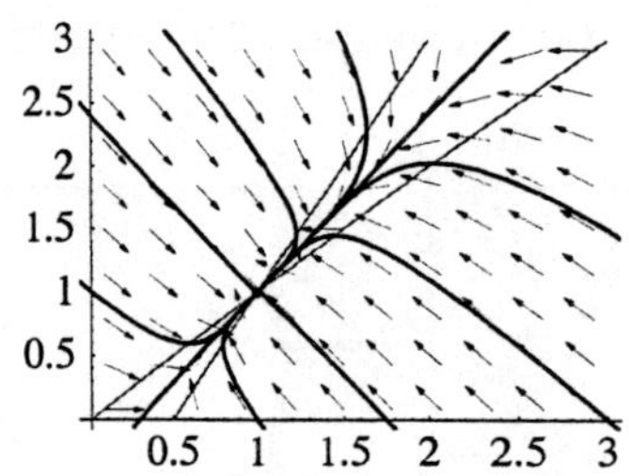

(a) Notice the equilibrium point at (1, 1) and the "straight-line" orbits that approach it as $t \to \infty$. Find the slopes of those lines by setting $A_2 = mA_1$ in the *slope equation*

$$\frac{dA_2}{dA_1} = \frac{2A_1 - 2A_2}{1 - 2A_1 + A_2}\,.$$

In the context of the two-tank mixing problem, give an interpretation of the solutions corresponding to the straight-line orbits.

(b) Use `NDSolve` to find the solution with initial values $A(0) = 0$, $A_2(0) = 1$, and plot the resulting $A_1(t)$ and $A_2(t)$ together versus t. Discuss the correspondence between these graphs and the orbit beginning at (0, 1) in the above phase portrait.

(c) Repeat (b) with initial values $A_1(0) = 3$, $A_2(0) = 0$, and the orbit beginning at (3, 0) in the above phase portrait.

2. Let $R_{\text{in}} = R_2 = 1$, $R_1 = R_{\text{out}} = 2$, and $V_1(0) = V_2(0) = 10$. Then

$$V_1(t) = 10 \text{ and } V_2(t) = 10 - t \text{ for } 0 \le t \le 10.$$

Suppose also that $k_{\text{in}} = 1$ and $A_1(0) = A_2(0) = 0$. Use `NDSolve` to help create a plot of $A_1(t)$ and $A_2(t)$ on the interval [0, 10]. Also, use `ParametricPlot` to plot the orbit, and then discuss the correspondence between the two plots.

3. Let $R_{\text{in}} = R_1 = R_{\text{out}} = 2$, $R_2 = 1$, each in gallons per hour, and suppose that initially the first tank is empty and the second tank contains 10 gallons of water in which $A_2(0) = 1$ lb of salt is dissolved. Note that the volumes are

$$V_1(t) = t \text{ and } V_2(t) = 10 - t \text{ for } 0 \le t \le 10.$$

Suppose that pure water is pumped into the first tank, i.e., $k_{\text{in}} = 0$. Note also that since $V_1(0) = 0$, we must have $A_1(0) = 0$. Obtain plots of $A_1(t)$ and $A_2(t)$ as well as the concentrations $A_1(t)/V_1(t)$ and $A_2(t)/V_2(t)$. What is the approximate salt concentration in each tank as $t \to 10^-$? (*Suggestion*: To avoid $t = 0$ and $t = 10$, where terms in the differential equations are undefined, use `NDSolve` on the interval $\epsilon \le t \le 10 - \epsilon$, where ϵ is some small positive number such as 10^{-10}.)

5.5.2 Mixing in a Draining Tank

Consider a mixing problem involving a cylindrical tank with cross-sectional area 1 and height H. At time $t = 0$ a salt solution with concentration k_{in} begins to flow in at a rate R_{in}, while the well-stirred contents of the tank drains out of a hole in the bottom at a rate given by Toricelli's law:

$$R_{\text{out}} = \rho\sqrt{y(t)} = \rho\sqrt{V(t)},$$

where $y(t)$ is the depth and $V(t)$ is the volume of the liquid in the tank. For given values of k_{in}, R_{in}, ρ, A_0 and V_0, the initial-value problem of interest is

$$A' = k_{\text{in}} R_{\text{in}} - \frac{\rho\sqrt{V}}{V} A, \quad A(0) = A_0,$$

$$V' = R_{\text{in}} - \rho\sqrt{V}, \qquad V(0) = V_0.$$

1. Let $k_{\text{in}} = 2$, $R_{\text{in}} = 0.5$, and $\rho = 1$. Use `PhasePlot` to make a phase portrait on the region $0 \le A \le 2$, $0.01 \le V \le 2$. Include in it orbits starting at initial points (0.01, 0.01), (2, 0.01), (0, 1), (1, 1), (2, 1), (0, 0.25), and (2, 0.25). Are there any straight-line orbits? In the context of this mixing problem, to what sort of behavior do they correspond?

2. Again with $k_{\text{in}} = 2$, $R_{\text{in}} = 0.5$, and $\rho = 1$, suppose that $V_0 = 3$ and $A_0 = 0$. Use `NDSolve` to help create a plot showing the graphs of the amount $A(t)$ and the concentration $A(t)/V(t)$. Estimate the maximum amount. How does the concentration behave?

Chapter 6

Second-Order Equations and Applications

6.1 Plotting Solutions

We begin here a study of second-order initial-value problems

$$y'' = g(t, y, y'),$$
$$y(t_0) = y_0, \quad y'(t_0) = v_0,$$

where g is continuous function on some region in $\mathbb{R}^3$ that contains the point (t_0, y_0, v_0). In any such problem we refer to the vector $(y, y')^T$ as the *state vector*. With $v = y'$, the problem is equivalent to the following first-order problem for the state vector:

$$\begin{pmatrix} y \\ v \end{pmatrix}' = \begin{pmatrix} v \\ g(t, y, v) \end{pmatrix}, \quad \begin{pmatrix} y(0) \\ v(0) \end{pmatrix} = \begin{pmatrix} y_0 \\ v_0 \end{pmatrix}.$$

Example 6.1.1

Consider the problem

$$y'' = (1 - y^2)y' - y, \quad y(0) = 2, \quad y'(0) = 0,$$

of which the corresponding first-order system is

$$\begin{pmatrix} y \\ v \end{pmatrix}' = \begin{pmatrix} v \\ (1 - y^2)v - y \end{pmatrix}, \quad \begin{pmatrix} y(0) \\ v(0) \end{pmatrix} = \begin{pmatrix} 2 \\ 0 \end{pmatrix}.$$

The differential equation here is *van der Pol's equation* (with particular parameter values), which arose in van der Pol's study of triode vacuum tube circuits in the 1920s.

We'll use `NDSolve` to compute a numerical solution.

```
tmax = 15; soln = {y[t], v[t]} /.
  NDSolve[{y'[t] == v[t], v'[t] == (1 - y[t]^2) y'[t] - y[t], y[0] == 2,
     v[0] == 0}, {y[t], v[t]}, {t, 0, tmax}, MaxSteps → ∞][[1]]
```

```
{InterpolatingFunction[{{0., 15.}}, <>][t],
 InterpolatingFunction[{{0., 15.}}, <>][t]}
```

We'll now view the solution three different ways. First we plot y and y' versus t.

```
Plot[Evaluate[soln], {t, 0, tmax}, PlotRange → All];
```

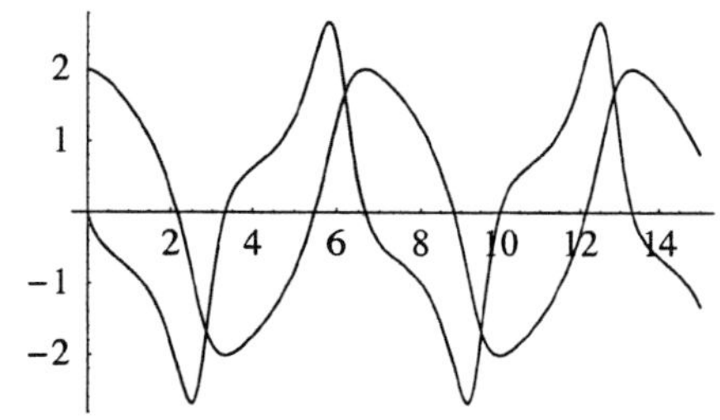

Next we make a parametric plot in the (y, y')-plane, which confirms that the solution is indeed periodic.

```
ParametricPlot[Evaluate[soln], {t, 0, tmax}, PlotRange → All];
```

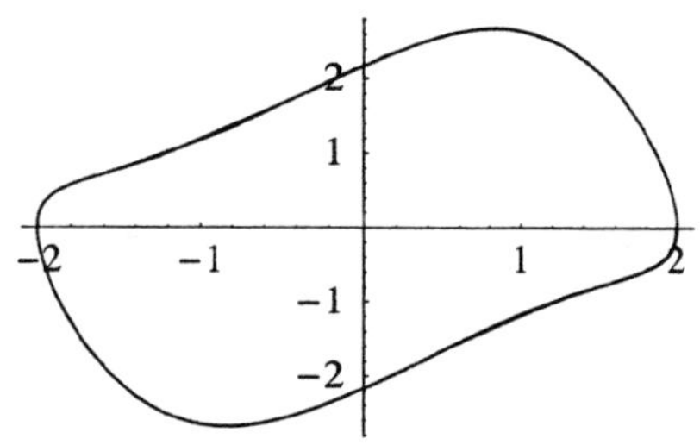

Our third plot will be of the "*time-state trajectory*". This can be done easily with the `TimeStatePlot` function in the `DEGraphics` package.

```
<< DiffEqs`DEGraphics`;
TimeStatePlot[Evaluate[soln], {t, 0, tmax}, PlotPoints → 400];
```

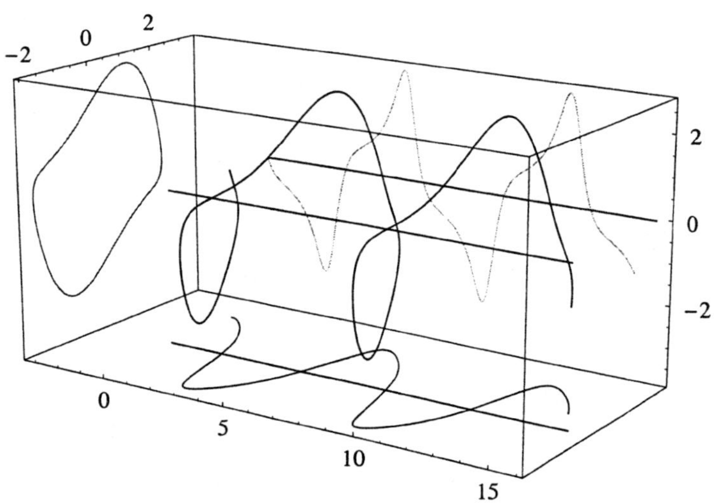

Notice that this plot includes the previous two. The graphs of y and y' are projections onto the bottom and back of the box, respectively, and the parametric plot in the (y, y')-plane is the projection onto the left end of the box.

Example 6.1.2

Consider the *linear* problem

$$y'' + y' + t^2 y = 2\cos 2t, \quad y(0) = 1, \quad y'(0) = 0.$$

The corresponding first-order system is

$$\begin{pmatrix} y \\ v \end{pmatrix}' = \begin{pmatrix} v \\ 2\cos 2t - v - t^2 y \end{pmatrix}, \quad \begin{pmatrix} y(0) \\ v(0) \end{pmatrix} = \begin{pmatrix} 1 \\ 0 \end{pmatrix}.$$

We compute a numerical solution,

```
tmax = 10; soln = {y[t], v[t]} /.
  NDSolve[{y'[t] == v[t], v'[t] == 2 Cos[2 t] - v[t] - t^2 y[t],
     y[0] == 1, v[0] == 0}, {y[t], v[t]}, {t, 0, tmax}][[1]]
```

```
{InterpolatingFunction[{{0., 10.}}, <>][t],
 InterpolatingFunction[{{0., 10.}}, <>][t]}
```

and skip directly to the time-state plot.

```
TimeStatePlot[Evaluate[soln], {t, 0, tmax}, PlotPoints → 200];
```

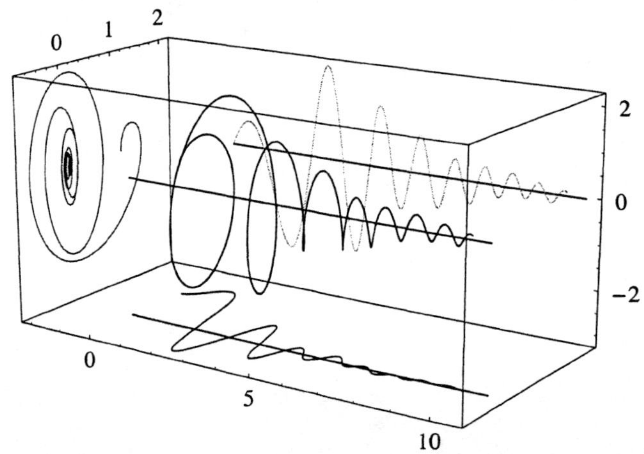

About TimeStatePlot

As always, the question mark reveals information on a *Mathematica* object.

```
? TimeStatePlot
```

```
TimeStatePlot[{x[t],y[t]}, {t,tmin,tmax},
  options]. Creates a 3D plot of the time-state trajectory (
  t,x,y) with projections onto the tx, ty, and xy planes.
```

The following shows the available options and their defaults. Of these, `BoxRatios` and `ViewPoint` are perhaps the most useful.

```
Options[TimeStatePlot]
```

```
{ShadowColors → {RGBColor[1, 0, 0], RGBColor[0, 1, 0], RGBColor[0, 0, 1]},
 ShadowPositions → {-0.67, 0.8, -0.67}, BoxRatios → {2, 1, 1},
 PlotPoints → Automatic, ViewPoint → {2, -3, 1}, PlotStyle → {},
 PlotRange → All, Axes → True, AxesLabel → None, Ticks → Automatic}
```

Exercises

Simple spring-mass systems (i.e. with linear characteristics) are generally governed by equations of the form

$$m\,y'' = -r\,y' - k\,y + F.$$

Here m is the mass, $r\,y'$ is the damping force of friction, $k\,y$ is the force exerted by the spring (by Hooke's law), and F accounts for all external forces. It is customary to rewrite the equation as

$$m\,y'' + r\,y' + k\,y = F,$$

or

$$y'' + \frac{r}{m}\,y' + \frac{k}{m}\,y = \frac{1}{m}\,F.$$

Converting to a pair of first-order equations gives us

$$y' = v \quad \text{and} \quad v' = -\frac{r}{m}\,v - \frac{k}{m}\,y + \frac{1}{m}\,F.$$

1. Consider an unforced spring-mass system with unit mass and unit spring stiffness:

$$y'' + r\,y' + y = 0.$$

With initial conditions $y(0) = 1$ and $y'(0) = 0$, create a time-state plot for each of $r = 0,\ 1,\ 2,$ and 3. In each case, describe the behavior of the mass and how it relates to the time-state trajectory.

2. Consider an unforced spring-mass system with unit mass and friction coefficient $r = 2$:

$$y'' + 2y' + k\,y = 0.$$

Using initial conditions $y(0) = 1$ and $y'(0) = 0$, create a time-state plot for each of $k = \frac{1}{2},\ 1,$ and 2. In each case, describe the behavior of the mass and how it relates to the time-state trajectory.

3. Consider an unforced, undamped spring-mass system with nonlinear spring stiffness governed by

$$y'' + y\,e^{-y} = 0.$$

(a) Plot the function $-y\,e^{-y}$ for $-1.5 \le y \le 3$. What does the graph tell you about the spring?

(b) Create a time-state plot with initial conditions $y(0) = y_0$, $y'(0) = 0$, for each of $y_0 = 1, 2, 4,$ and 6. In each case, describe the behavior of the mass and how it relates to the time-state trajectory.

4. Consider a spring-mass system as depicted in the figure below, with unit mass and unit stiffness, in which the mass slides along a surface that is perfectly smooth where $y < .5$ and rough where $y \ge .5$.

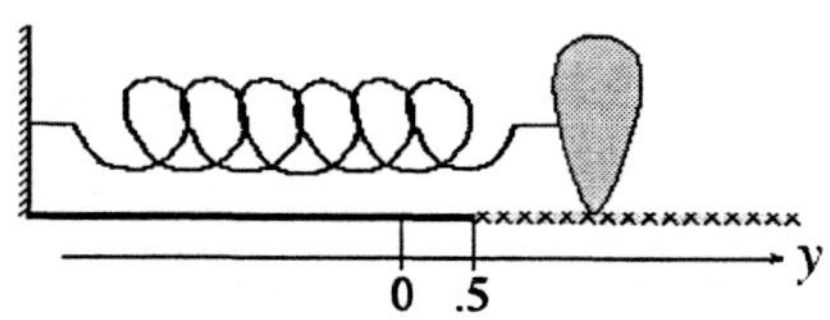

Suppose that the rough part of the surface causes a drag force of 1 dyne per 1 cm/sec of velocity, while the smooth part causes no drag at all. The motion is then governed by

$$y'' + r(y)\,y' + y = 0, \text{ where } r(y) = \begin{cases} 0, & \text{if } y < .5 \\ 1, & \text{if } y \geq .5 \end{cases}$$

(a) Plot the subsequent motion and velocity of the mass as functions of time (up to $t = 30$) for the initial conditions $y(0) = 2$, $y'(0) = 0$. Also plot the phase plane orbit and the time-state trajectory.

(b) Suppose that the rough part of the surface begins at $y = 0$ instead of $y = .5$. Repeat (a), and describe the qualitative differences between this motion and that seen in (a).

5. Consider a spring-mass system, as shown in the figure below, with mass $m = 1$ and where each spring has stiffness $k = 1$.

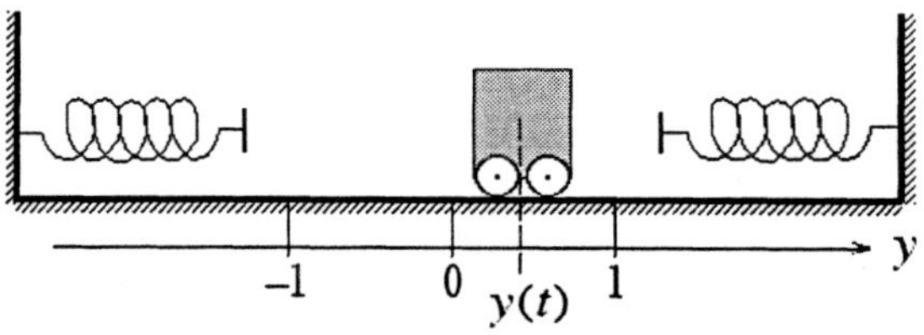

Neither spring is attached to the cart, so each spring loses contact with the cart when it becomes fully elongated. When there is no damping nor external force, the motion is then governed by

$$y'' + k(y)\,y = 0, \text{ where } k(y) = \begin{cases} 1, & \text{if } y \leq -1 \\ 0, & \text{if } -1 < y < 1 \\ -1, & \text{if } y \geq 1 \end{cases}$$

(a) Plot the subsequent motion and velocity of the mass as functions of time (up to $t = 30$) for the initial conditions $y(0) = 2$, $y'(0) = 0$. Also plot the phase plane orbit and the time-state trajectory.

(b) Suppose that there is a damping force given by $-0.2\,y'$. Repeat (a), and describe the observed motion.

(c) Suppose that the cart has an engine that propels it in the negative direction with a constant force of -0.1 dynes (and no change in mass). Repeat (b), and describe the observed motion.

6. A mass is allowed to move without friction in a straight line between $y = 0$ and $y = 1$. Located at each endpoint is a magnetic device that *repells* the mass with an "inverse-square-law" force. Thus the equation of motion is

$$y'' = \frac{k_1}{y^2} - \frac{k_2}{(1-y)^2},$$

where k_1 and k_2 are positive constants.

(a) If the equilibrium position of the mass is $y = \frac{1}{3}$, find the ratio k_1 / k_2.

(b) If the mass is gently released from position $y = \frac{1}{2}$, it is observed to oscillate with a period of 1.35 seconds. Use this information, along with your answer to (a) to experimentally estimate k_1 and k_2 to three decimal places.

(c) Using your values of k_1 and k_2 from (b), plot a time-state trajectory corresponding to initial conditions $y(0) = 0.1$, $y'(0) = 0$. Then experimentally estimate the period of the motion.

(d) Repeat (c) first with $y(0) = 0.9$, $y'(0) = 0$, and then with $y(0) = 0.99$, $y'(0) = 0$. Explain how and why the motion of the mass depends upon the initial position.

6.2 The Simple Pendulum

The equation of motion for the simple pendulum with arm length L is

$$\theta'' = -\frac{g}{L}\sin\theta,$$

which is equivalent to

$$\begin{pmatrix}\theta\\ \omega\end{pmatrix}' = \begin{pmatrix}\omega\\ -\frac{g}{L}\sin\theta\end{pmatrix},$$

Here g is the usual gravitational acceleration constant, θ is the angle formed by the pendulum with the downward vertical, and $\omega = \theta'$ is the *angular velocity*.

Example 6.2.1

Suppose that the arm of our pendulum has length $L = 1$ m and that it is gently released from a position where $\theta = \frac{\pi}{6}$. A numerical solution is computed by `NDSolve` as follows:

```
tmax = 5; L = 1; θ0 = π / 6; soln = {θ[t], ω[t]} /.
  NDSolve[{θ'[t] == ω[t], ω'[t] == -9.8 L^-1 Sin[θ[t]], θ[0] == θ0,
     ω[0] == 0}, {θ[t], ω[t]}, {t, 0, tmax}, MaxSteps → ∞][[1]]

{InterpolatingFunction[{{0., 5.}}, <>][t],
 InterpolatingFunction[{{0., 5.}}, <>][t]}
```

A plot of the angle and the angular velocity versus t shows that the solution is nearly sinusoidal.

```
Plot[Evaluate[soln], {t, 0, tmax}, Ticks → {Automatic, {-1, 1}}];
```

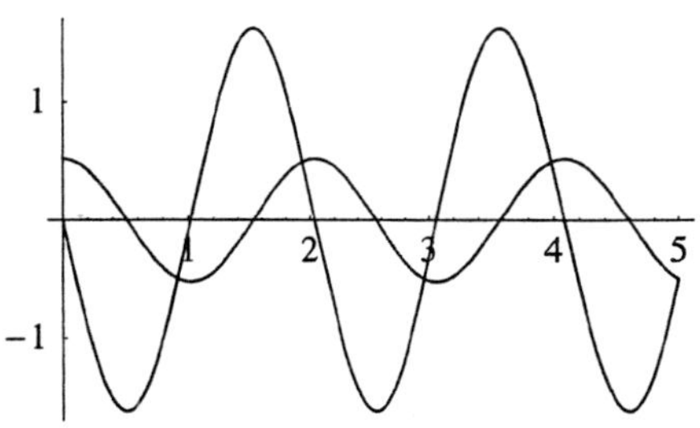

A parametric plot shows a nearly elliptical orbit.

```
ParametricPlot[Evaluate[soln],
  {t, 0, tmax}, Ticks → {Automatic, {-1, 1}}];
```

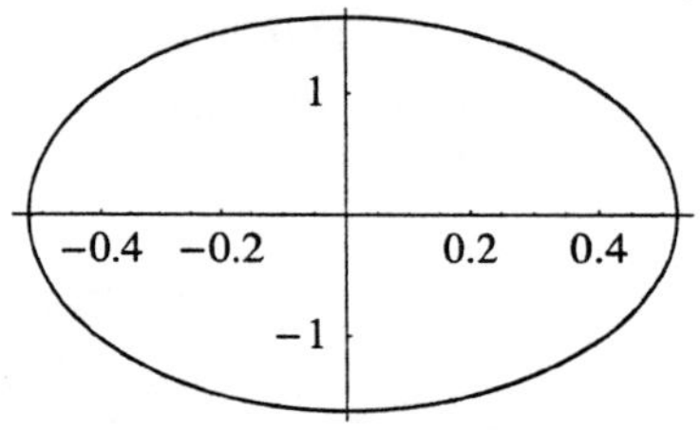

The following is the time-state trajectory of the solution.

```
TimeStatePlot[Evaluate[soln], {t, 0, tmax}];
```

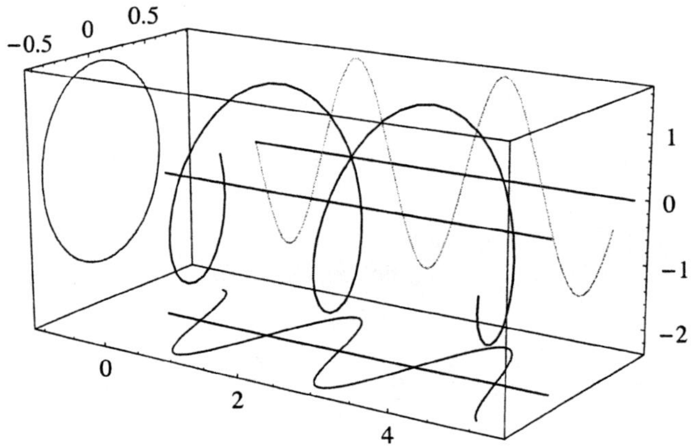

Example 6.2.2

Consider the same pendulum as in Example 6.2.1, but this time suppose that the initial angle is $\frac{5\pi}{6}$. The following plots show that the solution is far from being sinusoidal.

```
tmax = 10; L = 1; θ0 = 5 π / 6; soln = {θ[t], ω[t]} /.
  NDSolve[{θ'[t] == ω[t], ω'[t] == -9.8 L^-1 Sin[θ[t]], θ[0] == θ0,
    ω[0] == 0}, {θ[t], ω[t]}, {t, 0, tmax}, MaxSteps → ∞][[1]];
```

```
Plot[Evaluate[soln], {t, 0, tmax}];
```

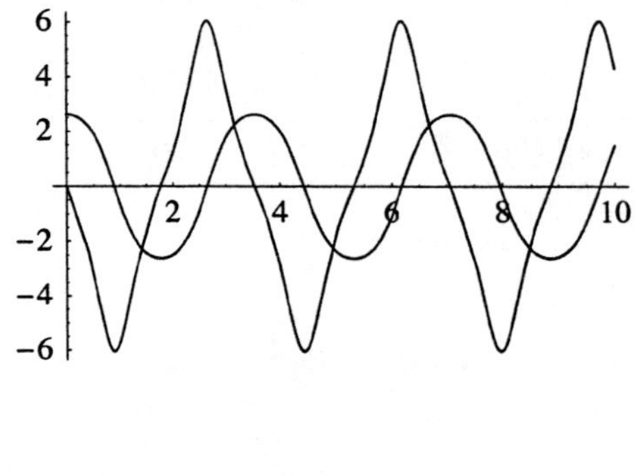

```
ParametricPlot[Evaluate[soln], {t, 0, tmax}];
```

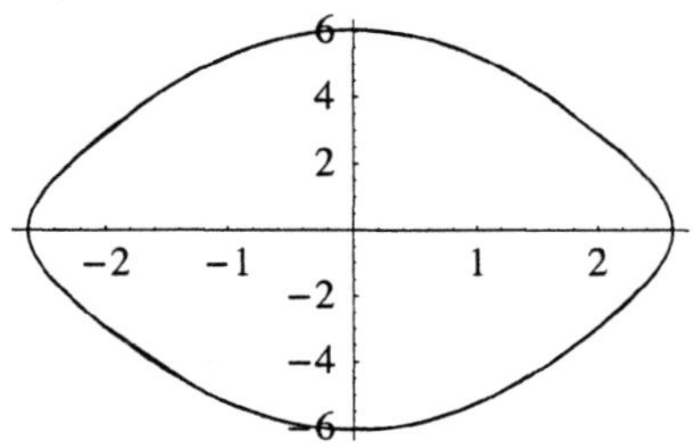


```
TimeStatePlot[Evaluate[soln], {t, 0, tmax}];
```

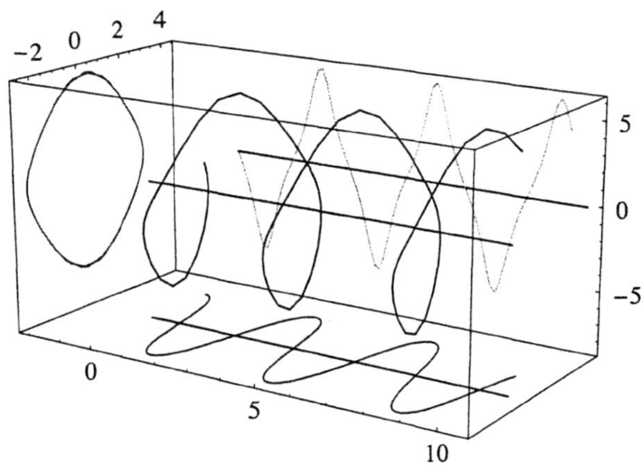

■ Animating the Pendulum

We're now going to have some fun by creating an animation that demonstrates the motion of pendulum. First we need to define some graphics *primitives* for the pieces of the pendulum. These are defined as functions of the angle θ and/or the arm length L as follows:

```
arm[θ_, L_] :=
  {Thickness[.01], GrayLevel[.5], Line[{{0, 0}, L {Sin[θ], -Cos[θ]}}]};
pivot[L_] := {Disk[{0, 0}, .06], {GrayLevel[.9], Disk[{0, 0}, .025]}};
bob[θ_, L_] := {Hue[.7], Disk[L {Sin[θ], -Cos[θ]}, .1]}
```

Now we put these together to form the pendulum.

```
pendulum[θ_, L_] := Graphics[{arm[θ, L], pivot[L], bob[θ, L]}]
```

Next we define a function to produce the frames of the animation.

```
PendulumMovie[{θ0_, ω0_}, L_, tmax_, nframes_] :=
 Module[{soln}, soln[t_] =
   θ[t] /. NDSolve[{θ''[t] == -9.8 L^-1 Sin[θ[t]], θ[0] == θ0, θ'[0] == ω0},
      θ[t], {t, 0, tmax}][[1]];
  Do[Show[pendulum[soln[t], L], AspectRatio → Automatic,
    PlotRange → 1.15 L {{-1, 1}, {-1, 1}}, Frame → True,
    FrameTicks → None], {t, 0, tmax, tmax / (nframes - 1)}]]
```

For illustration, the following creates five frames corresponding to $0 \le t \le 1.5$ with $\theta_0 = \frac{5\pi}{6}$, $\omega_0 = 0$, $L = 1$.

```
PendulumMovie[{5 π / 6, 0}, 1, 1.5, 5];
```

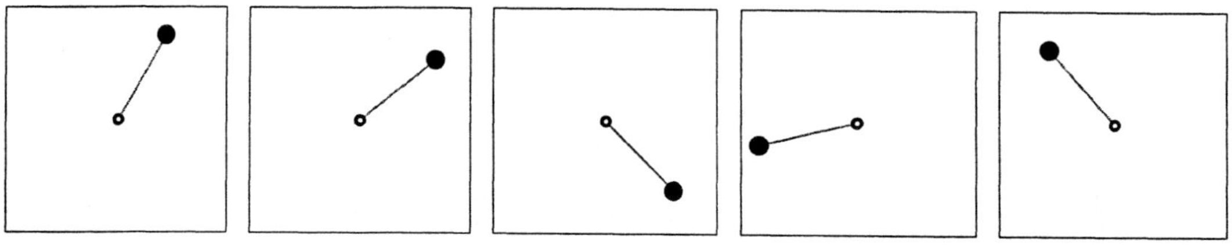

Exercises

1. For each of the following sets of initial conditions, create a time-state plot for the simple pendulum governed by $\theta'' = -\sin\theta$.

   (a) $\theta(0) = \frac{\pi}{4}$, $\theta'(0) = 0$ (b) $\theta(0) = \frac{\pi}{2}$, $\theta'(0) = 0$

   (c) $\theta(0) = \frac{2\pi}{3}$, $\theta'(0) = 0$ (d) $\theta(0) = \frac{11\pi}{12}$, $\theta'(0) = 0$

2. Use `PendulumMovie` to animate the motion for each of (a)–(d) in Exercise 1.

3. Consider a pendulum governed by $\theta'' = -\sin\theta$, and let the initial angular velocity be $\theta'(0) = 0$. Using initial angles of $\theta(0) = 2.8,\ 3,\ 3.1$, and 3.14, experimentally observe how the behavior of the motion changes as the initial position approaches π from below.

4. Consider a pendulum governed by $\theta'' = -\sin\theta$, and let the initial position be $\theta(0) = 0$. Experimentally approximate the (positive) initial angular velocity $\theta'(0)$ at which the resulting motion switches from "back-and-forth" to "round-and-round."

5. The equation $\theta'' + \frac{1}{2}\theta' = -\sin\theta$ models the motion of a certain *damped* pendulum. With initial conditions $\theta(0) = 3.14$ and $\theta'(0) = 0$, create both a time-state plot and an animation of the motion.

6.3 Linear Equations: The Green's Function

From here on in this chapter we will focus on *linear* second-order initial-value problems:

$$y'' + p\,y' + q\,y = f,$$
$$y(t_0) = y_0,\quad y'(t_0) = m_0,$$

where p, q, and f are continuous functions on an interval that contains t_0. Recall that for any two *linearly independent* solutions u and v, every solution of the homogeneous equation $y'' + p\,y' + q\,y = 0$ is of the form

$$y = c_1\,u + c_2\,v$$

for some constants c_1 and c_2. In addition, if $\tilde{y}$ is *any* solution of the nonhomogeneous equation $y'' + p\,y' + q\,y = f$, then *every* solution of that equation is of the form

$$y = c_1\,u + c_2\,v + \tilde{y}$$

for some constants c_1 and c_2.

Let T be the operator defined by $T\,y = y'' + p\,y' + q y$, so that the differential equation can be written as $T\,y = f$. Here's a *Mathematica* definition of T:

```
T[y_] := ∂t,t y + p[t] ∂t y + q[t] y
```

Given a pair of linearly independent solutions u, v of $T\,y = 0$, the Green's function associated with T is

```
              u[t] v[s] - u[s] v[t]
G[t_, s_] := ---------------------
             u'[s] v[s] - u[s] v'[s]
```

The *rest solution* (i.e., the solution satisfying $y(t_0) = y'(t_0) = 0$) is given by

```
yRest[t_] := ∫_t0^t G[t, s] f[s] ds
```

To verify this, we first check that the "rest" initial conditions are satisfied.

```
{yRest[t0], yRest'[t0]}
{0, 0}
```

Next we check that the differential equation is satisfied. The following expression should equal 0:

```
T[yRest[t]] - f[t]
```

$$\int_{t0}^{t} \frac{f[s]\,(v[s]\,u''[t] - u[s]\,v''[t])}{v[s]\,u'[s] - u[s]\,v'[s]}\,ds + \left(\int_{t0}^{t} \frac{f[s]\,(v[s]\,u'[t] - u[s]\,v'[t])}{v[s]\,u'[s] - u[s]\,v'[s]}\,ds\right) p[t\ \left(\int_{t0}^{t} \frac{f[s]\,(u[t]\,v[s] - u[s]\,v[t])}{v[s]\,u'[s] - u[s]\,v'[s]}\,ds\right) q[t]$$

The three integrals in this expression can be combined into the single expression

$$\int_{t_0}^{t} \frac{v[s]\;Tu[t] - u[s]\;Tv[t]}{v[s]\,u'[s] - u[s]\,v'[s]}\;f[s]\,ds,$$

which is clearly zero, provided that $T u = 0$ and $T v = 0$, which we assume.

Let's now define an operator $\mathcal{G}$ as follows (type ESC scG ESC):

```
𝒢[φ_][t_] := ∫_t0^t G[t, s] (φ /. t → s) ds
```

Note that $\mathcal{G}$ takes an expression $f(t)$ and returns the corresponding rest solution of the differential equation.

```
𝒢[f[t]][t]
```

$$\int_{t0}^{t} \frac{f[s]\,(u[t]\,v[s] - u[s]\,v[t])}{v[s]\,u'[s] - u[s]\,v'[s]}\,ds$$

Therefore, the preceding discussion tells us that $T\mathcal{G}f = f$ for any continuous function f and $\mathcal{G}T y = y$ for any twice-differentiable function y satisfying $y(t_0) = y(t_0) = 0$. So if we define the domain of T to be all twice-differentiable functions y satisfying $y(t_0) = y(t_0) = 0$, then T and $\mathcal{G}$ are in fact *inverses* of each other. What this all comes down to is that

$$\begin{matrix} Ty = f \\ y(t_0) = y(t_0) = 0 \end{matrix} \quad \text{if and only if } y = \mathcal{G}f.$$

Example 6.3.1

Consider the equation $y'' + y = f$. The operator here is

```
T[y_] := ∂t,t y + y
```

A pair of linearly independent solutions of the homogeneous equation $Ty = 0$ is given by

```
u[t_] := Cos[t]; v[t_] := Sin[t];
```

which we check (as if it were necessary) by entering

```
{T[u[t]], T[v[t]]}
{0, 0}
```

Thus the Green's function (as defined above) is

```
G[t, s] // Simplify
-Sin[s - t]
```

and the rest solution of $Ty = f$ (for $t_0 = 0$, with $\mathcal{G}$ as defined above) is

```
t0 = 0; 𝒢[f[t]][t] // Simplify
```

$$-\int_0^t f[s]\,\mathrm{Sin}[s-t]\,ds$$

The inverse properties of T and $\mathcal{G}$ are easy to check.

```
T[𝒢[f[t]][t]] // Simplify
f[t]
𝒢[T[y[t]]][t] // Simplify
-Cos[t] y[0] + y[t] - Sin[t] y'[0]
```

Note that the second of these does not reduce to $y(t)$ unless y satisfies the rest initial conditions.

```
𝒢[T[y[t]]][t] /. {y[0] → 0, y'[0] → 0} // Simplify
y[t]
```

For a specific instance, suppose we want to find the rest solution of $Ty = e^{-t}\cos t$. We get

```
yRest[t_] = 𝒢[e^-t Cos[t]][t] // Simplify
```

$$\frac{1}{5} e^{-t} \left(-\left(-1 + e^t\right) \mathrm{Cos}[t] + \left(-2 + 3\,e^t\right) \mathrm{Sin}[t]\right)$$

which is easily verified as follows:

```
T[yRest[t]] // Simplify
```

e^{-t} Cos[t]

```
{yRest[0], yRest'[0]}
```

{0, 0}

Example 6.3.2

Consider the operator T defined by $Ty = y'' - \frac{2}{t}\, y' + \frac{2}{t^2}\, y$.

$$T[y\_] := \partial_{t,t}\, y - \frac{2}{t}\, \partial_t\, y + \frac{2}{t^2}\, y$$

We first need a pair of solutions of the homogeneous equation $Ty = 0$. Let's try looking for them in the form of t^p.

```
T[t^p] // Simplify
```

$(2 - 3\,p + p^2)\; t^{-2+p}$

So we see that $T(t^p) = 0$ if $2 - 3\,p + p^2 = 0$. This quadratic equation is easy to solve and gives solutions $p = 1, 2$. Therefore, a pair of linearly independent solutions of $Ty = 0$ on $(0, \infty)$ is given by

```
u[t_] := t; v[t_] := t^2;
```

So the Green's function is

```
G[t, s] // Simplify
```

$$\frac{t\,(-s + t)}{s}$$

and the operator $\mathcal{G}$ (for $t_0 > 0$) is

```
Clear[t0]; 𝒢[f[t]][t] // Simplify
```

$$-t \int_{t0}^{t} \frac{(s - t)\; f[s]}{s}\, ds$$

■ Generating the Operator from a Given Pair of Solutions

Finding the general solution of a homogeneous equation $y'' + p\,y' + q\,y = 0$ essentially comes down to the following problem:

Given functions p and q, find linearly solutions u and v.

The point is that we can think of u and v as depending upon p and q. An associated *inverse problem* is:

Given linearly independent solutions u and v, find the coefficients p and q.

This inverse problem is really just an algebraic problem that is quite simple to solve. Suppose that u and v are given, linearly independent solutions of $y'' + p\,y' + q\,y = 0$ on an interval I. Then the coefficients p and q are given in terms of u and v as follows:

```
Solve[{u''[t] + p u'[t] + q u[t] == 0, v''[t] + p v'[t] + q v[t] == 0}, {p, q}]
```

$$\left\{\left\{p \to -\frac{v[t]\,u''[t] - u[t]\,v''[t]}{v[t]\,u'[t] - u[t]\,v'[t]},\ q \to -\frac{-v'[t]\,u''[t] + u'[t]\,v''[t]}{v[t]\,u'[t] - u[t]\,v'[t]}\right\}\right\}$$

Notice something important here: The denominators are both $u'v - uv'$, and so this quantity *must be nonzero throughout the interval I*. But recall that $u'v - uv'$ is precisely the *Wronskian* of u and v, and theory tells us that it is never zero on I, provided that the solutions u and v are linearly independent on I.

Let's use what we've found to define the operator T associated with the solutions u and v.

$$\texttt{T[u\_, v\_][y\_] :=} \ \partial_{t,t}\,y + \frac{u\,\partial_{t,t}\,v - v\,\partial_{t,t}\,u}{v\,\partial_t\,u - u\,\partial_t\,v}\,\partial_t\,y + \frac{\partial_t\,v\,\partial_{t,t}\,u - \partial_t\,u\,\partial_{t,t}\,v}{v\,\partial_t\,u - u\,\partial_t\,v}\,y$$

Now, for instance, $u = e^{-t}$ and $v = e^{-2t}$ satisfy $Ty = 0$, where T is defined by

```
T[e^-t, e^-2t][y[t]]
2 y[t] + 3 y'[t] + y''[t]
```

The functions $u = t$ and $v = t^2$ (cf. Example 6.3.2) satisfy the equation $Ty = 0$, where T is defined by

```
T[t, t^2][y[t]]
```

$$\frac{2\,y[t]}{t^2} - \frac{2\,y'[t]}{t} + y''[t]$$

Note that here the interval I must not contain 0.

Exercises

1. Let T be the operator defined by $Ty = y'' + y$. Use the Green's function (see Example 6.3.1) to compute the rest solution (at $t_0 = 0$) of $Ty = f$ where:

   (a) $f(t) = t\cos t$ (b) $f(t) = \cos^2 t$

   (c) $f(t) = e^{-t/3}\sin 2t$ (d) $f(t) = \cos 3t \sin 2t$

2. Let T be the operator defined by $Ty = y'' - y$, and let $t_0 = 0$.

   (a) Use $u = e^{t}$ and $v = e^{-t}$ to compute the Green's function.

   (b) Find the rest solution of $Ty = \sin t$.

   (c) Find the rest solution of $Ty = t\,e^{-t}$.

3. Let T be the operator defined by $Ty = y'' + 2y' + 2y$, and let $t_0 = 0$.

   (a) Use $u = e^{-t}\cos t$ and $v = e^{-t}\sin t$ to compute the Green's function.

   (b) Find the rest solution of $Ty = \sin t$.

   (c) Find the rest solution of $Ty = e^{-t}\sin t$.

4. Suppose that T is an operator defined by $Ty = y'' + py' + qy$, where p and q are continuous on an interval I containing $t_0 = 0$. Suppose further that $u = \cos t$ and $v = t \cos t$ are solutions of $Ty = 0$ on I.
   (a) What is the largest interval I on which these solutions can be valid?
   (b) Find (and simplify) the operator T (i.e., what's Ty?).
   (c) Compute the Green's function.
   (d) Find the rest solution of $Ty = \cos t$.
   (e) Find the rest solution of $Ty = t \cos t$.
5. Suppose that T is an operator defined by $Ty = y'' + py' + qy$, where p and q are continuous on an interval I containing $t_0 = 0$. Suppose further that $u = e^t$ and $v = t$ are solutions of $Ty = 0$ on I.
   (a) What is the largest interval containing 0 on which these solutions can be valid?
   (b) Find (and simplify) the operator T (i.e., what's Ty?).
   (c) Compute the Green's function.
   (d) Find the rest solution of $Ty = (t+1)\,e^{-t}$.
   (e) Find the rest solution of $Ty = (t+1)^{-1}$.

6.4 Constant Coefficients: Beats, Resonance, and Frequency Response

The following linear, second-order equations arise in models of spring-mass systems and RLC circuits:

Spring-mass system:	$my'' + ry' + ky = kf,$	y = position of the mass
RLC circuit:	$LV'' + RV' + C^{-1}V = C^{-1}f,$	V = voltage across the capacitor

In each case, we think of the "forcing function" f as an *input* to the system. In the case of the spring-mass system, f may viewed as the position of the other end of the spring. In an RLC circuit, f is a source voltage. We consider here a generic problem of this type, which we write in the form

$$y'' + py' + qy = qf,$$

where *the coefficients p and q are constants* with $p \geq 0$ and $q > 0$. The rest solution (i.e., the solution satisfying zero initial conditions) is the *response* to the input f.

We will concentrate on simple sinusoidal inputs of the form $\cos \omega t$—in particular on the effect of the angular frequency ω (i.e., the *frequency response*)—since many types of inputs can be at least approximated by a sum of sines and cosines with various frequencies.

Undamped Forced Response

Let's first consider the case where $p = 0$. With $\mu = \sqrt{q}$, the equation becomes

$$y'' + \mu^2 y = \mu^2 \cos \omega t.$$

This corresponds to an undamped spring-mass system or an LC circuit. Note that μ is the angular frequency of unforced vibrations, i.e., the sinusoidal solutions of the homogeneous equation $y'' + \mu^2 y = 0$. So μ is called the *natural* angular frequency of the system.

We start by defining the operator T for which the differential equation has the form $Ty = F$.

```
T[y_] := ∂t,t y + μ² y
```

Next we'll define the Green's function corresponding to T.

```
u[t_] := Cos[μ t]; v[t_] := Sin[μ t]

G[t_, s_] = (u[s] v[t] - u[t] v[s])/(u[s] v'[s] - u'[s] v[s]) // Simplify

-Sin[(s - t) μ]/μ
```

The response to the input $f(t) = \cos \omega t$ is

```
response[μ_, ω_][t_] = ∫_0^t G[t, s] μ² Cos[ω s] ds // Simplify

μ² (-Cos[t μ] + Cos[t ω])/((μ - ω) (μ + ω))
```

Note that this is valid only when $\omega^2 \neq \mu^2$. Also, the response can be expressed as a *product* of sines with different frequencies.

```
response[μ, ω][t] // TrigFactor

2 μ² Sin[tμ/2 - tω/2] Sin[tμ/2 + tω/2]/((μ - ω) (μ + ω))
```

So we can view the response as the product of the higher frequency factor $\sin(\frac{\mu+\omega}{2}\, t)$ with a lower frequency *envelope*:

```
envelope[μ_, ω_][t_] := (2 μ²)/(μ² - ω²) Sin[(μ - ω)/2 t]
```

The following plot illustrates the phenomenon of *beats,* which is noticeable when $\mu - \omega$ is small relative to $\mu + \omega$.

```
Plot[{response[5, 4][t], envelope[5, 4][t], -envelope[5, 4][t]},
  {t, 0, 6 π}, PlotStyle → {{}, Hue[0], Hue[0]},
  MaxBend → 1, Ticks → None];
```

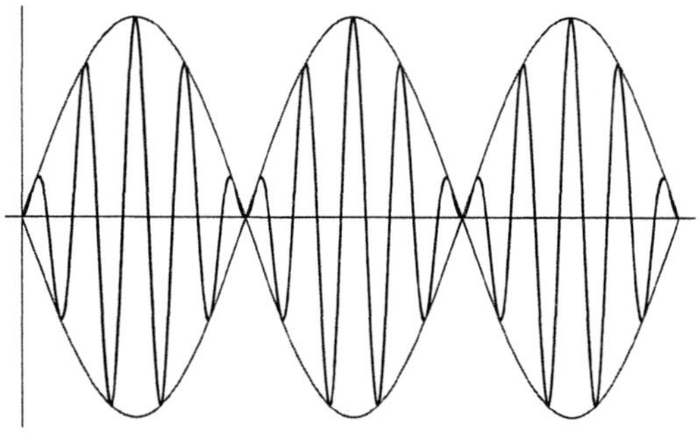

Note that arbitrarily large amplitudes are attained for input frequencies near the natural frequency of the system, i.e., for $\omega \approx \mu$. So what happens when the input frequency is the same as the natural frequency of the system? Let's suppose that $\omega = \mu$ and find out.

We look for a solution of $y'' + \mu^2 y = \mu^2 \cos \mu t$ in the form

$$\tilde{y} = g(t)\cos(\mu t - \phi).$$

Substitution of this into the left side of the equation yields the following:

```
T[y_] := ∂t,t y + μ² y

Lhs = T[g[t] Cos[μ t + φ]] // TrigExpand

-2 μ Cos[φ] Sin[t μ] g′[t] - 2 μ Cos[t μ] Sin[φ] g′[t] +
 Cos[t μ] Cos[φ] g″[t] - Sin[t μ] Sin[φ] g″[t]
```

We want the coefficient of $\cos \mu t$ in this expression to equal μ^2 and the coefficient of $\sin \mu t$ to equal zero.

```
eqn1 = Coefficient[Lhs, Cos[μ t]] == μ²

2 μ Sin[φ] g′[t] + Cos[φ] g″[t] == μ²

eqn2 = Coefficient[Lhs, Sin[μ t]] == 0

-2 μ Cos[φ] g′[t] + Sin[φ] g″[t] == 0
```

These equations suggest that a solution may be found in which g has the form $g(t) = a\,t$.

```
{eqn1, eqn2} /. {g′[t] → a, g″[t] → 0}

{2 a μ Sin[φ] == μ², -2 a μ Cos[φ] == 0}

solns = Solve[%, {a, φ}] // FullSimplify
```

$$\left\{\left\{a \to -\frac{\mu}{2}, \phi \to -\frac{\pi}{2}\right\}, \left\{a \to -\frac{\mu}{2}, \phi \to \frac{\pi}{2}\right\}, \left\{a \to \frac{\mu}{2}, \phi \to -\frac{\pi}{2}\right\}, \left\{a \to \frac{\mu}{2}, \phi \to \frac{\pi}{2}\right\}\right\}$$

The first and last of these yield the same desired solution—the *resonant* response

$$y = \frac{\mu}{2}\, t \cos(\mu t - \tfrac{\pi}{2}) = -\frac{\mu}{2}\, t \sin \mu t.$$

The following is a typical plot.

```
Plot[{t, -t, -t Sin[2 t]}, {t, 0, 8 π}, Ticks → None];
```

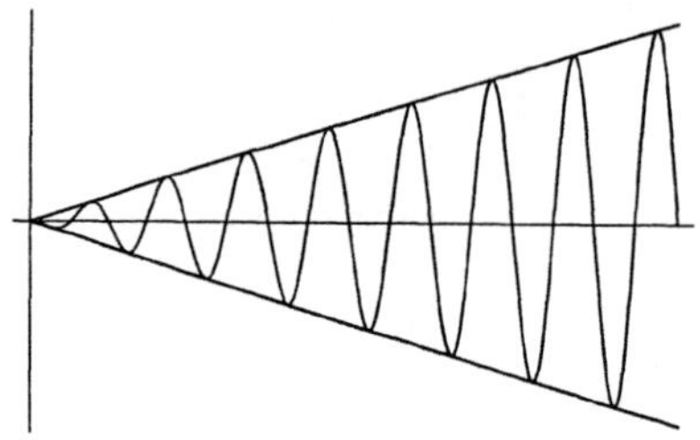

Damped Forced Response

Now let's consider the case where $p > 0$:

$$y'' + p\,y' + q\,y = q\cos\omega t.$$

In this case, solutions of the homogeneous equation decay exponentially to zero. So to study the system's response to the input f, we will simply focus on finding *a* particular solution. We start by defining the operator T.

```
T[y_] := ∂t,t y + p ∂t y + q y
```

Next we will look for a solution of the form

$$\tilde{y} = g_\omega \cos(\omega t - \phi_\omega),\ \text{ where } g_\omega > 0 \text{ and } 0 \le \phi_\omega \le \pi.$$

If there is such a solution, we will call the coefficient g_ω the *gain* of the system and ϕ_ω the *phase angle*. So let's compute $T\tilde{y}$ and expand to bring out factors $\cos\omega t$ and $\sin\omega t$:

```
Lhs = T[g Cos[ω t - φ]] // TrigExpand
```

$$g\,q\,\mathrm{Cos}[\phi]\,\mathrm{Cos}[t\,\omega] - g\,\omega^2\,\mathrm{Cos}[\phi]\,\mathrm{Cos}[t\,\omega] + g\,p\,\omega\,\mathrm{Cos}[t\,\omega]\,\mathrm{Sin}[\phi] - g\,p\,\omega\,\mathrm{Cos}[\phi]\,\mathrm{Sin}[t\,\omega] + g\,q\,\mathrm{Sin}[\phi]\,\mathrm{Sin}[t\,\omega] - g\,\omega^2\,\mathrm{Sin}[\phi]\,\mathrm{Sin}[t\,\omega]$$

Since this is supposed to equal $q\cos\omega t$, we need to solve the following two equations:

```
eqn1 = Coefficient[Lhs, Cos[ω t]] == q
```

$$g\,q\,\mathrm{Cos}[\phi] - g\,\omega^2\,\mathrm{Cos}[\phi] + g\,p\,\omega\,\mathrm{Sin}[\phi] == q$$

```
eqn2 = Coefficient[Lhs, Sin[ω t]] == 0
```

$$-g\,p\,\omega\,\mathrm{Cos}[\phi] + g\,q\,\mathrm{Sin}[\phi] - g\,\omega^2\,\mathrm{Sin}[\phi] == 0$$

```
solns = Solve[{eqn1, eqn2}, {g, φ}] // FullSimplify
```

$$\Big\{\Big\{g \to -\frac{q\,(q-\omega^2)}{\sqrt{(q-\omega^2)^2}\,\sqrt{p^2\,\omega^2 + (q-\omega^2)^2}},\ \phi \to -\mathrm{ArcCos}\Big[-\frac{\sqrt{(q-\omega^2)^2}}{\sqrt{p^2\,\omega^2 + (q-\omega^2)^2}}\Big]\Big\},$$

$$\Big\{g \to -\frac{q\,(q-\omega^2)}{\sqrt{(q-\omega^2)^2}\,\sqrt{p^2\,\omega^2 + (q-\omega^2)^2}},\ \phi \to \mathrm{ArcCos}\Big[-\frac{\sqrt{(q-\omega^2)^2}}{\sqrt{p^2\,\omega^2 + (q-\omega^2)^2}}\Big]\Big\},$$

$$\Big\{g \to \frac{q\,(q-\omega^2)}{\sqrt{(q-\omega^2)^2}\,\sqrt{p^2\,\omega^2 + (q-\omega^2)^2}},\ \phi \to -\mathrm{ArcCos}\Big[\frac{\sqrt{(q-\omega^2)^2}}{\sqrt{p^2\,\omega^2 + (q-\omega^2)^2}}\Big]\Big\},$$

$$\Big\{g \to \frac{q\,(q-\omega^2)}{\sqrt{(q-\omega^2)^2}\,\sqrt{p^2\,\omega^2 + (q-\omega^2)^2}},\ \phi \to \mathrm{ArcCos}\Big[\frac{\sqrt{(q-\omega^2)^2}}{\sqrt{p^2\,\omega^2 + (q-\omega^2)^2}}\Big]\Big\}\Big\}$$

Since we want $0 \le \phi_\omega \le \pi$, we dismiss the first and third of these. Since we want $g_\omega > 0$, we take the second solution when $q - \omega^2 < 0$ and the fourth when $q - \omega^2 > 0$. When $q - \omega^2 = 0$, we get $\phi_\omega = \arccos 0 = \frac{\pi}{2}$ and $g_\omega = \frac{q}{\sqrt{p^2\,\omega^2}} = \frac{\omega}{p}$. However, it turns out that the following formulas work in all cases:

```
gain[p_, q_, ω_] := q / Sqrt[p² ω² + (q - ω²)²];

ϕ[p_, q_, ω_] := ArcCos[(q - ω²) / Sqrt[p² ω² + (q - ω²)²]];
```

The following are typical plots (with $p = 1$, $q = 2$) of the gain and phase angle versus ω.

```
Plot[gain[1, 2, ω], {ω, 0, 5}, Ticks → {Automatic, {1}}];
```

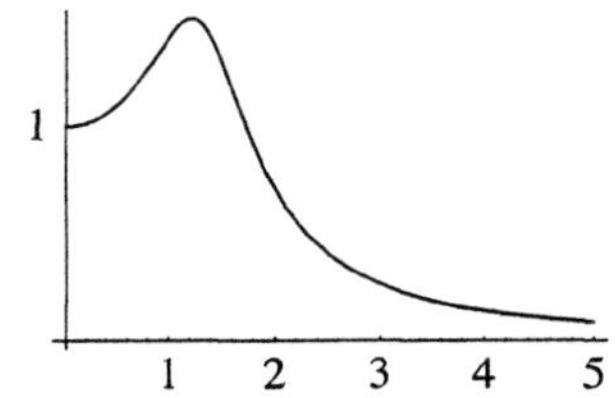

```
Plot[ϕ[1, 2, ω], {ω, 0, 5},
  PlotRange → {0, π}, Ticks → {Automatic, {π / 2, π}}];
```

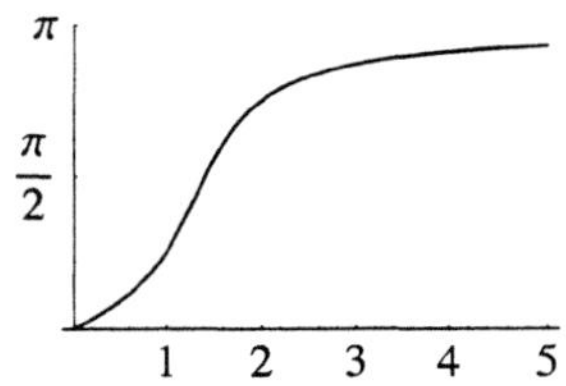

The solution corresponding to $p = 1$, $q = 2$, and $\omega = 1$ has the following gain and phase angle. (Notice these values in the preceding plots.)

```
gain[1, 2, 1]
```

$\sqrt{2}$

```
ϕ[1, 2, 1]
```

$\frac{\pi}{4}$

Here is a graph of that solution along with the forcing function. The phase angle $\frac{\pi}{4}$ shifts the graph of the response by $\frac{1}{8}$ of one cycle (since $\frac{\pi}{4} = \frac{1}{8}\,2\pi$).

```
Plot[{Cos[t], √2 Cos[t - π / 4]}, {t, 0, 4 π}];
```

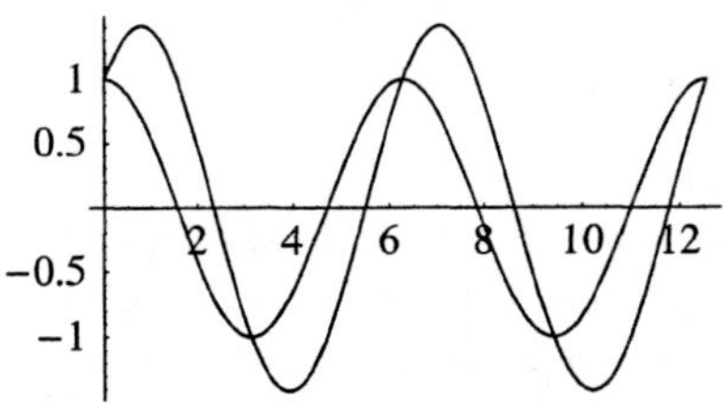

Bode Plots. Engineers are accustomed to thinking of the phase angle in degrees and the gain in *decibels* (dB), which is defined as $20 \log_{10} g_\omega$. A pair of plots showing graphs these quantities versus $\log_{10}\omega$ is called a *Bode plot* and describes the frequency response of the system. The following creates a Bode plot for parameter values $p = 0.5$ and $q = 5$.

```
gainplot[p_, q_] :=
  Plot[20 Log[10, gain[p, q, 10^x]], {x, -1, 1}, PlotRange → All,
   AxesLabel → {"log10ω", "Gain (dB)"}, DisplayFunction → Identity];

angleplot[p_, q_] := Plot[ϕ[p, q, 10^x] / Degree, {x, -1, 1},
   PlotRange → All, AxesLabel → {"log10ω", "Phase Angle (deg)"},
   Ticks → {Automatic, 30 Range[1, 6]}, DisplayFunction → Identity];

BodePlot[p_, q_] :=
  Show[GraphicsArray[{gainplot[p, q], angleplot[p, q]}], PlotLabel →
    "Bode Plot (p=" <> ToString[p] <> ", q=" <> ToString[q] <> ")",
   Frame → True, DisplayFunction → $DisplayFunction];

BodePlot[.5, 5];
```

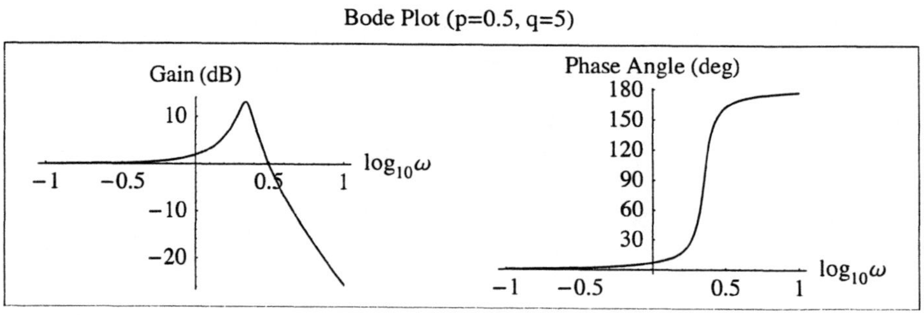

■ Animating a Spring-Mass System

In order to create an animation of a spring-mass system, we first need to build some graphics elements. For this we'll make use of the standard package `Shapes`.

```
<< Graphics`Shapes`
```

Using `Sphere` from the `Shapes` package, we define the mass—in terms of the spring's unstretched length L and the mass's position y—as a ball with radius 1. (Down is the positive direction.)

```
ball[L_, y_] := TranslateShape[
  Graphics3D[{EdgeForm[{}], Sphere[1]} ], {0, 0, -y - L - 1}]
```

We'll build the spring out of a parametric plot of a helix with radius r that makes 5 "loops" between its ends located at y and z. A bit of tweaking is done to bring the ends of the spring in to the center axis. (Notice the strange $(\sin\pi t)^{1/7}$ factor in the second coordinate.)

```
spring[L_, r_, y_, z_] :=
 ParametricPlot3D[{r Sin[10 π t], r Sin[π t]^(1/7) Cos[10 π t],
   -z (1 - t) - (L + y) t, {Thickness[.02], GrayLevel[.3]}},
  {t, 0, 1}, PlotPoints → Automatic, DisplayFunction → Identity]
```

The top end of the spring will be attached to a "handle," for which we'll just make a thick line segment.

```
handle[z_] := ParametricPlot3D[
  {x / 5, x, z, {Thickness[.04], RGBColor[.5, .2, .2]}},
  {x, -1.5, 2}, DisplayFunction → Identity]
```

Now we put our elements together in a frame of the animation.

```
frame[L_, r_, y_, z_] := Show[ball[L, y], handle[-z],
  spring[L, r, y, z], Boxed → False, ViewPoint -> {.5, -3, 1.25},
  PlotRange → {{-2, 2}, {-2, 2}, {-1.8 L, 1.8}},
  PlotRegion → {{0, 1}, {0, 1}}, DisplayFunction → Identity]
```

The following plot shows a spring with unstretched length 4 and radius 1 in compressed, unstretched, and stretched configurations.

```
Show[GraphicsArray[
   {frame[4, 1, -1, 1], frame[4, 1, 0, 0], frame[4, 1, 1, -1]}],
  DisplayFunction → $DisplayFunction];
```

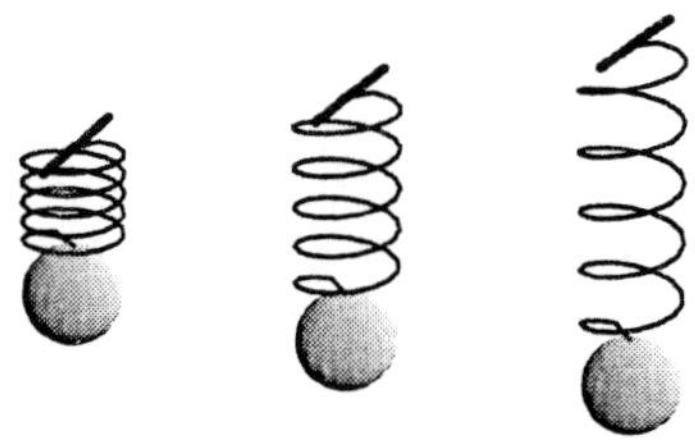

Finally, to set the system in motion, we create a sequence of frames based upon a solution of the system's differential equation. For example, if the natural angular frequency is $\mu = 2$ and motion is undamped and forced by $\cos t$, then the response is $y = \frac{1}{3}(\cos t - \cos 2t)$. The following creates 20 frames with $0 \leq t < 2\pi$. Since this covers one full period of the motion, the motion will seem continuous if you allow the animation to "loop." The picture that follows shows every other frame.

```
Do[Show[frame[4, 1, 1/3 (Cos[t] - Cos[2 t]), Cos[t]],
  DisplayFunction → $DisplayFunction], {t, 0, 1.9 π, .1 π}]
```

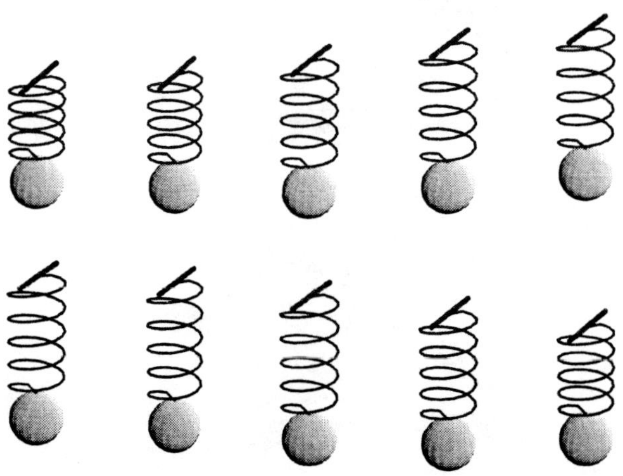

■ Playing With Sound

Beats

The phenomenon of beats is best illustrated with sound. The following produces a sound that consists of two pure sinusoidal tones with frequencies of 275 Hz and 280 Hz. You'll hear audible 5 Hz beats.

```
Play[Sin[275*2πt] + Sin[280*2πt], {t, 0, 3}];
```

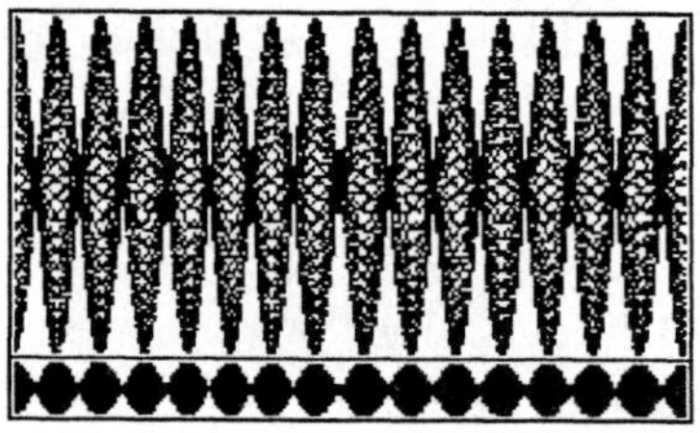

In the following, the beats will have a frequency of 2 Hz.

```
Play[Sin[275*2πt] + Sin[277*2πt], {t, 0, 3}];
```

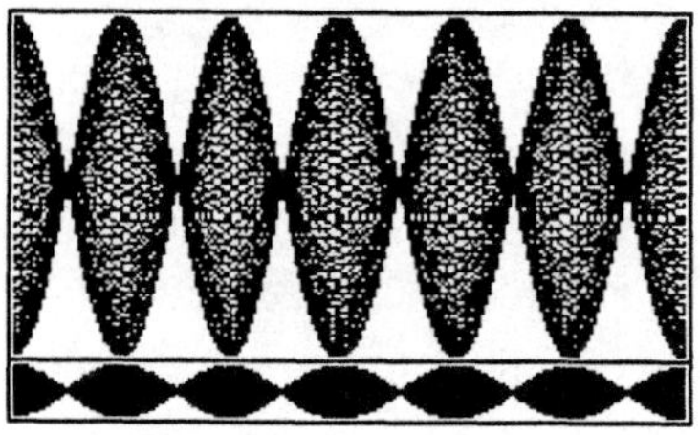

As the component frequencies become closer to one another, the frequency of the beats gets smaller.

```
Play[Sin[275*2πt] + Sin[276*2πt], {t, 0, 3}];
```

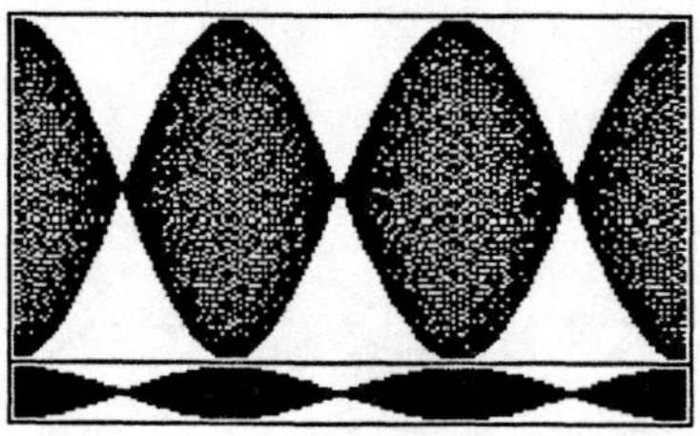

```
Play[Sin[275*2πt] + Sin[275.5*2πt], {t, 0, 3}];
```

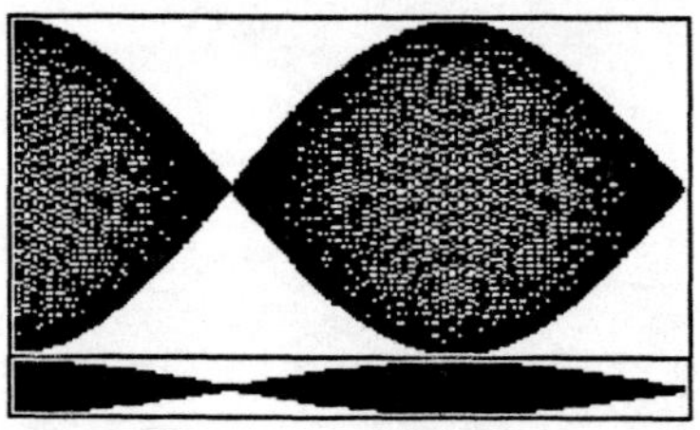

As the component frequencies merge, the beats go away.

```
Play[Sin[275 * 2 π t] + Sin[(275 + 5 e^-t) * 2 π t], {t, 0, 7}];
```

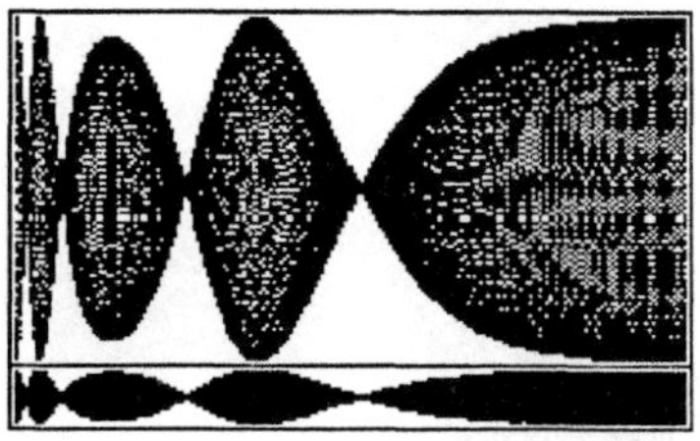

Music

The standard package `Music` allows us to play combinations of tones.

```
Needs["Miscellaneous`Music`"]
```

Here is a function that creates a chord based on a matrix of tones and intensities.

```
Chord[tones_] := Sum[tones[[i, 2]] Sin[2 π tones[[i, 1]] t], {i, 1, Length[tones]}]
```

$$\text{Chord[tones\_]} := \sum_{i=1}^{\text{Length[tones]}} \text{tones[[i, 2]] Sin[2}\pi\text{ tones[[i, 1]] t]}$$

The following plays a *D-9th* chord:

$$\text{Play}\left[\text{Chord}\left[\begin{pmatrix} \text{D2} & 3 \\ \text{Gflat2} & 2 \\ \text{A3} & 1 \\ \text{C3} & 2 \\ \text{E3} & 3 \end{pmatrix}\right], \{t, 0, 1\}\right];$$

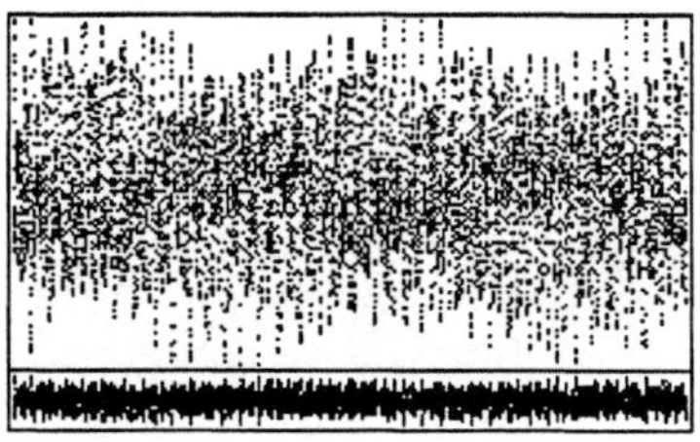

This is just random:

$$\text{Play}\left[\text{Chord}\left[\begin{pmatrix} \text{Bflat2} & 2 \\ \text{Fsharp2} & 3 \\ \text{B3} & 2 \\ \text{C3} & 2 \\ \text{Aflat4} & 1 \end{pmatrix}\right], \{t, 0, 1\}\right];$$

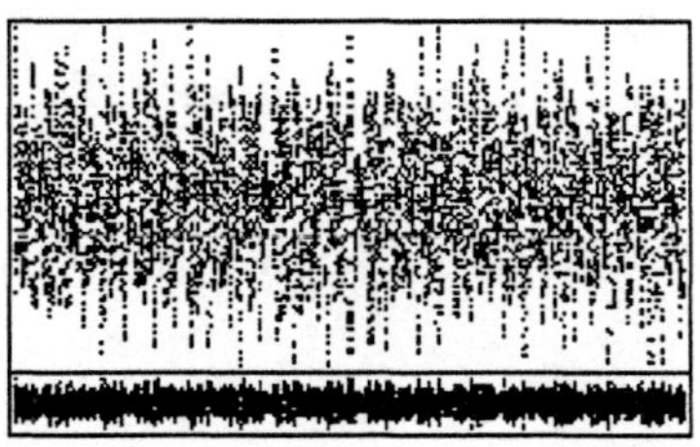

■ Exercises

In Exercises 1–4, solve the initial-value problem (with DSolve), and use the solution to create a spring-mass animation.

1. $y'' + \frac{2}{3}y' + \frac{10}{9}y = 0,\ y(0) = 1,\ y'(0) = 0$
2. $y'' + y' + \frac{1}{4}y = 0,\ y(0) = 1,\ y'(0) = -1$
3. $y'' + 4y = \frac{1}{2}\cos(2.1\,t),\ y(0) = 0,\ y'(0) = 0$
4. $y'' + 2y' + 2y = \sin t,\ y(0) = 0,\ y'(0) = 0$

In Exercises 5–10, solve the initial-value problem (with DSolve), and *listen* to the solution.

5. $y'' + (220\,j\pi)^2\,y = 0,\ y(0) = 1,\ y'(0) = 0,$ for each of $j = 1, 2, 4, 8, 16$
6. $y'' + (220\,j\pi)^2\,y = 0,\ y(0) = 1,\ y'(0) = 0,$ for each of $j = 2, 3, 4, 5, 6, 7, 8$
7. $y'' + 4.0\,y + (880\,\pi)^2 y = 0,\ y(0) = 1,\ y'(0) = 0$ (Set PlayRange→All.)
8. $y'' + 2\,y + (440\,\pi)^2 y = (440\,\pi)^2 \sin(660\,\pi\,t),\ y(0) = 10,\ y'(0) = 0$
9. $y'' + 2\,y + (880\,\pi)^2 y = (880\,\pi)^2 \sin(550\,\pi\,t),\ y(0) = 10,\ y'(0) = 0$
10. $y'' + 2\,y + (440\,\pi)^2 y = (440\,\pi)^2 \sin(587\,\pi\,t),\ y(0) = 10,\ y'(0) = 0$

In Exercises 11 and 12, plot 2 cycles of each function, and listen (for 2 or 3 seconds) to each one, noticing the difference in the way they sound.

11. $\sin(330\cdot 2\pi\,t),\ \sin^3(330\cdot 2\pi\,t),\ \sin^{11}(330\cdot 2\pi\,t)$
12. $f(t)$ and $\max\{f(t), k\}$ for $k = -\frac{1}{2}, 0, \frac{1}{2}, \frac{3}{4}$, where $f(t) = \sin(330\cdot 2\pi\,t)$

6.5 Laplace Transforms

The functions LaplaceTransform and InverseLaplaceTransform, formerly provided by the LaplaceTransform package, were implemented in the kernel beginning with *Mathematica* 4.0. (So if you're using an older version of *Mathematica*, you'll need to load that package.)

Each of these functions takes three arguments. For instance, to compute the Laplace transform of $e^{-2t}\cos 3\,t$ and have the result as a function of s, we enter

```
F[s_] = LaplaceTransform[e^(-2 t) Cos[3 t], t, s]
```

$$\frac{2 + s}{13 + 4\,s + s^2}$$

`InverseLaplaceTransform` will now recover the original function if we enter

```
invtrans = InverseLaplaceTransform[F[s], s, t]
```

$$\frac{1}{2} e^{(-2-3 i) t} (1 + e^{6 i t})$$

To simplify this, use `ComplexExpand`,

```
invtrans // ComplexExpand // Simplify
```

$e^{-2 t}$ `Cos[3 t]`

or `ExpToTrig`.

```
invtrans // ExpToTrig // Simplify
Cos[3 t] (Cosh[2 t] - Sinh[2 t])
```

`FullSimplify` also gives the function in terms of cosh and sinh.

```
invtrans // FullSimplify
Cos[3 t] (Cosh[2 t] - Sinh[2 t])
```

Suppose we want to solve an initial-value problem of the type

$$y'' + py' + qy = f(t),\ y(0) = y_0,\ y'(0) = v_0,$$

where the coefficients p and q are constants. We start by defining a symbol for the differential equation.

```
Clear[f];
DE = y''[t] + p y'[t] + q y[t] == f[t]
q y[t] + p y'[t] + y''[t] == f[t]
```

Next we apply `LaplaceTransform` to the differential equation.

```
transDE = LaplaceTransform[DE, t, s]
q LaplaceTransform[y[t], t, s] + s^2 LaplaceTransform[y[t], t, s] +
  p (s LaplaceTransform[y[t], t, s] - y[0]) - s y[0] - y'[0] ==
 LaplaceTransform[f[t], t, s]
```

For convenience, let's define a simple symbol for `LaplaceTransform[y[t],t,s]`.

```
Y[s] := LaplaceTransform[y[t], t, s]
```

Now we solve the algebraic transformed equation for the transform of the solution of the differential equation.

```
Y[s] = Y[s] /. Solve[transDE, Y[s]][[1]]
```

$$\frac{\text{LaplaceTransform}[f[t], t, s] + p\,y[0] + s\,y[0] + y'[0]}{q + p\,s + s^2}$$

Notice that this is a completely general result, provided that p and q are constants. We will use this in the examples that follow.

Example 6.5.1

Consider the initial-value problem

$$y'' + 2y' + 2y = e^{-t}\cos t,\ y(0) = 1,\ y'(0) = -1.$$

With `Y[s]` as defined above, the transform of the solution is

```
transy = Y[s] /. {p → 2, q → 2, f[t] → e^-t Cos[t], y[0] → 1, y'[0] → -1} // Apa
```

$$\frac{1+s}{(2+2s+s^2)^2} + \frac{1+s}{2+2s+s^2}$$

The solution is

```
InverseLaplaceTransform[transy, s, t] // FullSimplify
```

$$\frac{1}{2} e^{-t} (2\,\mathrm{Cos}[t] + t\,\mathrm{Sin}[t])$$

■ Piecewise-Defined Inputs

Recall that the Heaviside unit step function is defined by

$$h(t) = \begin{cases} 1, & \text{if } t \geq 0, \\ 0, & \text{if } t \leq 0. \end{cases}$$

Mathematica's version of this is named `UnitStep` and provides an important way of defining piecewise-defined functions. The following are two examples.

```
Plot[UnitStep[t - 1] - UnitStep[t - 2] + UnitStep[t - 3] - UnitStep[t - 4],
  {t, 0, 5}, AspectRatio → Automatic, Ticks → {Automatic, {1}}];
```

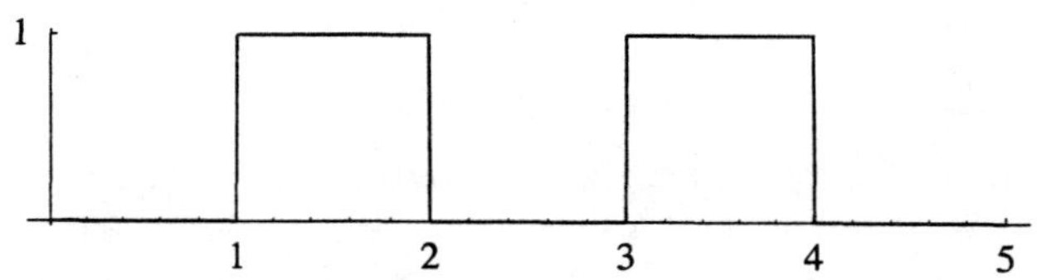


```
Plot[t UnitStep[t] + (1 - t) UnitStep[t - 1], {t, -1, 2},
  AspectRatio → Automatic, Ticks → {{-1, 1, 2}, {1}}];
```

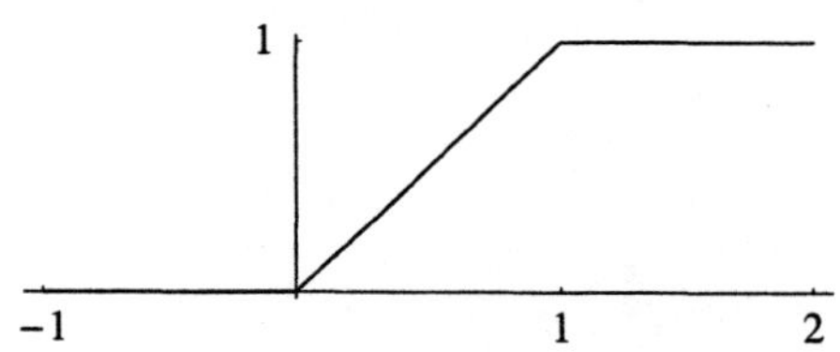

The Laplace transform of $h(t)$ is the same as the Laplace transform of 1, since those functions are the same for all $t \geq 0$.

```
LaplaceTransform[UnitStep[t], t, s]
```

$$\frac{1}{s}$$

The transform of $h(t-c)$ (i.e., $h(t)$ "shifted" by c) is the transform of 1 multiplied by e^{-cs}.

```
LaplaceTransform[UnitStep[t - c], t, s]
```

$$\frac{e^{-c\,s}}{s}$$

The same notion applies in general. The transform of $h(t-c)\,f(t-c)$ (i.e., $f(t)$ "shifted" by c and made zero for $t < c$) is the transform of $f(t)$ multiplied by e^{-cs}. For instance, the graph and the transform of $h(t-3)\cos(t-3)$ are as follows:

```
Plot[UnitStep[t - 3] Cos[t - 3], {t, 0, 4 π + 3}];
```

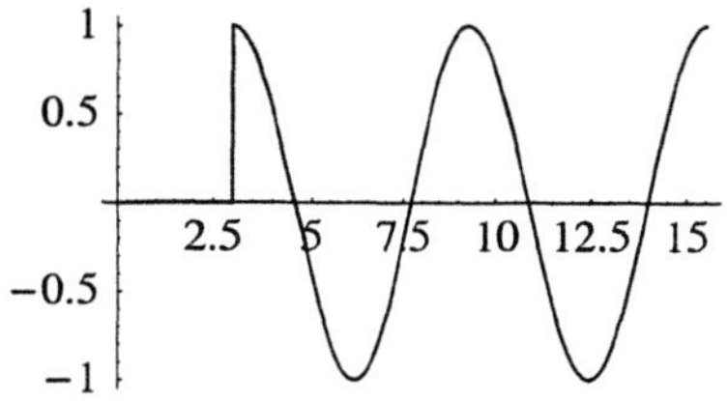


```
LaplaceTransform[UnitStep[t - 3] Cos[t - 3], t, s] // Simplify
```

$$\frac{e^{-3\,s}\,s}{1+s^2}$$

Example 6.5.2

Consider the initial-value problem

$$y'' + y' + \tfrac{5}{4}y = \tfrac{1}{2}\,h(t-2\pi)\sin(t-2\pi),\;\; y(0)=1,\;\; y'(0)=0.$$

With `Y[s]` as defined above, the transform of the solution is

```
transy = Y[s] /.
  {p → 1, q → 5/4, f[t] → 1/2 UnitStep[t - 2 π] Sin[t - 2 π], y[0] → 1, y'[0] →
```

$$\frac{1+s+\frac{e^{-2\pi s}}{2\,(1+s^2)}}{\frac{5}{4}+s+s^2}$$

Before giving this to `InverseLaplaceTransform`, it is a good idea to do a partial fraction decomposition.

```
transy = Apart[transy]
```

$$\frac{4\,(1+s)}{5+4\,s+4\,s^2}+\frac{2\,e^{-2\pi s}}{(1+s^2)\,(5+4\,s+4\,s^2)}$$

Finally, the solution of the initial-value problem is

```
soln =
 InverseLaplaceTransform[transy, s, t] // ComplexExpand // Simplify
```

$$\frac{1}{68}\,e^{-t/2}\,(34\,(2\,\text{Cos}[t]+\text{Sin}[t])+ \\ 8\,(4\,(e^{\pi}-e^{t/2})\,\text{Cos}[t]+(e^{\pi}+e^{t/2})\,\text{Sin}[t])\,\text{UnitStep}[-2\pi+t])$$

The following plot shows the solution along with the forcing function.

```
Plot[{soln, 1/2 UnitStep[t - 2 π] Sin[t - 2 π]}, {t, 0, 6 π},
  Ticks → {π Range[1, 6], {1}}, PlotStyle → {Thickness[.008], Hue[0]}]
```

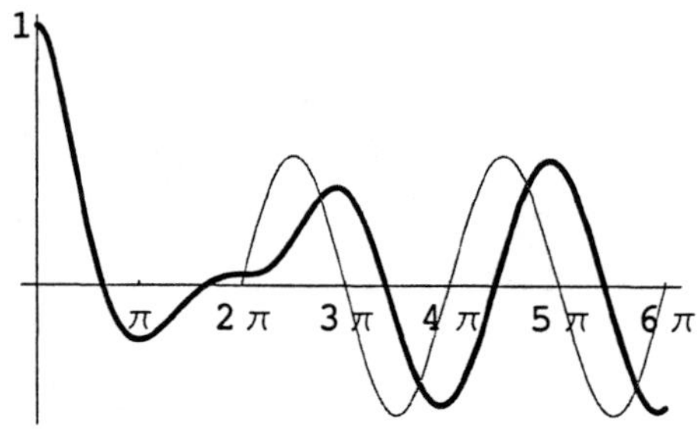

■ Periodic Inputs

Suppose that $f(t)$ is a periodic function with period ρ, and let $f_0(t) = f(t)\,(h(t) - h(t - \rho))$. Then f is the *ρ-periodic extension* of f_0. Note that $f_0(t) = f(t)$ for all t in $[0, \rho)$ and $f_0(t) = 0$ for all t outside of $[0, \rho)$. If $F_0(s)$ is the transform of f_0, then the transform of f is

$$F(s) = \frac{F_0(s)}{1-e^{-\rho s}}.$$

Example 6.5.3

Consider the function $f(t) = t - \lfloor t \rfloor$, where $\lfloor \cdot \rfloor$ denotes the *floor* (or greatest integer) function.

```
Plot[t - Floor[t], {t, 0, 3},
  AspectRatio → Automatic, Ticks → {{1, 2, 3}, {1}}];
```

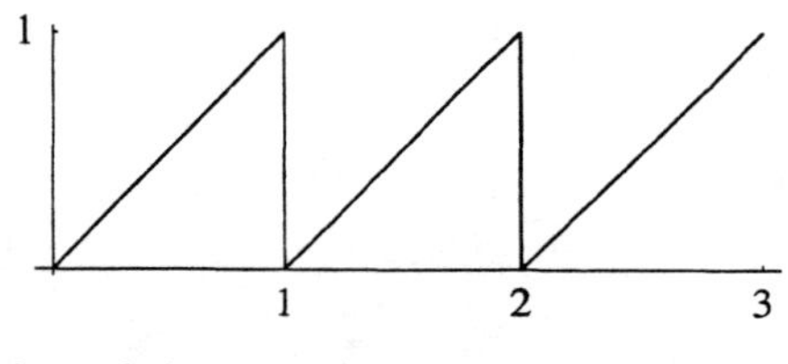

This f is the 1-periodic extension of $f_0(t) = t\,(h(t) - h(t - 1))$.

```
Plot[t (UnitStep[t] - UnitStep[t - 1]), {t, -.5, 3},
  AspectRatio → Automatic, Ticks → {{1, 2, 3}, {1}}];
```

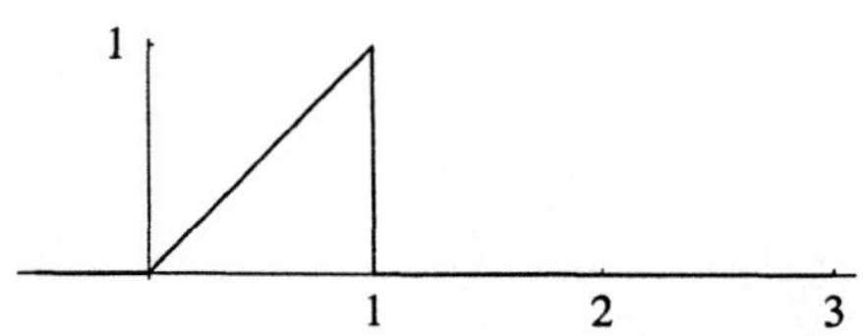

The transform of f is therefore

```
LaplaceTransform[t (UnitStep[t] - UnitStep[t - 1]), t, s]
---------------------------------------------------------- // Simplify
                        1 - e^-s
```

$$\frac{-1 + e^{s} - s}{(-1 + e^{s})\, s^{2}}$$

Finding the Inverse Transform. Recovering a periodic function such as the one in Example 6.5.3 from its transform is a delicate matter. For instance, note that `InverseLaplace-Transform` balks at such a problem

```
InverseLaplaceTransform[(-1 + e^s - s)/((-1 + e^s) s^2), s, t]
```

```
InverseLaplaceTransform[(-1 + e^s - s)/((-1 + e^s) s^2), s, t]
```

One remedy is to define a new rule for `InverseLaplaceTransform`. To do this we must first "unprotect" it.

```
Unprotect[InverseLaplaceTransform];
```

Now we use a partial sum of the series expansion

$$\frac{1}{-1+e^{a\,s}} = -\sum_{j=0}^{\infty} e^{j\,a\,s}$$

to tell `InverseLaplaceTransform` what to do with an expression of the form $\frac{F_0(s)}{-1+e^{as}}$.

```
InverseLaplaceTransform[η_ : 1/(-1 + e^((a_ : 1) s_)), s_, t_, n_] :=
 InverseLaplaceTransform[Expand[If[a < 0, -η Sum[e^(j a s), {j, 0, n-1}], η e^(-a s) Sum[e^(-j a s), {j, 0, n-1}]]], s, t]
```

Now to get three terms of the Heaviside expansion of the inverse transform of $\frac{-1+e^s-s}{(-1+e^s)s^2}$, valid for $t < 5$, we enter

```
invtrans = InverseLaplaceTransform[(-1 + e^s - s)/((-1 + e^s) s^2), s, t, 5] // Simplify
```

```
t + 4 UnitStep[-5 + t] - t UnitStep[-5 + t] - UnitStep[-4 + t] -
 UnitStep[-3 + t] - UnitStep[-2 + t] - UnitStep[-1 + t]
```

```
Plot[invtrans, {t, 0, 5},
 AspectRatio → Automatic, Ticks → {{1, 2, 3, 4, 5}, {1}}];
```

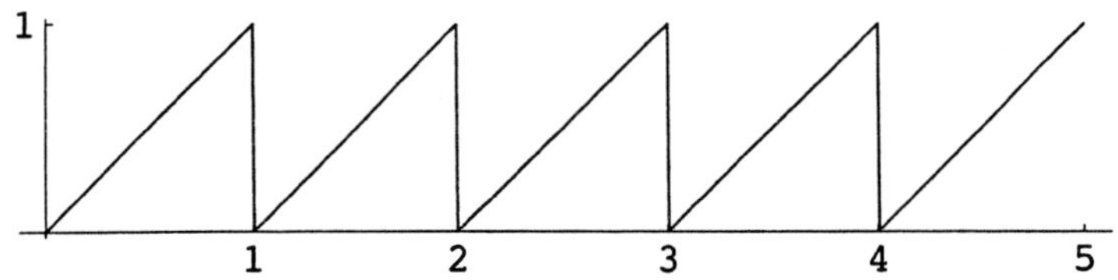

Note that unless the fourth argument is supplied, the new rule has no effect.

```
InverseLaplaceTransform[(-1 + e^s - s)/((-1 + e^s) s^2), s, t]
```

```
InverseLaplaceTransform[(-1 + e^s - s)/((-1 + e^s) s^2), s, t]
```

Example 6.5.4

Consider the initial-value problem

$$y'' + 4\pi^2 y = 4\pi^2(t - \lfloor t \rfloor),\ \ y(0) = 0,\ \ y'(0) = 0.$$

With `Y[s]` as defined previously, and with the transform found in Example 6.5.3, the transform of the solution is

```
transy = Y[s] /. {p → 0, q → 4 π², y[0] → 0, y'[0] → 0} /.
  LaplaceTransform[f[t], t, s] → 4 π² (-1 + ⅇ^s - s)/((-1 + ⅇ^s) s²)
```

$$\frac{4\pi^2 (-1 + e^s - s)}{(-1 + e^s)\, s^2\, (4\pi^2 + s^2)}$$

Four terms of the solution's Heaviside expansion are

```
soln = InverseLaplaceTransform[transy, s, t, 4] //
   ComplexExpand // Simplify
```

$$-\frac{1}{2\pi}\big(-2\pi t + \text{Sin}[2\pi t] - (6\pi - 2\pi t + 2\pi\,\text{Cos}[2\pi t] + \text{Sin}[2\pi t])\,\text{UnitStep}[-4 + t] + 4\pi\,\text{Sin}[\pi t]^2\,\text{UnitStep}[-3 + t] + 4\pi\,\text{Sin}[\pi t]^2\,\text{UnitStep}[-2 + t] + 4\pi\,\text{Sin}[\pi t]^2\,\text{UnitStep}[-1 + t]\big)$$

The following plot shows the solution along with the forcing function.

```
Plot[{soln, t - Floor[t]}, {t, 0, 4},
  PlotStyle → {Thickness[.008], Hue[0]}, PlotRange → All];
```

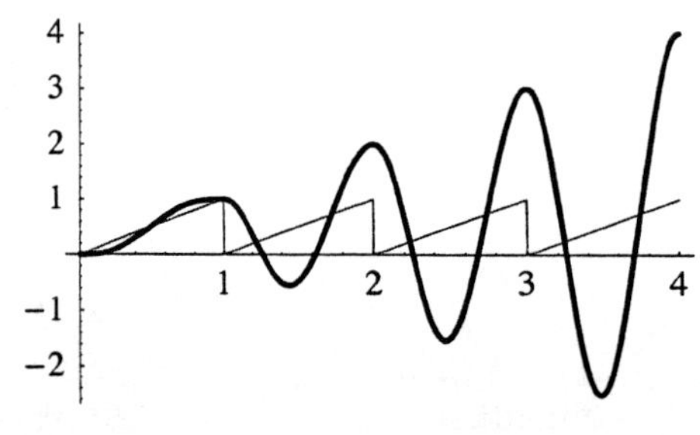

What we are seeing here is resonance. This is a response to a frequency component of the forcing function that matches system's natural frequency. For instance, notice how the function plotted below, which contains the term $\sin 2\pi t$, mimics the forcing function.

```
Plot[1/2 - Sin[2 π t]/π - Sin[4 π t]/(2 π) - Sin[6 π t]/(3 π) - Sin[8 π t]/(4 π), {t, 0, 3}];
```

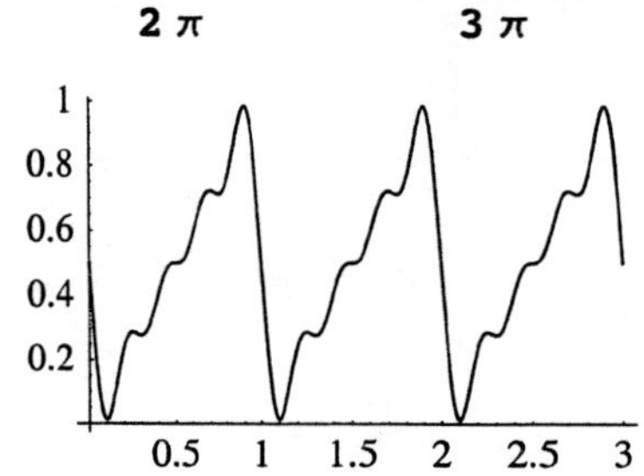

(Notice the pattern in the terms here. Add a few more terms, and observe the improvement in the approximation.)

■ Impulses

Consider a spring-mass system that is either at rest or undergoing unforced motion. Suppose that the mass is struck with a hammer in such a way that the mass experiences an instantaneous change in momentum. Such an instantaneous force—called an *impulse*—can be modeled with the *Dirac distribution* $\delta(t)$, a.k.a., the *Dirac delta function*. The Dirac distribution is defined by the following properties:

$$\text{(i) } \delta(t-t_1)=0 \text{ for all } t\neq t_1, \text{ and (ii) } \int_a^b \delta(t-t_1)\, f(t)\, dt = f(t_1) \text{ if } a<t_1<b.$$

In particular, note that (ii) implies that $\int_{-\infty}^{\infty} \delta(t-t_1)\, dt = 1$ for any t_1. So we may think of $\delta(t-t_1)$ as an impulse that occurs at time t_1 and imparts a unit change in momentum.

Mathematica's Dirac distribution is named `DiracDelta`.

```
∫_{-∞}^{∞} DiracDelta[t] dt
```

```
1
```

The Laplace transform of $\delta(t-c)$ is e^{-cs}; for instance:

```
LaplaceTransform[DiracDelta[t - 2], t, s]
```

$$e^{-2 s}$$

Example 6.5.5

Consider the initial-value problem

$$y'' + y' + \frac{5}{4} y = -\delta(t-3), \quad y(0)=1, \quad y'(0)=0,$$

which we can think of as a model of a damped spring-mass system that is struck with a hammer at time $t=3$. With `Y[s]` as defined previously,

```
Y[s] := (LaplaceTransform[f[t], t, s] + p y[0] + s y[0] + y'[0]) / (q + p s + s^2)
```

the transform of the solution is

```
transy = Y[s] /.
  {p → 1, q → 5 / 4, f[t] → DiracDelta[t - 3], y[0] → 1, y'[0] → 0} // Apart
```

$$\frac{4\, e^{-3 s}}{5+4 s+4 s^2} + \frac{4\,(1+s)}{5+4 s+4 s^2}$$

The solution is

```
soln =
 InverseLaplaceTransform[transy, s, t] // ComplexExpand // Simplify
```

$$\frac{1}{2} e^{-t/2} \left(2 \operatorname{Cos}[t] + \operatorname{Sin}[t] - 2 e^{3/2} \operatorname{Sin}[3-t] \operatorname{UnitStep}[-3+t]\right)$$

```
Plot[soln, {t, 0, 10}];
```

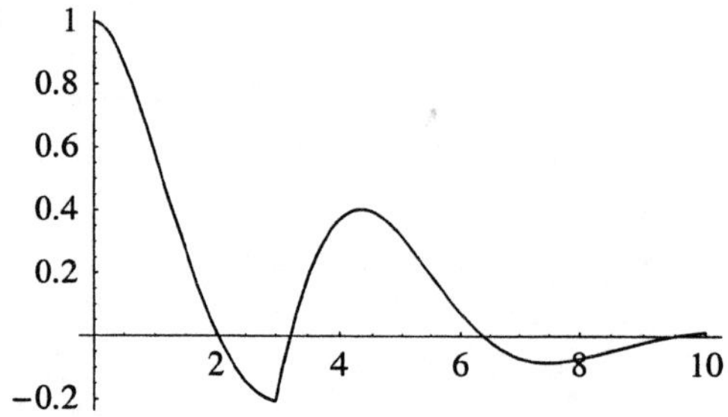

■ Exercises

Use Laplace transforms to solve the initial-value problems in Exercises 1--4.

1. $y'' + 4y' + 4y = e^{-2t}$, $y(0) = 1$, $y'(0) = 0$

2. $y'' + 4y' + 3y = \cos t$, $y(0) = 1$, $y'(0) = -1$

3. $y'' + 9y = e^{-t} \cos 3t$, $y(0) = 1$, $y'(0) = 0$

4. $y'' + 4y' + 13y = e^{-2t} \cos 3t$, $y(0) = 0$, $y'(0) = 0$

5. Let $f(t) = t - 2\left\lfloor \frac{t-1}{2} \right\rfloor - 2$.

```
Plot[t - 2 - 2 Floor[(t - 1) / 2], {t, 0, 6}, AspectRatio → Automatic];
```

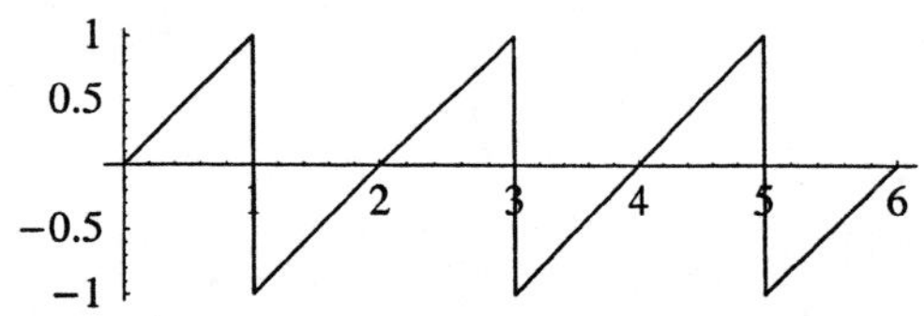

Observe that f is the 2-periodic extension of

$$f_0(t) = t\,h(t) - 2\,h(t-1) - (t-2)\,h(t-2).$$

```
Plot[t UnitStep[t] - 2 UnitStep[t - 1] + (2 - t) UnitStep[t - 2],
  {t, 0, 3}, AspectRatio → Automatic];
```

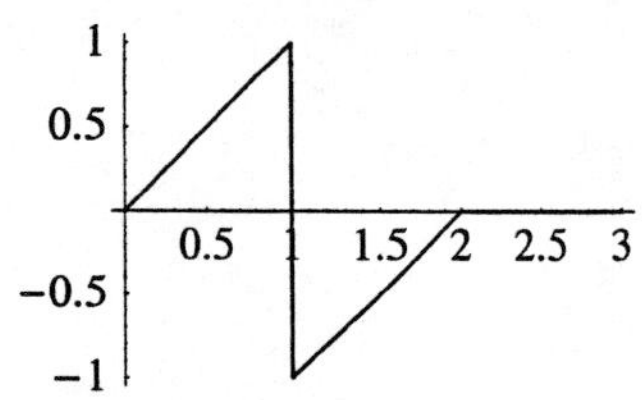

(a) Find the Laplace transform of f_0 and then f.

(b) Find the solution of

$$y'' + \pi^2 y = \pi^2 f(t),\ y(0) = 0,\ y'(0) = 0,$$

for $0 \le t < 5$. Plot the solution on that interval along with the function f.

(c) Find the solution of

$$y'' + 4\pi^2 y = 4\pi^2 f(t), \ y(0) = 0, \ y'(0) = 0,$$

for $0 \le t < 5$. Plot the solution on that interval along with the function f.

6. Let $f(t) = (-1)^{\lfloor t \rfloor}$.

```
Plot[(-1)^Floor[t], {t, 0, 6}, AspectRatio → Automatic];
```

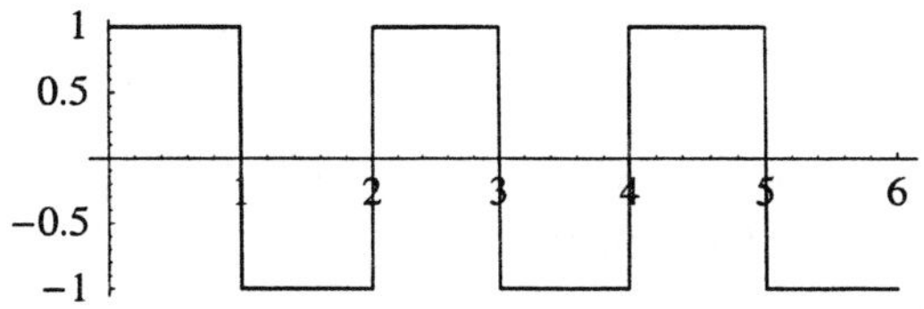

Observe that f is the 2-periodic extension of

$$f_0(t) = h(t) - 2\,h(t-1) - h(t-2).$$

```
Plot[UnitStep[t] - 2 UnitStep[t - 1] + UnitStep[t - 2],
  {t, 0, 3}, AspectRatio → Automatic];
```

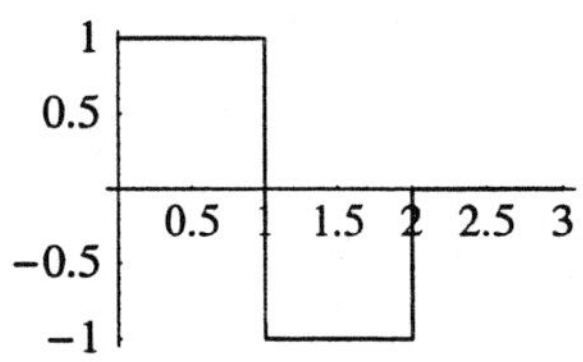

(a) Find the Laplace transform of f_0 and then f.

(b) Find the solution of

$$y'' + 4\pi^2 y = 4\pi^2 f(t), \ y(0) = 0, \ y'(0) = 0,$$

for $0 \le t < 5$. Plot the solution on that interval along with the function f.

(c) Find the solution of

$$y'' + 2\pi y' + \pi^2 y = \pi^2 f(t), \ y(0) = 0, \ y'(0) = 0,$$

for $0 \le t < 5$. Plot the solution on that interval along with the function f.

7. Let $f(t) = \max\{\sin t, 0\}$.

```
Plot[Max[{Sin[t], 0}], {t, 0, 5 π}];
```

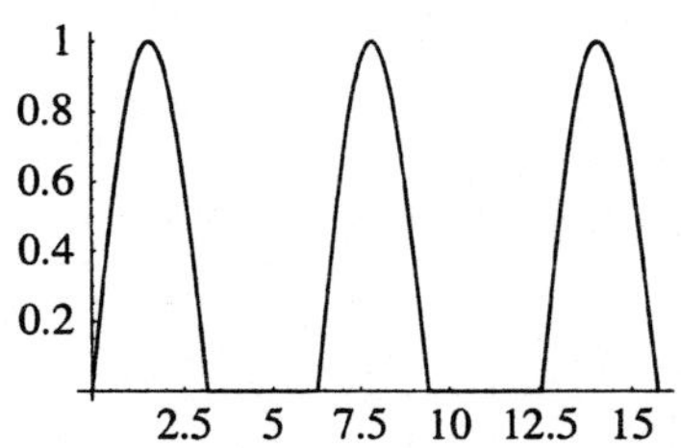

Observe that f is the 2π-periodic extension of

$$f_0(t) = (h(t) - h(t - \pi)) \sin t\ .$$

```
Plot[(UnitStep[t] - UnitStep[t - π]) Sin[t], {t, 0, 10}];
```

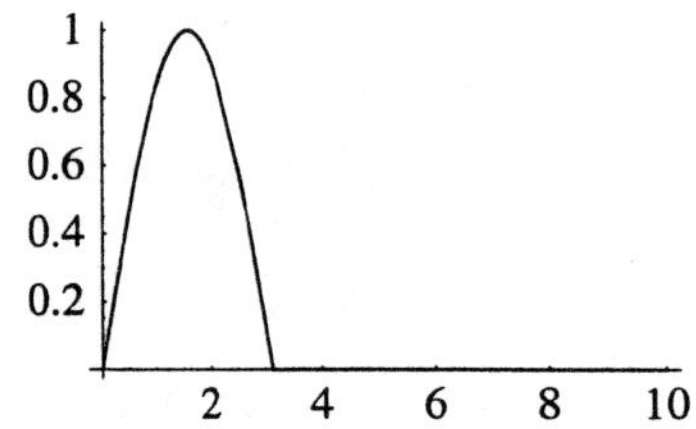

(a) Find the Laplace transform of f_0 and then f.

(b) Find the solution of

$$y'' + y = f(t),\ y(0) = 0,\ y'(0) = 0,$$

for $0 \le t < 5\pi$. Plot the solution on that interval along with the function f.

(c) Find the solution of

$$y'' + 2y' + 2y = 2f(t),\ y(0) = 0,\ y'(0) = 0,$$

for $0 \le t < 5\pi$. Plot the solution on that interval along with the function f.

In Exercises 8–11, find and plot the solution.

8. $y'' + 4y' + 13y = \delta(t-3),\ y(0) = 0,\ y'(0) = 2$

9. $y'' + 2y' + y = 2\,\delta(t-3),\ y(0) = 1,\ y'(0) = 0$

10. $y'' + \frac{1}{4}\pi^2 y = -\frac{1}{2}\pi\,\delta(t-3),\ y(0) = 1,\ y'(0) = 0$

11. $y'' + 4\pi^2 y = \pi^2(4\cos(2\pi t) + 6\,\delta(t-3) + 6\,\delta(t-6)),\ y(0) = 0,\ y'(0) = 0$

6.6 Series Solutions

■ Finding the Recurrence Formula

Let T be the operator defined by $Ty = y'' + p\,y' + qy$, so that the differential equation

$$y'' + p\,y' + qy = 0$$

can be written as $Ty = 0$. To find a power series solution $y = \sum_{k=0}^{\infty} a_k\, t^k$, we compute

$$Ty = T \sum_{k=0}^{\infty} a_k\, t^k = \sum_{n=0}^{\infty} b_n\, t^n$$

and choose $a_0, a_1, a_2, \ldots$ so that $b_n = 0$ for all n. When the coefficients p and q are polynomials of small degree, b_n can be expressed in terms of a small number of the a_k's, typically $a_{n-i}, \ldots, a_n, \ldots, a_{n+j}$ for some i, j. Setting that expression to zero gives rise to a recurrence formula for $a_0, a_1, a_2, \ldots$. Thus we need only compute $T \sum_{k=n-i}^{n+j} a_k t^k$ to determine the recurrence formula. Appropriate values of i and j can be inferred from $T t^n$, as we will demonstrate in the following example.

Example 6.6.1

Consider the equation

$$y'' + t^2 y' + y = 0$$

We first define the operator T and apply it to t^n.

```
T[y_] := ∂t,t y + t² ∂t y + y

T[tⁿ]

(-1 + n) n t^(-2+n) + t^n + n t^(1+n)
```

From this we can tell that t^n will appear (only) in each of $T t^{n-1}, \ldots, T t^{n+2}$. In fact we can verify this as follows:

```
Table[T[t^k], {k, n - 2, n + 3}] // TableForm

(-3 + n) (-2 + n) t^(-4+n) + t^(-2+n) + (-2 + n) t^(-1+n)
(-2 + n) (-1 + n) t^(-3+n) + t^(-1+n) + (-1 + n) t^n
(-1 + n) n t^(-2+n) + t^n + n t^(1+n)
n (1 + n) t^(-1+n) + t^(1+n) + (1 + n) t^(2+n)
(1 + n) (2 + n) t^n + t^(2+n) + (2 + n) t^(3+n)
(2 + n) (3 + n) t^(1+n) + t^(3+n) + (3 + n) t^(4+n)
```

So in order to determine the coefficient of t^n in $T \sum_{k=0}^{\infty} a_k t^k$, we need only include powers t^k for $k = n - 1, \ldots, n + 2$.

```
terms = T[ Sum[a[j] t^j, {j, n-1, n+2}] ] // Expand

2 t^(-3+n) a[-1 + n] - 3 n t^(-3+n) a[-1 + n] + n^2 t^(-3+n) a[-1 + n] + t^(-1+n) a[-1 + n] - t^n a[-1 + n] +
  n t^n a[-1 + n] - n t^(-2+n) a[n] + n^2 t^(-2+n) a[n] + t^n a[n] + n t^(1+n) a[n] + n t^(-1+n) a[1 + n] +
  n^2 t^(-1+n) a[1 + n] + t^(1+n) a[1 + n] + t^(2+n) a[1 + n] + n t^(2+n) a[1 + n] + 2 t^n a[2 + n] +
  3 n t^n a[2 + n] + n^2 t^n a[2 + n] + t^(2+n) a[2 + n] + 2 t^(3+n) a[2 + n] + n t^(3+n) a[2 + n]

b[n] = Coefficient[terms, t^n] // Simplify

(-1 + n) a[-1 + n] + a[n] + (2 + 3 n + n^2) a[2 + n]
```

Therefore the (three-term) recurrence formula is

```
Solve[b[n] == 0, a[n + 2]] // Simplify

{{a[2 + n] -> -((-1 + n) a[-1 + n] + a[n])/(2 + 3 n + n^2)}}
```

This will be more useful if we shift the index to obtain an expression for a_n.

```
(-((-1 + n) a[-1 + n] + a[n])/(2 + 3 n + n^2) /. n -> n - 2) // Simplify
```

$$-\frac{(-3+n)\,a[-3+n]+a[-2+n]}{(-1+n)\,n}$$

Now we enter a recursive definition of a_n.

```
a[n_] := -((-3 + n) a[-3 + n] + a[-2 + n])/((-1 + n) n)
```

For the fundamental solution satisfying $y(0) = 1$ and $y'(0) = 0$, we provide starting values as follows:

```
a[-1] := 0; a[0] := 1; a[1] := 0;
```

The resulting series up to the degree-11 term is

```
u[t_] = Sum[a[n] t^n, {n, 0, 11}]
```

$$1-\frac{t^2}{2}+\frac{t^4}{24}+\frac{t^5}{20}-\frac{t^6}{720}-\frac{13\,t^7}{2520}-\frac{179\,t^8}{40320}+\frac{17\,t^9}{90720}+\frac{109\,t^{10}}{241920}+\frac{641\,t^{11}}{1995840}$$

For the fundamental solution satisfying $y(0) = 0$ and $y'(0) = 1$, we define

```
a[-1] := 0; a[0] := 0; a[1] := 1;
```

The resulting series up to the degree-11 term is

```
v[t_] = Sum[a[n] t^n, {n, 0, 11}]
```

$$t-\frac{t^3}{6}-\frac{t^4}{12}+\frac{t^5}{120}+\frac{7\,t^6}{360}+\frac{13\,t^7}{1680}-\frac{11\,t^8}{10080}-\frac{209\,t^9}{120960}-\frac{107\,t^{10}}{181440}+\frac{23\,t^{11}}{241920}$$

Finally, a plot of u and v:

```
Plot[{u[t], v[t]}, {t, -1, 1}];
```

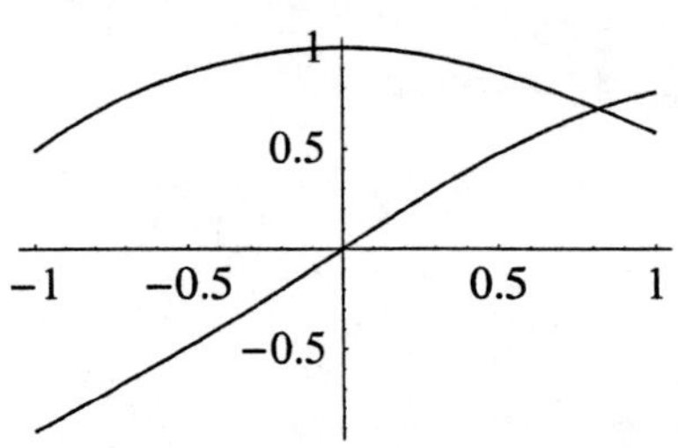

■ The First n Terms Without the Recurrence Formula

Another approach to finding the first several terms of a series solution is to construct a system of equations for the coefficients and solve the system directly (instead of using the recurrence formula). The system of equations is constructed by substituting a partial sum $\sum_{k=0}^{n} a_k t^k$ to the differential equation and then setting all the coefficients to zero. A little finessing is required, which we will describe in the following example.

Example 6.6.2

Consider the equation

$$y'' + y' + t\,y = 0.$$

Suppose we want to find fundamental series solutions with terms up through degree 9. We begin by substituting $\sum_{k=0}^{9} a_k t^k$ into the left side of the equation.

```
Clear[a]; T[y_] := ∂t,t y + ∂t y + t y;
```

```
terms = T[Sum[a[n] t^n, {n, 0, 9}]] // Simplify
```

```
a[1] + 2 a[2] + t (a[0] + 2 a[2] + 6 a[3]) + t^2 (a[1] + 3 (a[3] + 4 a[4])) +
 t^3 (a[2] + 4 (a[4] + 5 a[5])) + t^4 (a[3] + 5 (a[5] + 6 a[6])) +
 t^5 (a[4] + 6 (a[6] + 7 a[7])) + t^9 a[8] + t^6 (a[5] + 7 (a[7] + 8 a[8])) +
 t^10 a[9] + t^8 (a[7] + 9 a[9]) + t^7 (a[6] + 8 (a[8] + 9 a[9]))
```

Next we peel off the coefficients.

```
coeffs = CoefficientList[terms, t]
```

```
{a[1] + 2 a[2], a[0] + 2 a[2] + 6 a[3], a[1] + 3 (a[3] + 4 a[4]),
 a[2] + 4 (a[4] + 5 a[5]), a[3] + 5 (a[5] + 6 a[6]), a[4] + 6 (a[6] + 7 a[7]),
 a[5] + 7 (a[7] + 8 a[8]), a[6] + 8 (a[8] + 9 a[9]), a[7] + 9 a[9], a[8], a[9]}
```

The last two of these are no good, so we throw them away.

```
coeffs = Drop[coeffs, -2]
```

```
{a[1] + 2 a[2], a[0] + 2 a[2] + 6 a[3], a[1] + 3 (a[3] + 4 a[4]),
 a[2] + 4 (a[4] + 5 a[5]), a[3] + 5 (a[5] + 6 a[6]), a[4] + 6 (a[6] + 7 a[7]),
 a[5] + 7 (a[7] + 8 a[8]), a[6] + 8 (a[8] + 9 a[9]), a[7] + 9 a[9]}
```

Now we turn the list of coefficients into a list of equations.

```
eqns = Map[(# == 0) &, coeffs]
```

```
{a[1] + 2 a[2] == 0, a[0] + 2 a[2] + 6 a[3] == 0, a[1] + 3 (a[3] + 4 a[4]) == 0,
 a[2] + 4 (a[4] + 5 a[5]) == 0, a[3] + 5 (a[5] + 6 a[6]) == 0, a[4] + 6 (a[6] + 7 a[7]) == 0,
 a[5] + 7 (a[7] + 8 a[8]) == 0, a[6] + 8 (a[8] + 9 a[9]) == 0, a[7] + 9 a[9] == 0}
```

At this point we have 8 equations in the 10 unknowns $a_0, a_1, \ldots, a_9$. Next we add two equations, $a_0 = 1$ and $a_1 = 0$, to the list, solve the equations, and apply the resulting rules to the sum.

```
Sum[a[n] t^n, {n, 0, 9}] /. Solve[Join[{a[0] == 1, a[1] == 0}, eqns], Table[a[i], {i, 0, 9}]]
```

$$\left\{1 - \frac{t^3}{6} + \frac{t^4}{24} - \frac{t^5}{120} + \frac{t^6}{144} - \frac{t^7}{504} + \frac{t^8}{2520} - \frac{17\,t^9}{120960}\right\}$$

That gives us the series expansion of the solution satisfying $y(0) = 1$, $y'(0) = 0$. Now to find the other fundamental solution, we just make a slight modification the preceding command.

```
 9
 ∑ a[n] tⁿ /. Solve[Join[{a[0] == 0, a[1] == 1}, eqns], Table[a[i], {i, 0, 9}]]
n=0
```

$$\left\{t - \frac{t^2}{2} + \frac{t^3}{6} - \frac{t^4}{8} + \frac{t^5}{20} - \frac{t^6}{72} + \frac{5\,t^7}{1008} - \frac{61\,t^8}{40320} + \frac{131\,t^9}{362880}\right\}$$

■ Frobenius-Series Solutions

Here we consider equations of the form

$$p_2\,y'' + p_1\,y' + p_0\,y = 0,$$

where $p_2(0) = 0$. Such "degenerate" equations typically fail to have a pair of linearly independent power-series solutions about $t = 0$. In certain cases, however, such an equation may have a *Frobenius-series* solution. By that we mean a solution of the form

$$y = t^\mu \sum_{k=0}^{\infty} a_k t^k,$$

where μ is neither 0 nor 1. (In those cases y is an ordinary power series.) We may assume with no loss of generality that $a_0 = 1$ and thus write y as

$$y = t^\mu + \sum_{k=1}^{\infty} a_k t^{k+\mu}.$$

Example 6.6.3

Consider *Bessel's equation of order* $1/2$:

$$t^2\,y'' + t\,y' + \left(t^2 - \left(\tfrac{1}{2}\right)^2\right) y = 0.$$

We will attempt to find the first several terms of a Frobenius-series solution. We proceed in a way similar to what we did in Example 6.6.2. First we substitute the partial sum $t^\mu + \sum_{k=1}^{9} a_k t^{k+\mu}$ into the left side of the equation (after multiplying by 4).

```
T[y_] := 4 t² ∂t,t y + 4 t ∂t y + (4 t² - 1) y;
              9
psum = t^μ + ∑ a[n] t^(n+μ);
             n=1
terms = T[psum] // Expand
```

$-t^{\mu} + 4\,t^{2+\mu} + 4\,t^{\mu}\,\mu^2 + 3\,t^{1+\mu}\,a[1] + 4\,t^{3+\mu}\,a[1] + 8\,t^{1+\mu}\,\mu\,a[1] + 4\,t^{1+\mu}\,\mu^2\,a[1] + 15\,t^{2+\mu}\,a[2] + 4\,t^{4+\mu}\,a[2] + 16\,t^{2+\mu}\,\mu\,a[2] + 4\,t^{2+\mu}\,\mu^2\,a[2] + 35\,t^{3+\mu}\,a[3] + 4\,t^{5+\mu}\,a[3] + 24\,t^{3+\mu}\,\mu\,a[3] + 4\,t^{3+\mu}\,\mu^2\,a[3] + 63\,t^{4+\mu}\,a[4] + 4\,t^{6+\mu}\,a[4] + 32\,t^{4+\mu}\,\mu\,a[4] + 4\,t^{4+\mu}\,\mu^2\,a[4] + 99\,t^{5+\mu}\,a[5] + 4\,t^{7+\mu}\,a[5] + 40\,t^{5+\mu}\,\mu\,a[5] + 4\,t^{5+\mu}\,\mu^2\,a[5] + 143\,t^{6+\mu}\,a[6] + 4\,t^{8+\mu}\,a[6] + 48\,t^{6+\mu}\,\mu\,a[6] + 4\,t^{6+\mu}\,\mu^2\,a[6] + 195\,t^{7+\mu}\,a[7] + 4\,t^{9+\mu}\,a[7] + 56\,t^{7+\mu}\,\mu\,a[7] + 4\,t^{7+\mu}\,\mu^2\,a[7] + 255\,t^{8+\mu}\,a[8] + 4\,t^{10+\mu}\,a[8] + 64\,t^{8+\mu}\,\mu\,a[8] + 4\,t^{8+\mu}\,\mu^2\,a[8] + 323\,t^{9+\mu}\,a[9] + 4\,t^{11+\mu}\,a[9] + 72\,t^{9+\mu}\,\mu\,a[9] + 4\,t^{9+\mu}\,\mu^2\,a[9]$

Since we require this to be zero, we may as well multiply by $t^{-\mu}$ in order to obtain a polynomial in t. We'll define the result as a function of μ and t.

```
zero[μ_][t_] = t^-μ terms // Expand // Simplify
```

$$-1+4\mu^2+t\,(3+8\mu+4\mu^2)\,a[1]+t^2\,(4+(15+16\mu+4\mu^2)\,a[2])+t^3\,(4\,a[1]+(35+24\mu+4\mu^2)\,a[3])+t^4\,(4\,a[2]+(63+32\mu+4\mu^2)\,a[4])+t^5\,(4\,a[3]+(99+40\mu+4\mu^2)\,a[5])+t^6\,(4\,a[4]+(143+48\mu+4\mu^2)\,a[6])+t^7\,(4\,a[5]+(195+56\mu+4\mu^2)\,a[7])+4\,t^{10}\,a[8]+t^8\,(4\,a[6]+(255+64\mu+4\mu^2)\,a[8])+4\,t^{11}\,a[9]+t^9\,(4\,a[7]+(323+72\mu+4\mu^2)\,a[9])$$

Here we observe that the constant term determines μ. In this case μ must satisfy $4\mu^2 - 1 = 0$, which requires that $\mu = \pm 1/2$. Let's first handle the case $\mu = 1/2$. We start by extracting the coefficients from the above polynomial.

```
coeffs = CoefficientList[zero[1 / 2][t], t]
```

```
{0, 8 a[1], 4 + 24 a[2], 4 a[1] + 48 a[3], 4 a[2] + 80 a[4], 4 a[3] + 120 a[5],
 4 a[4] + 168 a[6], 4 a[5] + 224 a[7], 4 a[6] + 288 a[8], 4 a[7] + 360 a[9], 4 a[8], 4 a[9]}
```

Then we throw away the first and the last two entries and then turn the list of coefficients into a list of equations.

```
coeffs = Drop[Drop[coeffs, 1], -2]
```

```
{8 a[1], 4 + 24 a[2], 4 a[1] + 48 a[3], 4 a[2] + 80 a[4], 4 a[3] + 120 a[5],
 4 a[4] + 168 a[6], 4 a[5] + 224 a[7], 4 a[6] + 288 a[8], 4 a[7] + 360 a[9]}
```

```
eqns = Map[(# == 0) &, coeffs]
```

```
{8 a[1] == 0, 4 + 24 a[2] == 0, 4 a[1] + 48 a[3] == 0,
 4 a[2] + 80 a[4] == 0, 4 a[3] + 120 a[5] == 0, 4 a[4] + 168 a[6] == 0,
 4 a[5] + 224 a[7] == 0, 4 a[6] + 288 a[8] == 0, 4 a[7] + 360 a[9] == 0}
```

At this point we have 9 equations for the 9 unknowns $a_1, \ldots, a_9$. The last step is to solve these equations and apply the resulting rules.

```
(psum /. μ → 1 / 2) /. Solve[eqns, Table[a[i], {i, 1, 9}]]
```

$$\left\{\sqrt{t}-\frac{t^{5/2}}{6}+\frac{t^{9/2}}{120}-\frac{t^{13/2}}{5040}+\frac{t^{17/2}}{362880}\right\}$$

Now we proceed similarly with $\mu = -1/2$.

```
coeffs = CoefficientList[zero[-1 / 2][t], t]
```

```
{0, 0, 4 + 8 a[2], 4 a[1] + 24 a[3], 4 a[2] + 48 a[4], 4 a[3] + 80 a[5], 4 a[4] + 120 a[6],
 4 a[5] + 168 a[7], 4 a[6] + 224 a[8], 4 a[7] + 288 a[9], 4 a[8], 4 a[9]}
```

Notice here that the first *two* entries are zero. So our pruned list is

```
coeffs = Drop[Drop[coeffs, 2], -2]
```

```
{4 + 8 a[2], 4 a[1] + 24 a[3], 4 a[2] + 48 a[4], 4 a[3] + 80 a[5],
 4 a[4] + 120 a[6], 4 a[5] + 168 a[7], 4 a[6] + 224 a[8], 4 a[7] + 288 a[9]}
```

```
eqns = Map[(# == 0) &, coeffs]
```

```
{4 + 8 a[2] == 0, 4 a[1] + 24 a[3] == 0,
 4 a[2] + 48 a[4] == 0, 4 a[3] + 80 a[5] == 0, 4 a[4] + 120 a[6] == 0,
 4 a[5] + 168 a[7] == 0, 4 a[6] + 224 a[8] == 0, 4 a[7] + 288 a[9] == 0}
```

Here there are only 8 equations for the 9 unknowns $a_1, \ldots, a_9$. So we reduce the number of unknowns (or increase the number of equations) by setting $a_1 = 0$.

```
eqns = Prepend[eqns, a[1] == 0]
```

```
{a[1] == 0, 4 + 8 a[2] == 0, 4 a[1] + 24 a[3] == 0,
 4 a[2] + 48 a[4] == 0, 4 a[3] + 80 a[5] == 0, 4 a[4] + 120 a[6] == 0,
 4 a[5] + 168 a[7] == 0, 4 a[6] + 224 a[8] == 0, 4 a[7] + 288 a[9] == 0}
```

The result is

```
(psum /. μ → -1 / 2) /. Solve[eqns, Table[a[i], {i, 1, 9}]]
```

$$\left\{\frac{1}{\sqrt{t}} - \frac{t^{3/2}}{2} + \frac{t^{7/2}}{24} - \frac{t^{11/2}}{720} + \frac{t^{15/2}}{40320}\right\}$$

The alert reader will notice that the two partial sums we have found point to the fact that

$$\frac{1}{\sqrt{t}}\cos t \quad \text{and} \quad \frac{1}{\sqrt{t}}\sin t$$

are solutions of Bessel's equation of order $1/2$.

■ SeriesDSolve

The `DETools` package contains a function called `SeriesDSolve` that computes partial sums of power series solutions.

```
<< Diffeqs`DETools`; ? SeriesDSolve
```

```
SeriesDSolve[eqn, y[t], {t, t0, n}] returns a pair of independent
  series solutions of a linear 2nd-order differential equation.
  The series expansions are about t=t0, with terms up to degree n.
```

`SeriesDSolve` has the nice feature of being able to handle nonpolynomial coefficients. For instance, we get terms up through degree 10 of the fundamental power-series solutions of

$$y'' + t\,y' + \cos(t)\,y = 0$$

as follows:

```
SeriesDSolve[y''[t] + t y'[t] + Cos[t] y[t] == 0, y[t], {t, 0, 10}]
```

$$\left\{1 - \frac{t^2}{2} + \frac{t^4}{6} - \frac{3\,t^6}{80} + \frac{53\,t^8}{8064} - \frac{23\,t^{10}}{24192},\ t - \frac{t^3}{3} + \frac{11\,t^5}{120} - \frac{13\,t^7}{720} + \frac{37\,t^9}{12960}\right\}$$

`SeriesDSolve` will also compute the first n terms of the rest solution of a nonhomogeneous equation. For instance, consider the problem

$$y'' + \frac{1}{1+t^2}\,y' + y = e^t, \quad y(0) = y'(0) = 0.$$

The terms up through degree 11 in the power series expansion of the solution are

```
SeriesDSolve[y''[t] + 1/(1 + t^2) y'[t] + y[t] == e^t, y[t], {t, 0, 11}]
```

$$\frac{t^2}{2} + \frac{7\,t^5}{120} - \frac{t^6}{120} - \frac{t^7}{42} + \frac{337\,t^8}{40320} + \frac{127\,t^9}{10080} - \frac{1949\,t^{10}}{302400} - \frac{60331\,t^{11}}{7983360}$$

In addition, `SeriesDSolve` will compute partial sums of Frobenius-series solutions.

```
SeriesDSolve[t² y''[t] + t y'[t] + (t³ - 1/9) y[t] == 0, y[t], {t, 0, 12}]
```

$$\left\{\frac{1}{t^{1/3}} - \frac{t^{8/3}}{7} + \frac{t^{17/3}}{224} - \frac{t^{26/3}}{16800} + \frac{t^{35/3}}{2284800},\ t^{1/3} - \frac{t^{10/3}}{11} + \frac{t^{19/3}}{440} - \frac{t^{28/3}}{38280} + \frac{t^{37/3}}{5818560}\right\}$$

■ Exercises

In Exercises 1–6, find the recurrence formula for the equation, and use it to generate partial sums of fundamental power series solutions containing terms up through degree 12. Use `Series-DSolve` to check, and then plot the results on $[-1, 1]$.

1. $(1+t^2)\,y'' + y = 0$
2. $(1-t^2)\,y'' + y = 0$
3. $y'' + t\,y' + y = 0$
4. $y'' - t\,y' + y = 0$
5. $(t^2+1)\,y'' + t\,y' + y = 0$
6. $y'' + t^3\,y' + t^2\,y = 0$

7–12. For each equation in Exercises 1–6, use `DSolve` to find the fundamental solutions (i.e., solutions u and v satisfying $u(0) = 1,\ u'(0) = 0$, and $v(0) = 0,\ v'(0) = 1$.) In terms of what *special function* (if any) is the solution given?

In Exercises 13 and 14, find—with the help of `SeriesDSolve`—the sum of terms up through degree 12 of a power-series solution of the initial-value problem. Plot the result on a sensible interval.

13. $y'' + t\,y' + 9y = 0,\ y(0) = 1,\ y(0) = -1$

14. $y'' + \cos(t)\,y' + y = \sin(t),\ y(0) = 1,\ y(0) = 2.$

15. An instance of *Chebyshev's equation* is

$$(1-t^2)\,y'' - t\,y' + 9y = 0.$$

Use `SeriesDSolve` to find terms up through degree 10 of the equation's fundamental power series solutions. What interesting fact do you notice about one of the solutions?

16. *Bessel's equation of order n* is

$$t^2\,y'' + t\,y' + (t^2 - n^2)\,y = 0.$$

For each of $n = 0, 1, 2$, and 3, use `SeriesDSolve` to find a degree-40 partial sum of a power-series solution (about $t = 0$). Name these Taylor polynomials `j0`, `j1`, `j2`, and `j3`, respectively. Then create a plot containing the graphs of `j0`, …, `j3` for $0 \le t \le 16$.

17. Find the recurrence formula for *Gauss's hypergeometric equation*,

$$t\,(1-t)\,y'' + (c - (a+b+1)\,t)\,y' - a\,b\,y = 0,$$

where a, b, and c are constant parameters.

18. The equation

$$2\,t\,y'' + (1 - 2\,t)\,y' - y = 0$$

has an ordinary series solution and a Frobenius series solution. Walk through the process of Example 6.6.3 to find their terms up through degree 7. Then use `SeriesDSolve` to check your results. Identify the ordinary series solution as a familiar elementary function.

19. The equation

$$t\,y'' + 2y' + t\,y = 0$$

has a pair of Frobenius series solutions. Walk through the process of Example 6.6.3 to find their terms up through degree 9. Then use `SeriesDSolve` to check your results. Identify each solution as a familiar elementary function.

20. Equations of the form

$$t^2 y'' + b\,t\,y' + c\,y = 0$$

are known as *Cauchy-Euler equations* or *equidimensional equations*. Use `SeriesDSolve` to find a pair of solutions in powers of t. Under what conditions on b and c are the exponents real? What *real solutions* arise when the exponents are not real? (Suggestion: As a first step, look at the special case $b = c = 1$, and determine the real and imaginary parts of the complex solutions.)

Chapter 7

Linear Differential Systems

7.1 Vectors and Matrices in *Mathematica*

Our goal in this prelminary section is to become acquainted with how algebraic matrix and vector computations are done with *Mathematica*. We begin with a brief discussion of the way vectors and matrices are represented.

Vectors are simply lists. *Mathematica* does not distinguish between row vectors and column vectors.

```
vec := {1, 2, 3}
```

A matrix is a list of lists. Each inner list is a row of the matrix.

```
mat = {{1, 2, 3}, {3, 1, 2}, {2, 3, 1}}
{{1, 2, 3}, {3, 1, 2}, {2, 3, 1}}
```

`MatrixForm` displays a matrix in the traditional form.

```
mat // MatrixForm
```

$$\begin{pmatrix} 1 & 2 & 3 \\ 3 & 1 & 2 \\ 2 & 3 & 1 \end{pmatrix}$$

`MatrixForm` also displays a vector as a matrix with one column.

```
vec // MatrixForm
```

$$\begin{pmatrix} 1 \\ 2 \\ 3 \end{pmatrix}$$

💡 One should never use `MatrixForm` as part of a definition. For instance, if we define

```
bad = MatrixForm[{{1, 2}, {3, 4}}]
```

$$\begin{pmatrix} 1 & 2 \\ 3 & 4 \end{pmatrix}$$

then *Mathematica* cannot evaluate a product such as

```
2 bad
```

$$2 \begin{pmatrix} 1 & 2 \\ 3 & 4 \end{pmatrix}$$

However, `FullForm` reveals that the matrix can be recovered by extracting the first entry.

```
FullForm[bad]
MatrixForm[List[List[1, 2], List[3, 4]]]
```

```
First[bad]
{{1, 2}, {3, 4}}
```

Also note that a matrix may be entered in two-dimensional form, but it still represented as a list of lists.

$$\texttt{okay} = \begin{pmatrix} 1 & 2 \\ 2 & 1 \end{pmatrix}$$

```
{{1, 2}, {2, 1}}
```

To build a matrix this way, first click on 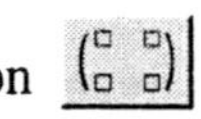 in the `BasicTypesetting` palette. Use the tab key to move from placeholder to placeholder. To add a row, click to the right of any entry and press CTRL-return. To add a column, click to the right of any entry and press CTRL-comma. For more information, bring up the Help Browser and look under Other Information : 2D Expression Input ▷ Entering 2D Expressions ▷ Matrices.

Matrix Algebra

Matrix-vector multiplication requires `Dot` (a period).

```
mat.vec // MatrixForm
```

$$\begin{pmatrix} 14 \\ 11 \\ 11 \end{pmatrix}$$

Notice the result if `Dot` is not used. Each row of the matrix is multiplied by the corresponding entry of the vector.

```
mat vec // MatrixForm
```

$$\begin{pmatrix} 1 & 2 & 3 \\ 6 & 2 & 4 \\ 6 & 9 & 3 \end{pmatrix}$$

We can define *vector-valued functions* and *matrix-valued functions*.

```
u[t_] := {Sin[t], Cos[t]};
```

$$\texttt{Amat[t\_]} := \begin{pmatrix} t & t^2 \\ e^{-t} & -t \end{pmatrix}$$

```
Amat[t].u[t] // MatrixForm
```

$$\begin{pmatrix} t^2\,\texttt{Cos[t]} + t\,\texttt{Sin[t]} \\ -t\,\texttt{Cos[t]} + e^{-t}\,\texttt{Sin[t]} \end{pmatrix}$$

The derivative of a vector (or matrix)-valued function is computed automatically.

```
u'[t]
```

```
{Cos[t], -Sin[t]}
```

```
∂t Amat[t]
```

$\{\{1, 2t\}, \{-e^{-t}, -1\}\}$

Scalar multiplication and matrix addition/subtraction are done in the natural way.

$5\begin{pmatrix} 3 & -2 \\ -2 & 3 \end{pmatrix}$ `// MatrixForm`

$\begin{pmatrix} 15 & -10 \\ -10 & 15 \end{pmatrix}$

$\begin{pmatrix} 1 & 2 \\ 3 & 4 \end{pmatrix} - \begin{pmatrix} 3 & -2 \\ -2 & 3 \end{pmatrix}$ `// MatrixForm`

$\begin{pmatrix} -2 & 4 \\ 5 & 1 \end{pmatrix}$

Matrix-matrix multiplication requires the use of `Dot`.

$\begin{pmatrix} 1 & 2 \\ 3 & 4 \end{pmatrix} . \begin{pmatrix} 3 & -2 \\ -2 & 3 \end{pmatrix}$ `// MatrixForm`

$\begin{pmatrix} -1 & 4 \\ 1 & 6 \end{pmatrix}$

Multiplication without `Dot` is done entry by entry. This is rarely useful.

$\begin{pmatrix} 1 & 2 \\ 3 & 4 \end{pmatrix} \begin{pmatrix} 3 & -2 \\ -2 & 3 \end{pmatrix}$ `// MatrixForm`

$\begin{pmatrix} 3 & -4 \\ -6 & 12 \end{pmatrix}$

A related issue is that to compute a matrix power, we need to use `MatrixPower` rather than an exponent.

`MatrixPower[`$\begin{pmatrix} 1 & -2 \\ 3 & -1 \end{pmatrix}$`, 3] // MatrixForm`

$\begin{pmatrix} -5 & 10 \\ -15 & 5 \end{pmatrix}$

Determinants and Inverses

`Det` returns the determinant of a square matrix.

`Det[`$\begin{pmatrix} a & b \\ c & d \end{pmatrix}$`]`

```
-b c + a d
```

```
Det[(1  -2   1
     3   1  -1
     2  -3   1)]
```

-3

If a square matrix has a nonzero determinant, we can compute its inverse.

```
A = (1   3   1
     1   1  -1
     2  -3   1);
invA = Inverse[A]; MatrixForm[invA]
```

$$\begin{pmatrix} \frac{1}{8} & \frac{3}{8} & \frac{1}{4} \\ \frac{3}{16} & \frac{1}{16} & -\frac{1}{8} \\ \frac{5}{16} & -\frac{9}{16} & \frac{1}{8} \end{pmatrix}$$

Recall that the product of a matrix with its inverse is the *identity matrix*.

```
A.invA // MatrixForm
```

$$\begin{pmatrix} 1 & 0 & 0 \\ 0 & 1 & 0 \\ 0 & 0 & 1 \end{pmatrix}$$

Matrix Construction

If there is a formula for the ij^{th} entry of a matrix, then `Table` can be used to build the matrix. For instance, if a 3×3 matrix A has $a_{ij} = i - j$, then A is

```
Table[i - j, {i, 1, 3}, {j, 1, 3}] // MatrixForm
```

$$\begin{pmatrix} 0 & -1 & -2 \\ 1 & 0 & -1 \\ 2 & 1 & 0 \end{pmatrix}$$

A diagonal matrix can be constructed from a list of its diagonal entries with `DiagonalMatrix`.

```
DiagonalMatrix[{a, b, c}] // MatrixForm
```

$$\begin{pmatrix} a & 0 & 0 \\ 0 & b & 0 \\ 0 & 0 & c \end{pmatrix}$$

The $n \times n$ identity matrix is `IdentityMatrix[n]`.

```
IdentityMatrix[2] // MatrixForm
```

$$\begin{pmatrix} 1 & 0 \\ 0 & 1 \end{pmatrix}$$

The `MatrixTools` Package

The `MatrixTools` package contains several functions that are handy for manipulating the columns of a matrix. These are `TakeColumn`, `DropColumn`, `ReplaceColumn`, `AppendColumn`, `PrependColumn`, and `InsertColumn`. Their usage is illustrated as follows:

```
<< DiffEqs`MatrixTools`
```

$$A := \begin{pmatrix} 1 & 2 \\ 3 & 4 \end{pmatrix}$$

```
TakeColumn[A, 2] // MatrixForm
```

$$\begin{pmatrix} 2 \\ 4 \end{pmatrix}$$

```
DropColumn[A, 2] // MatrixForm
```

$$\begin{pmatrix} 1 \\ 3 \end{pmatrix}$$

```
ReplaceColumn[A, {0, 0}, 2] // MatrixForm
```

$$\begin{pmatrix} 1 & 0 \\ 3 & 0 \end{pmatrix}$$

```
AppendColumn[A, {0, 0}] // MatrixForm
```

$$\begin{pmatrix} 1 & 2 & 0 \\ 3 & 4 & 0 \end{pmatrix}$$

```
PrependColumn[A, {0, 0}] // MatrixForm
```

$$\begin{pmatrix} 0 & 1 & 2 \\ 0 & 3 & 4 \end{pmatrix}$$

```
InsertColumn[A, {0, 0}, 2] // MatrixForm
```

$$\begin{pmatrix} 1 & 0 & 2 \\ 3 & 0 & 4 \end{pmatrix}$$

7.2 Linear Differential Systems

A pair of linear differential equations,

$$y_1' = a_{11}\, y_1 + a_{12}\, y_2\,,$$
$$y_2' = a_{21}\, y_1 + a_{22}\, y_2\,,$$

can be written in vector form as

$$\begin{pmatrix} y_1 \\ y_2 \end{pmatrix}' = \begin{pmatrix} a_{11} & a_{12} \\ a_{21} & a_{22} \end{pmatrix} \begin{pmatrix} y_1 \\ y_2 \end{pmatrix}.$$

We can similarly construct such a system in *Mathematica*. To do so, we will first give `Equal` the `Listable` attribute

```
SetAttributes[Equal, Listable]
```

and then enter

```
∂t (y1[t]; y2[t]) == (a11 a12; a21 a22).(y1[t]; y2[t]) // Flatten
{y1'[t] == a11 y1[t] + a12 y2[t], y2'[t] == a21 y1[t] + a22 y2[t]}
```

Also notice that the same can be accomplished as follows, where `Hold` is used to prevent the derivative from being evaluated until after the rule is applied.

```
Hold[∂t y[t] == (a11 a12; a21 a22).y[t]] /. y[t] → {y1[t], y2[t]} // Release
{y1'[t] == a11 y1[t] + a12 y2[t], y2'[t] == a21 y1[t] + a22 y2[t]}
```

■ Solution Matrices and Fundamental Matrices

Consider an $n \times n$, homogeneous, linear, differential system

$$\mathbf{y}' = A\mathbf{y},$$

where the entries of A are continuous (perhaps constant) on an interval I. A matrix $Y(t)$ whose columns are solutions of the system on I is said to be a *solution matrix.* An $n \times n$ solution matrix whose columns are linearly independent functions on I is called a *fundamental matrix.* An $n \times n$ solution matrix $Y(t)$ is a fundamental matrix if and only if the determinant of $Y(t)$ is never zero on I.

Example 7.2.1

Consider the system $\mathbf{y}' = A\mathbf{y}$, where the matrix A is

```
A := 1/(1 + t^2) (-t -1; 1 -t)
```

We will verify that the following is a fundamental matrix on $(-\infty, \infty)$:

```
Y[t_] := 1/(1 + t^2) (t 1; -1 t)
```

First we enter a definition for the system, using. `Hold` to delay the evaluation of the derivative.

```
sys = Hold[∂t y[t] == A.y[t]]
Hold[∂t y[t] == A.y[t]]
```

Next we substitute the supposed solution matrix into the system, `Release`, and `Simplify`.

```
sys /. y[t] → Y[t] // Release // Simplify
{{True, True}, {True, True}}
```

So the matrix is indeed a solution matrix. Now we compute its determinant.

```
Det[Y[t]] // Simplify
```

$$\frac{1}{1+t^2}$$

This is never zero; hence the solution matrix is a fundamental matrix on $(-\infty, \infty)$.

■ Initial-Value Problems and Evolution Matrices

Consider an initial-value problem

$$\mathbf{y}' = A\mathbf{y}, \quad \mathbf{y}(t_0) = \mathbf{y}_0,$$

where A is an $n \times n$ matrix whose entries are continuous on an interval I containing t_0. Given any fundamental matrix $Y(t)$, the solution of the initial-value problem is

$$\mathbf{y}(t) = Y(t)\, Y(t_0)^{-1}\, \mathbf{y}_0.$$

This is a solution of the system because it is a linear combination of the columns of $Y(t)$. Furthermore, its value at $t = t_0$ is $\mathbf{y}(t_0) = Y(t_0)\, Y(t_0)^{-1}\, \mathbf{y}_0 = \mathbf{y}_0$ as required.

Consider the matrix $\Phi_{t_0}(t) = Y(t)\, Y(t_0)^{-1}$ in the above formula. This matrix—which can be computed from any fundamental matrix $Y(t)$—is itself a fundamental matrix, and it is the only fundamental matrix that equals the identity matrix at $t = t_0$, i.e., with the property that $\Phi_{t_0}(t_0) = I$. We call this matrix *the evolution matrix* generated by A at t_0.

Example 7.2.2

Consider the initial-value problem

$$\mathbf{y}' = \frac{1}{1+t^2}\begin{pmatrix} -t & -1 \\ 1 & -t \end{pmatrix}\mathbf{y}, \quad \mathbf{y}(0) = \begin{pmatrix} 2 \\ 1 \end{pmatrix}$$

In Example 7.1.1, we verified that a fundamental matrix on $(-\infty, \infty)$ is given by

```
Y[t_] := 1/(1 + t²) (t  1
                     -1 t)
```

The evolution matrix generated by A at 0 is therefore

```
Φ[t_] = Y[t].Inverse[Y[0]]; Φ[t] // MatrixForm
```

$$\begin{pmatrix} \frac{1}{1+t^2} & -\frac{t}{1+t^2} \\ \frac{t}{1+t^2} & \frac{1}{1+t^2} \end{pmatrix}$$

and the solution of the initial-value problem is

```
Φ[t].{2, 1} // Simplify
```

$$\left\{\frac{2-t}{1+t^2}, \frac{1+2t}{1+t^2}\right\}$$

■ The Variation of Constants Formula

Consider the nonhomogeneous system

$$\mathbf{y}' = A\mathbf{y} + \mathbf{f}(t), \quad \mathbf{y}(t_0) = \mathbf{y}_0,$$

where A is an $n \times n$ matrix and f is a vector-valued function, each of whose entries are continuous on an interval I containing t_0. Given the evolution matrix $\Phi_{t_0}(t)$, the solution of the initial-value problem is

$$\mathbf{y}(t) = \Phi_{t_0}(t)\left(\mathbf{y}_0 + \int_0^t \Phi_{t_0}(s)^{-1}\,\mathbf{f}(s)\,ds\right).$$

This is the *variation of constants formula* (a.k.a., the variation of parameters formula).

Example 7.2.3

Consider the nonhomogeneous initial-value problem

$$\mathbf{y}' = \frac{1}{1+t^2}\begin{pmatrix} -t & -1 \\ 1 & -t \end{pmatrix}\mathbf{y} + \begin{pmatrix} t \\ t^2 \end{pmatrix}, \quad \mathbf{y}(0) = \begin{pmatrix} 0 \\ 1 \end{pmatrix}$$

In Example 7.2.2, we found that the evolution matrix generated by A at 0 is

```
Φ[t_] := ( 1/(1+t²)   -t/(1+t²) )
         ( t/(1+t²)    1/(1+t²) )
```

Therefore, the solution of the initial-value problem is

```
Φ[t].({0, 1} + ∫₀ᵗ Inverse[Φ[s]].{s, s²} ⅆs) // Simplify
```

```
{t (-4 + 2 t + t³)/(4 (1 + t²)), (4 + 2 t³ + t⁵)/(4 + 4 t²)}
```

■ Exercises

1. Let $A = \frac{1}{1+t^2}\begin{pmatrix} t & 1 \\ -1 & t \end{pmatrix}$.

   a) Verify that $Y(t) = \begin{pmatrix} 1 & t \\ -t & 1 \end{pmatrix}$ is a fundamental matrix generated by A on $(-\infty, \infty)$.

   b) Verify that $Y(t)$ is also the evolution matrix generated by A at 0.

   c) Find the solution of $\mathbf{y}' = A\mathbf{y}$, $y(0) = \begin{pmatrix} -1 \\ 2 \end{pmatrix}$.

   d) Find the solution of $\mathbf{y}' = A\mathbf{y} + \begin{pmatrix} e^{-t} \\ \cos t \end{pmatrix}$, $\mathbf{y}(0) = \begin{pmatrix} 1 \\ 1 \end{pmatrix}$.

2. Let $A = \sec t\begin{pmatrix} -\sin t & 1 \\ 1 & -\sin t \end{pmatrix}$.

   a) Verify that $Y(t) = \begin{pmatrix} 1 & \sin t \\ \sin t & 1 \end{pmatrix}$ is a fundamental matrix generated by A on $(-\frac{\pi}{2}, \frac{\pi}{2})$.

b) Verify that $Y(t)$ is also the evolution matrix generated by A at 0.

c) Find the solution of $\mathbf{y}' = A\mathbf{y},\ \mathbf{y}(0) = \begin{pmatrix} -1 \\ 1 \end{pmatrix}$.

d) Find the solution of $\mathbf{y}' = A\mathbf{y} + \begin{pmatrix} t \\ \sin t \end{pmatrix},\ \mathbf{y}(0) = \begin{pmatrix} 0 \\ 0 \end{pmatrix}$.

3. Let $A = \dfrac{1}{1-t^2}\begin{pmatrix} -t & 1 \\ 1 & -t \end{pmatrix}$.

a) Verify that $Y(t) = \begin{pmatrix} t & 1 \\ 1 & t \end{pmatrix}$ is a fundamental matrix generated by A on $(-1, 1)$.

b) Find the evolution matrix generated by A at 0.

c) Find the solution of $\mathbf{y}' = A\mathbf{y},\ \mathbf{y}(0) = \begin{pmatrix} -1 \\ 1 \end{pmatrix}$.

d) Find the solution of $\mathbf{y}' = A\mathbf{y} + \begin{pmatrix} \cos t \\ \sin t \end{pmatrix},\ \mathbf{y}(0) = \begin{pmatrix} 0 \\ 0 \end{pmatrix}$.

4. Let $A = \dfrac{1}{e^t - t}\begin{pmatrix} t & -1 \\ e^t & -1 \end{pmatrix}$.

a) Verify that $Y(t) = \begin{pmatrix} 1 & e^{-t} \\ t & 1 \end{pmatrix}$ is a fundamental matrix generated by A on $(-\infty, \infty)$.

b) Find the evolution matrix generated by A at 0.

c) Find the solution of $\mathbf{y}' = A\mathbf{y},\ \mathbf{y}(0) = \begin{pmatrix} -1 \\ 1 \end{pmatrix}$.

d) Find the solution of $\mathbf{y}' = A\mathbf{y} + \begin{pmatrix} t\,e^{-t} \\ e^{-t} \end{pmatrix},\ \mathbf{y}(0) = \begin{pmatrix} 0 \\ 0 \end{pmatrix}$.

5. Let $A = \begin{pmatrix} -2 & 1 & 1 \\ 1 & -2 & 1 \\ 1 & 1 & -2 \end{pmatrix}$.

a) Verify that $Y(t) = \begin{pmatrix} 1 & -e^{-3t} & -e^{-3t} \\ 1 & e^{-3t} & 0 \\ 1 & 0 & e^{-3t} \end{pmatrix}$ is a fundamental matrix generated by A on $(-\infty, \infty)$.

b) Find the evolution matrix generated by A at 0.

c) Find the solution of $\mathbf{y}' = A\mathbf{y},\ \mathbf{y}(0) = \begin{pmatrix} -1 \\ 1 \\ 2 \end{pmatrix}$.

d) Find the solution of $\mathbf{y}' = A\mathbf{y} + \begin{pmatrix} t\,e^{-t} \\ e^{-t} \\ e^{-3t} \end{pmatrix},\ \mathbf{y}(0) = \begin{pmatrix} 0 \\ 0 \end{pmatrix}$.

7.3 Matrix Eigenvalues and Eigenvectors

Let A be a square matrix. Recall that a (complex) number λ is an *eigenvalue* of A if there is a nonzero vector $\boldsymbol{p}$ such that

$$A\boldsymbol{p} = \lambda \boldsymbol{p}.$$

The vector $\boldsymbol{p}$ is an *eigenvector* of A corresponding to the eigenvalue λ. The *Mathematica* function `Eigenvalues` computes the eigenvalues of a square matrix.

```
Eigenvalues[(a b
             c d)] // FullSimplify
```

$$\left\{\frac{1}{2}\left(a-\sqrt{4\,b\,c+(a-d)^2}+d\right),\ \frac{1}{2}\left(a+\sqrt{4\,b\,c+(a-d)^2}+d\right)\right\}$$

```
Eigenvalues[(-1 0  2
              2 1  2
             -2 3 -1)] // N
```

```
{-1.93012 + 2.07747 i, -1.93012 - 2.07747 i, 2.86025}
```

`Eigenvectors` returns the corresponding eigenvectors.

```
Eigenvectors[(-1 0  2
               2 1  2
              -2 3 -1)] // N
```

```
{{-0.359051 - 0.801955 i, -0.549408 + 0.157854 i, 1.},
 {-0.359051 + 0.801955 i, -0.549408 - 0.157854 i, 1.}, {0.518102, 1.63215, 1.}}
```

Note that the eigenvectors are returned as the *rows* of a matrix. We will find it useful later to have the matrix whose *columns* are the eigenvectors. For this, we need only apply `Transpose`.

```
evecs = Eigenvectors[(-2 3
                       4 2)]
```

```
{{-3, 2}, {1, 2}}
```

```
(evecmat = Transpose[evecs]) // MatrixForm
```

$$\begin{pmatrix} -3 & 1 \\ 2 & 2 \end{pmatrix}$$

`Eigensystem` returns both the eigenvalues and the eigenvectors of a square matrix.

```
esys = Eigensystem[(-1  2
                    -2 -1)] // FullSimplify // Expand
```

```
{{-1 - 2 i, -1 + 2 i}, {{i, 1}, {-i, 1}}}
```

The first entry in the result is a list of the eigenvalues, and the second entry is a list containing the eigenvectors. Transposing the result produces a list of the *eigenpairs* of the matrix.

```
esys // Transpose
```

```
{{-1 - 2 i, {i, 1}}, {-1 + 2 i, {-i, 1}}}
```

Example 7.3.1

Let A be the defined by

$$A = \begin{pmatrix} -1 & 2 \\ -2 & -1 \end{pmatrix};$$

The eigenvalues and eigenvectors of A are

```
evals = Eigenvalues[A]
{-1 - 2 i, -1 + 2 i}
evecs = Eigenvectors[A]
{{i, 1}, {-i, 1}}
```

Note that an eigenvector multiplied by A is the same as the eigenvector multiplied by the eigenvalue to which it corresponds.

```
A.evecs[[1]]
{2 - i, -1 - 2 i}
evals[[1]] evecs[[1]]
{2 - i, -1 - 2 i}
A.evecs[[2]]
{2 + i, -1 + 2 i}
evals[[2]] evecs[[2]]
{2 + i, -1 + 2 i}
```

If we transpose the list of eigenvectors, we obtain a matrix E in which the eigenvectors are columns.

```
evecmat = Transpose[evecs]; evecmat // MatrixForm
```

$$\begin{pmatrix} i & -i \\ 1 & 1 \end{pmatrix}$$

The product AE is the same as the product $E\Lambda$, where Λ is the diagonal matrix with the eigenvalues of A along its diagonal.

```
A.evecmat // MatrixForm
```

$$\begin{pmatrix} 2 - i & 2 + i \\ -1 - 2 i & -1 + 2 i \end{pmatrix}$$

```
evalsdiag = DiagonalMatrix[evals]; evalsdiag // MatrixForm
```

$$\begin{pmatrix} -1 - 2 i & 0 \\ 0 & -1 + 2 i \end{pmatrix}$$

```
Transpose[evecs].evalsdiag // MatrixForm
```

$$\begin{pmatrix} 2 - i & 2 + i \\ -1 - 2 i & -1 + 2 i \end{pmatrix}$$

As a result, if E is invertible, then $A = E\Lambda E^{-1}$ and $E^{-1}A\,E = \Lambda$.

```
evecmat.evalsdiag.Inverse[evecmat] // MatrixForm
```

$$\begin{pmatrix} -1 & 2 \\ -2 & -1 \end{pmatrix}$$

```
Inverse[evecmat].A.evecmat // MatrixForm
```

$$\begin{pmatrix} -1-2\,i & 0 \\ 0 & -1+2\,i \end{pmatrix}$$

■ Exercises

For each of the following matrices A,

(a) find the eigenvalues λ_i and eigenvectors p_i of A,

(b) construct the matrices Λ and E, where Λ is the diagonal matrix whose diagonal entries are $\lambda_1, \lambda_2, \ldots, \lambda_n$ and $E = (p_1 \;\; p_2 \;\; \cdots \;\; p_n)$. Then verify that $AE = E\Lambda$.

1. $\begin{pmatrix} -2 & 1 \\ 1 & -2 \end{pmatrix}$
2. $\begin{pmatrix} 0 & -1 \\ 1 & 0 \end{pmatrix}$
3. $\begin{pmatrix} 1 & -1 \\ 1 & 1 \end{pmatrix}$
4. $\begin{pmatrix} 1 & 2 \\ 3 & -4 \end{pmatrix}$
5. $\begin{pmatrix} -2 & 1 & 0 \\ 1 & -2 & 1 \\ 0 & 1 & -2 \end{pmatrix}$
6. $\begin{pmatrix} -2 & -1 & 0 \\ 1 & 0 & -1 \\ 0 & 1 & -2 \end{pmatrix}$
7. $\begin{pmatrix} -1 & -1 & -2 & 0 \\ 1 & -1 & 0 & -2 \\ 2 & 0 & -1 & -1 \\ 0 & 2 & 1 & -1 \end{pmatrix}$

7.4 Autonomous Systems

When the entries of A are constants, the system $\mathbf{y}' = A\mathbf{y}$ is said to be an *autonomous* system. In this case, if λ is an eigenvalue of A with corresponding eigenvector p, then the vector $e^{\lambda t} p$ is a solution of the system. Consequently, a solution matrix can be constructed from the eigenvalues and eigenvectors of A. If $\lambda_1, \lambda_2, \ldots, \lambda_k$ are the eigenvalues with corresponding eigenvectors $p_1, p_2, \ldots, p_k$, then a solution matrix is

$$Y(t) = (e^{\lambda_1 t} p_1 \;\; e^{\lambda_2 t} p_2 \;\; \cdots \;\; e^{\lambda_k t} p_k).$$

When A has n linearly independent eigenvectors—which will be the case when A has n distinct eigenvalues—this will be a fundamental matrix.

It is worth noting that

$$(e^{\lambda_1 t} p_1 \;\; e^{\lambda_2 t} p_2 \;\; \cdots \;\; e^{\lambda_k t} p_k) = PE(t),$$

where $P = (p_1 \;\; p_2 \;\; \cdots \;\; p_n)$ and $E(t) = \mathrm{diag}\,\{e^{\lambda_1 t}, e^{\lambda_2 t}, \ldots, e^{\lambda_n t}\}$.

The following examples illustrate the calculation of fundamental matrices by this method.

Example 7.4.1

Consider the system $\mathbf{y}' - A\mathbf{y} = 0$, where A is the following 2×2 matrix.

$$A := \begin{pmatrix} -2 & 1 \\ 1 & -2 \end{pmatrix}$$

Let's first compute the eigenvalues λ_1, λ_2 of A and form a diagonal matrix from $e^{\lambda_1 t}$, $e^{\lambda_2 t}$.

```
λ = Eigenvalues[A]
{-3, -1}
```

```
expλt = DiagonalMatrix[e^(λ t)]; expλt // MatrixForm
```

$$\begin{pmatrix} e^{-3t} & 0 \\ 0 & e^{-t} \end{pmatrix}$$

Next we form a matrix whose columns are the corresponding eigenvectors.

```
evecmat = Eigenvectors[A] // Transpose; MatrixForm[evecmat]
```

$$\begin{pmatrix} -1 & 1 \\ 1 & 1 \end{pmatrix}$$

The last step is to left-multiply the first of these matrices by the second. This produces a fundamental matrix.

```
solnmat = evecmat.expλt; MatrixForm[solnmat]
```

$$\begin{pmatrix} -e^{-3t} & e^{-t} \\ e^{-3t} & e^{-t} \end{pmatrix}$$

Finally, let's verify the result. The differential equation is satisfied, and the determinant of the matrix is never zero.

```
∂t solnmat - A.solnmat // MatrixForm
```

$$\begin{pmatrix} 0 & 0 \\ 0 & 0 \end{pmatrix}$$

```
Det[solnmat]
```

$-2\, e^{-4t}$

Example 7.4.2

Consider the system $\mathbf{y}' - A\mathbf{y} = 0$, where A is the following 2×2 matrix.

$$A = \begin{pmatrix} -1 & -2 \\ 2 & -1 \end{pmatrix};$$

The eigenvalues and eigenvectors of A are complex. Nevertheless, we proceed as before.

```
λ = Eigenvalues[A]
{-1 - 2 i, -1 + 2 i}
```

```
expλt = DiagonalMatrix[e^(λ t)]; expλt // MatrixForm
```

$$\begin{pmatrix} e^{(-1-2i)t} & 0 \\ 0 & e^{(-1+2i)t} \end{pmatrix}$$

```
evecmat = Eigenvectors[A] // Transpose; MatrixForm[evecmat]
```

$$\begin{pmatrix} -i & i \\ 1 & 1 \end{pmatrix}$$

```
solnmat = evecmat.expλt; MatrixForm[solnmat]
```

$$\begin{pmatrix} -i\,e^{(-1-2i)t} & i\,e^{(-1+2i)t} \\ e^{(-1-2i)t} & e^{(-1+2i)t} \end{pmatrix}$$

At this point we have a complex solution matrix.

```
∂t solnmat - A.solnmat // MatrixForm
```

$$\begin{pmatrix} 0 & 0 \\ 0 & 0 \end{pmatrix}$$

To obtain a *real* solution matrix, we extract the real and imaginary parts of either column.

```
Needs["DiffEqs`MatrixTools`"];

col1 = ComplexExpand[Re[TakeColumn[solnmat, 1]]]
```

$\{-e^{-t}\,\mathrm{Sin}[2\,t],\ e^{-t}\,\mathrm{Cos}[2\,t]\}$

```
col2 = ComplexExpand[Im[TakeColumn[solnmat, 1]]]
```

$\{-e^{-t}\,\mathrm{Cos}[2\,t],\ -e^{-t}\,\mathrm{Sin}[2\,t]\}$

```
realsolnmat = Transpose[{col1, col2}]; realsolnmat // MatrixForm
```

$$\begin{pmatrix} -e^{-t}\,\mathrm{Sin}[2\,t] & -e^{-t}\,\mathrm{Cos}[2\,t] \\ e^{-t}\,\mathrm{Cos}[2\,t] & -e^{-t}\,\mathrm{Sin}[2\,t] \end{pmatrix}$$

Finally we check the result.

```
∂t realsolnmat - A.realsolnmat // MatrixForm
```

$$\begin{pmatrix} 0 & 0 \\ 0 & 0 \end{pmatrix}$$

```
Det[realsolnmat] // Simplify
```

$e^{-2\,t}$

Example 7.4.3

Consider the system $\mathbf{y}' - A\mathbf{y} = 0$, where A is the following 3×3 matrix.

$$\mathbf{A} = \begin{pmatrix} -1 & 2 & -3 \\ 7 & 4 & 7 \\ -1 & -1 & 2 \end{pmatrix};$$

A has only two distinct eigenvalues.

```
λ = Eigenvalues[A]
{-3, 4, 4}
```

```
expλt = DiagonalMatrix[e^(λ t)]; expλt // MatrixForm
```

$$\begin{pmatrix} e^{-3t} & 0 & 0 \\ 0 & e^{4t} & 0 \\ 0 & 0 & e^{4t} \end{pmatrix}$$

More importantly, A does not have three linearly independent eigenvectors. The eigenvalue $\lambda = 4$ has a *deficiency of order* 1. *Mathematica* automatically puts the zero vector in place of the missing eigenvector.

```
evecmat = Eigenvectors[A] // Transpose; MatrixForm[evecmat]
```

$$\begin{pmatrix} -1 & -1 & 0 \\ 1 & -1 & 0 \\ 0 & 1 & 0 \end{pmatrix}$$

Consequently, we get a zero column in our solution matrix.

```
solnmat = evecmat.expλt; solnmat // MatrixForm
```

$$\begin{pmatrix} -e^{-3t} & -e^{4t} & 0 \\ e^{-3t} & -e^{4t} & 0 \\ 0 & e^{4t} & 0 \end{pmatrix}$$

We want to replace the zero column with a solution that is linearly independent from the first two columns. To do this we compute a *generalized eigenvector* as follows. First we form the matrix $A - 4I$.

```
Am4I = A - 4 IdentityMatrix[3]; Am4I // MatrixForm
```

$$\begin{pmatrix} -5 & 2 & -3 \\ 7 & 0 & 7 \\ -1 & -1 & -2 \end{pmatrix}$$

Then we find a nonzero solution of $(A - 4I)^2 x = \mathbf{0}$. `NullSpace` is handy for this.

```
genevec = First[NullSpace[MatrixPower[Am4I, 2]]];
MatrixForm[genevec]
```

$$\begin{pmatrix} -5 \\ 0 \\ 6 \end{pmatrix}$$

From this generalized eigenvector we get a third solution as follows:

```
soln3 = e^(4 t) (IdentityMatrix[3] + t Am4I).genevec // Simplify;
soln3 // MatrixForm
```

$$\begin{pmatrix} e^{4t}(-5 + 7t) \\ 7e^{4t}t \\ e^{4t}(6 - 7t) \end{pmatrix}$$

Next we want to replace the zero column in the solution matrix with this third solution. Doing so produces a fundamental matrix.

```
Needs["DiffEqs`MatrixTools`"];

solnmat = ReplaceColumn[solnmat, soln3, 3]; solnmat // MatrixForm
```

$$\begin{pmatrix} -e^{-3t} & -e^{4t} & e^{4t}(-5+7t) \\ e^{-3t} & -e^{4t} & 7e^{4t}t \\ 0 & e^{4t} & e^{4t}(6-7t) \end{pmatrix}$$

Finally, we verify the result.

```
∂t solnmat - A.solnmat // Simplify // MatrixForm
```

$$\begin{pmatrix} 0 & 0 & 0 \\ 0 & 0 & 0 \\ 0 & 0 & 0 \end{pmatrix}$$

```
Det[solnmat] // Simplify
```

$7e^{5t}$

Example 7.4.4

Consider the system $\mathbf{y}' - A\mathbf{y} = 0$, where A is the following 4×4 matrix.

$$\mathtt{A} = \begin{pmatrix} -1 & 1 & -1 & 0 \\ 1 & -1 & 0 & 0 \\ 10 & 0 & -1 & -1 \\ 0 & 0 & 1 & -1 \end{pmatrix};$$

The exact eigenvalues and eigenvectors of this matrix turn out to be very complicated expressions; so let's use numerical values.

```
λ = Eigenvalues[A] // N
```

{-1.31467, -0.685326, -1. - 3.1779 i, -1. + 3.1779 i}

```
evecs = Eigenvectors[A] // N // Transpose; MatrixForm[evecs]
```

$$\begin{pmatrix} 0.109902 & 0.109902 & -0.909902 & -0.909902 \\ -0.349257 & 0.349257 & 0.-0.286322\,i & 0.+0.286322\,i \\ -0.314674 & 0.314674 & 0.-3.1779\,i & 0.+3.1779\,i \\ 1. & 1. & 1. & 1. \end{pmatrix}$$

```
solnmat = evecs.DiagonalMatrix[e^(λ t)] // Simplify // Chop;
solnmat // MatrixForm
```

$$\begin{pmatrix} 0.109902\,e^{-1.31467\,t} & 0.109902\,e^{-0.685326\,t} & -0.909902\,e^{(-1.-3.1779\,i)\,t} & -0.909902\,e^{(-1.+3.1779\,i)\,t} \\ -0.349257\,e^{-1.31467\,t} & 0.349257\,e^{-0.685326\,t} & -0.286322\,i\,e^{(-1.-3.1779\,i)\,t} & 0.286322\,i\,e^{(-1.+3.1779\,i)\,t} \\ -0.314674\,e^{-1.31467\,t} & 0.314674\,e^{-0.685326\,t} & -3.1779\,i\,e^{(-1.-3.1779\,i)\,t} & 3.1779\,i\,e^{(-1.+3.1779\,i)\,t} \\ 1.\,e^{-1.31467\,t} & 1.\,e^{-0.685326\,t} & 1.\,e^{(-1.-3.1779\,i)\,t} & 1.\,e^{(-1.+3.1779\,i)\,t} \end{pmatrix}$$

(The use of `Chop` here is to rid the expression of insignificant terms that arise due to rounding error.) So at this point we have a complex solution matrix in which the third and fourth columns are complex conjugates. We will replace these columns with the real and imaginary parts of the third column.

```
<< DiffEqs`MatrixTools`

realsoln3 = ComplexExpand[Re[TakeColumn[solnmat, 3]]] // Simplify

{-0.909902 e^(-1. t) Cos[3.1779 t], -0.286322 e^(-1. t) Sin[3.1779 t],
 -3.1779 e^(-1. t) Sin[3.1779 t], 1. e^(-1. t) Cos[3.1779 t]}

realsoln4 = ComplexExpand[Im[TakeColumn[solnmat, 3]]] // Simplify

{0.909902 e^(-1. t) Sin[3.1779 t], -0.286322 e^(-1. t) Cos[3.1779 t],
 -3.1779 e^(-1. t) Cos[3.1779 t], -1. e^(-1. t) Sin[3.1779 t]}

solnmat = ReplaceColumn[solnmat, realsoln3, 3];
solnmat = ReplaceColumn[solnmat, realsoln4, 4];
```

Now we verify the result.

```
∂t solnmat - A.solnmat // Simplify // Chop

{{0, 0, 0, 0}, {0, 0, 0, 0}, {0, 0, 0, 0}, {0, 0, 0, 0}}

Det[solnmat /. t → 0] // Chop

2.08
```

Let's use this fundamental matrix to compute the evolution matrix generated by A at 0.

```
Φ[t_] = solnmat.Inverse[solnmat /. t → 0] // Simplify;
```

Now, for example, the solution that satisfies the initial condition $\mathbf{y}(0) = (2, 0, -1, 1)^T$ is computed and plotted as follows.

```
soln = Φ[t].{2, 0, -1, 1} // Simplify;
Plot[Evaluate[soln], {t, 0, 5}, PlotRange → All];
```

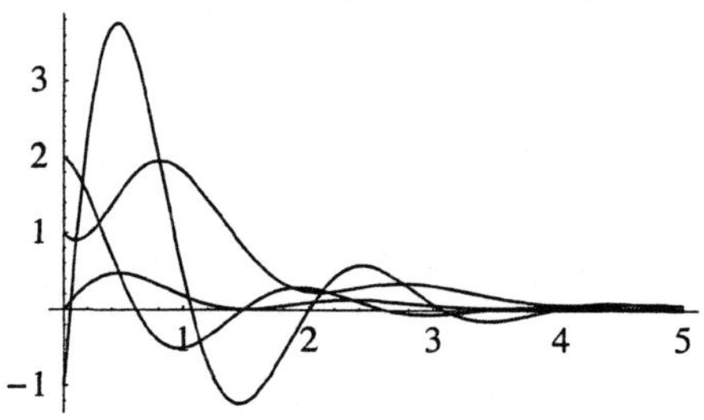

■ Exercises

For each of the following matrices A, find a fundamental matrix and then the evolution matrix generated by A.

1. $\begin{pmatrix} -2 & 1 \\ 1 & -2 \end{pmatrix}$

2. $\begin{pmatrix} 0 & -1 \\ 1 & 0 \end{pmatrix}$

3. $\begin{pmatrix} 1 & -1 \\ 1 & 1 \end{pmatrix}$

4. $\begin{pmatrix} 1 & 2 \\ 3 & -4 \end{pmatrix}$

5. $\begin{pmatrix} -2 & 1 & 0 \\ 1 & -2 & 1 \\ 0 & 1 & -2 \end{pmatrix}$

6. $\begin{pmatrix} -2 & -1 & 0 \\ 1 & 0 & -1 \\ 0 & 1 & -2 \end{pmatrix}$

7. $\begin{pmatrix} -1 & -1 & -2 & 0 \\ 1 & -1 & 0 & -2 \\ 2 & 0 & -1 & -1 \\ 0 & 2 & 1 & -1 \end{pmatrix}$

In Exercises 8 and 9, let A be the matrix in Exercise 6. Use the evolution matrix generated by A to solve the initial-value problem. Plot the components of the solution versus t. Then use `ParametricPlot3D` to plot the solution curve in $\mathbb{R}^3$.

8. $\mathbf{y}' - A\mathbf{y} = 0, \quad \mathbf{y}(0) = \begin{pmatrix} 1 \\ -1 \\ 1 \end{pmatrix}$

9. $\mathbf{y}' - A\mathbf{y} = \begin{pmatrix} e^{-t} \\ \sin t \\ \cos t \end{pmatrix}, \quad \mathbf{y}(0) = \begin{pmatrix} 0 \\ 0 \\ 0 \end{pmatrix}$

7.5 The Matrix Exponential

For an autonomous system $\mathbf{y}' = A\mathbf{y}$, the evolution matrix $\Phi_0(t)$ is denoted by e^{At} because of the following properties, which have direct analogies with e^{at}, where a is a real number.

$$\frac{d}{dt}\Phi_0(t) = A\Phi_0(t), \quad \Phi_0(0) = I;$$

$$\Phi_0(s)\,\Phi_0(t) = \Phi_0(s+t);$$

$$\Phi_0(t)^{-1} = \Phi_0(-t);$$

$$\Phi_0(t) = \sum_{k=1}^{\infty} \frac{1}{k!}\,(tA)^k.$$

Mathematica has a function named `MatrixExp` that computes e^M for any square matrix M.

```
MatrixExp[( 2  -1
           -1   2 )] // Simplify // MatrixForm
```

$$\begin{pmatrix} \frac{1}{2}(e + e^3) & \frac{1}{2}(e - e^3) \\ \frac{1}{2}(e - e^3) & \frac{1}{2}(e + e^3) \end{pmatrix}$$

Our only interest lies in computing matrix exponentials of the form e^{At}.

```
A = ( 2  -1
     -1   2 );
expA[t_] = MatrixExp[A t] // Simplify; expA[t] // MatrixForm
```

$$\begin{pmatrix} \frac{1}{2}e^t(1 + e^{2t}) & -\frac{1}{2}e^t(-1 + e^{2t}) \\ -\frac{1}{2}e^t(-1 + e^{2t}) & \frac{1}{2}e^t(1 + e^{2t}) \end{pmatrix}$$

Observe in this particular instance the following results:

```
∂t expA[t] == A.expA[t] // Simplify
True
expA[0] // MatrixForm
```

$$\begin{pmatrix} 1 & 0 \\ 0 & 1 \end{pmatrix}$$

```
expA[s].expA[t] == expA[s + t] // Simplify
True
expA[-t] == Inverse[expA[t]] // Simplify
True
```

In terms of the matrix exponential, the variation of constants formula gives the following:

The solution of $\begin{array}{l} \mathbf{y}' = A\mathbf{y} + \mathbf{f} \\ \mathbf{y}(0) = \mathbf{y}_0 \end{array}$ *is* $\mathbf{y} = e^{At}\left(\mathbf{y}_0 + \int_0^t e^{-As}\mathbf{f}(s)\,ds\right).$

Note that this formula has precisely the same form as the familiar variation of constants formula for a single linear first-order equation.

Example 7.4.1

Consider the 3×3 initial-value problem

$$\mathbf{y}' = A\mathbf{y} + \mathbf{f}(t),\ \mathbf{y}(0) = \mathbf{y}_0,$$

where

```
A := {{-3, 2, 1}, {1, -2, 1}, {0, 1, -1}}; f[t_] := {0, Sin[t], 0}; y0 := {2, 1, 0}
```

First we compute e^{At}:

```
expA[t_] = Simplify[MatrixExp[A t]]; expA[t] // MatrixForm
```

$$\begin{pmatrix} \frac{1}{8}(1 + 5e^{-4t} + 2e^{-2t}) & \frac{1}{8}(3 - 5e^{-4t} + 2e^{-2t}) & \frac{1}{2} - \frac{e^{-2t}}{2} \\ \frac{1}{8}(1 - 3e^{-4t} + 2e^{-2t}) & \frac{1}{8}(3 + 3e^{-4t} + 2e^{-2t}) & \frac{1}{2} - \frac{e^{-2t}}{2} \\ \frac{1}{8}e^{-4t}(-1 + e^{2t})^2 & \frac{1}{8}(3 - e^{-4t} - 2e^{-2t}) & \frac{1}{2}(1 + e^{-2t}) \end{pmatrix}$$

The variation of constants formula then gives the solution:

```
soln = expA[t].(y0 + ∫_0^t expA[-s].f[s] ds) // Simplify
```

$$\left\{\frac{1}{85}(85 + 50e^{-4t} + 68e^{-2t} - 33\,\mathrm{Cos}[t] - 4\,\mathrm{Sin}[t]),\ \frac{1}{85}(85 - 30e^{-4t} + 68e^{-2t} - 38\,\mathrm{Cos}[t] + 16\,\mathrm{Sin}[t]),\ \frac{1}{85}(85 + 10e^{-4t} - 68e^{-2t} - 27\,\mathrm{Cos}[t] - 11\,\mathrm{Sin}[t])\right\}$$

```
Plot[Evaluate[soln], {t, 0, 15}];
```

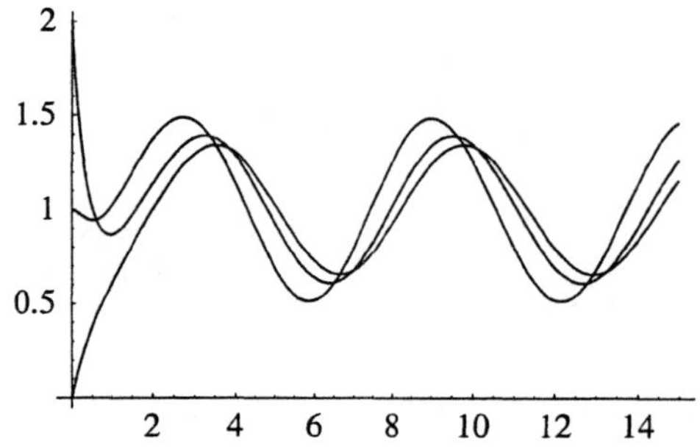

■ Exercises

1. Let $A = \begin{pmatrix} -1 & 1 & 1 \\ -1 & 0 & -1 \\ 1 & 1 & -1 \end{pmatrix}$. Use `MatrixExp` to solve $\mathbf{y}' - A\mathbf{y} = 0$, $\mathbf{y}(0) = \begin{pmatrix} 1 \\ -1 \\ 1 \end{pmatrix}$.
   (Apply `ComplexExpand` to the result.) Plot the components of the solution versus t, and then use `ParametricPlot3D` to plot the solution curve in $\mathbb{R}^3$.

2. Let A be the matrix in Exercise 1. Use `MatrixExp` and the variation of constants formula to solve
$$\mathbf{y}' - A\mathbf{y} = \begin{pmatrix} 2\,s\,e^{-s} \\ 0 \\ -s\,e^{-s} \end{pmatrix}, \quad \mathbf{y}(0) = \begin{pmatrix} 0 \\ 0 \\ 0 \end{pmatrix}.$$
   (Apply `ComplexExpand` to the result.) Plot the components of the solution versus t, and then use `ParametricPlot3D` to plot the solution curve in $\mathbb{R}^3$.

3. Let $A = \begin{pmatrix} -2 & 1 & 0 \\ -1 & 0 & -1 \\ 0 & 1 & -2 \end{pmatrix}$. Use `MatrixExp` and the variation of constants formula to solve
$$\mathbf{y}' - A\mathbf{y} = \begin{pmatrix} \cos t \\ 0 \\ \sin 2t \end{pmatrix}, \quad \mathbf{y}(0) = \begin{pmatrix} -1 \\ 0 \\ 1 \end{pmatrix}.$$
   (Apply `ComplexExpand` to the result.) Plot the components of the solution versus t, and then use `ParametricPlot3D` to plot the solution curve in $\mathbb{R}^3$.

7.6 Eigenvalues and Phase Portraits

The phase portraits of 2 × 2 autonomous linear systems $\mathbf{y}' = A\mathbf{y}$ — and likewise the character of the equilibrium point (0, 0) — can be categorized according to the eigenvalues of A. In the following we will present a simple example of each case in which the eigenvalues are distinct.

```
Needs["DiffEqs`DEGraphics`"]
```

Complex, negative real part

This matrix has complex eigenvalues whose real parts are negative.

```
A := (-1  2
      -2 -1); Eigenvalues[A]
```

```
{-1 - 2 i, -1 + 2 i}
```

The phase portrait shows that orbits spiral in toward the origin as $t \to \infty$. Thus the origin is an *(asymptotically) stable spiral point.*

```
PhasePlot[Evaluate[A.{x, y}], {t, -2, 4}, {x, -2, 2}, {y, -2, 2}];
```

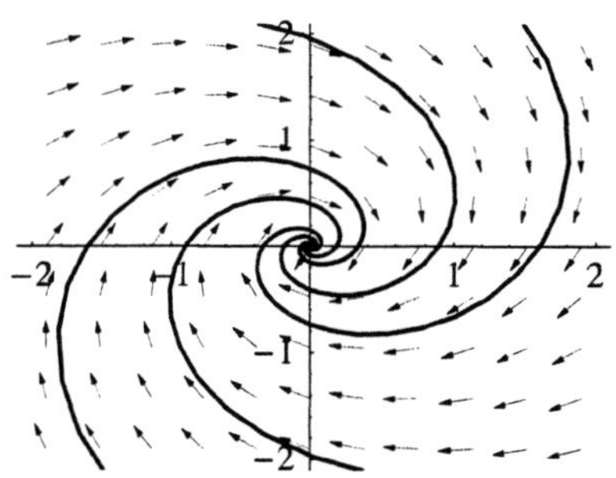

Imaginary

This matrix has imaginary eigenvalues.

```
A := ( 1  2
      -2 -1); Eigenvalues[A]
```

$\{-i\sqrt{3}, i\sqrt{3}\}$

The phase portrait shows that solutions are periodic. The origin is a *(neutrally stable) center.*

```
PhasePlot[Evaluate[A.{x, y}], {t, 0, 2}, {x, -2, 2}, {y, -2, 2}];
```

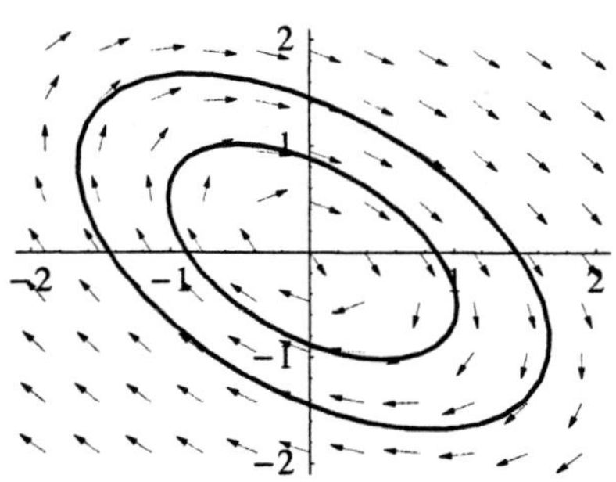

Complex, positive real part

This matrix has complex eigenvalues whose real parts are positive.

```
A := (1  2
      -2 1); Eigenvalues[A]
```

{1 - 2 i, 1 + 2 i}

The phase portrait shows that orbits spiral away from the origin as $t \to \infty$. Thus the origin is an *unstable spiral point.*

```
PhasePlot[Evaluate[A.{x, y}], {t, -4, 1}, {x, -2, 2}, {y, -2, 2}];
```

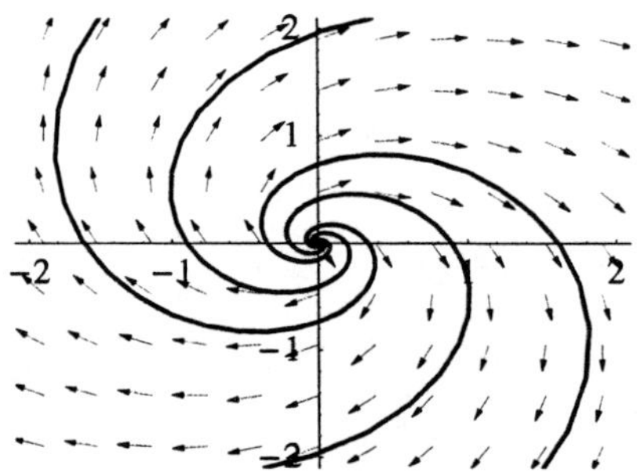

Real and positive

This matrix has real eigenvalues, both of which are positive.

```
A := (2  -1
      -1  2); Eigenvalues[A]
```

{1, 3}

The phase portrait shows that all orbits flow away from the origin as $t \to \infty$. The origin is an *unstable node.*

```
PhasePlot[Evaluate[A.{x, y}], {t, -3, 2},
  {x, -2, 2}, {y, -2, 2}, InitialPoints → {{-1, 0}, {1, 0},
    {0, -1}, {0, 1}, {1, 1}, {-1, -1}, {-1, 1}, {1, -1}}];
```

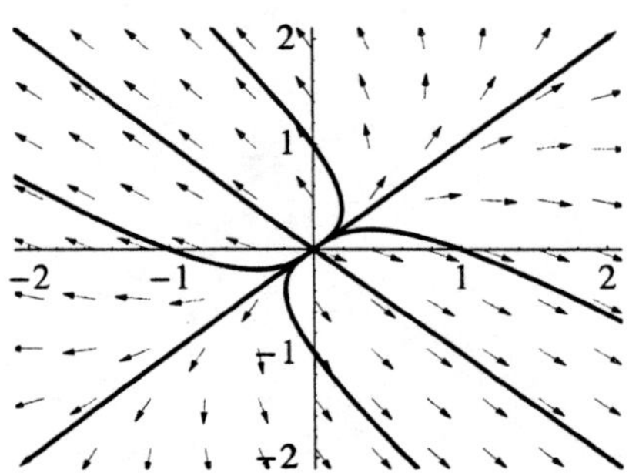

Real, opposite sign

This matrix has real eigenvalues with opposite signs.

$$\mathbf{A := \begin{pmatrix} 1 & -2 \\ -2 & 1 \end{pmatrix}; Eigenvalues[A]}$$

```
{-1, 3}
```

The phase portrait shows that orbits are *hyperbolic*. The origin is a *saddle point*.

```
PhasePlot[Evaluate[A.{x, y}], {t, -2, 2}, {x, -2, 2}, {y, -2, 2},
  InitialPoints → {{-.75, 0}, {.75, 0}, {0, -.75}, {0, .75}}];
```

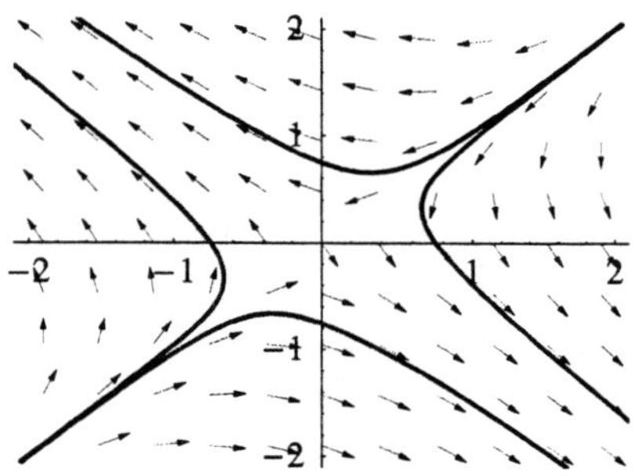

Real and negative

This matrix has real eigenvalues, both of which are negative.

$$\mathbf{A := \begin{pmatrix} -2 & 1 \\ 1 & -2 \end{pmatrix}; Eigenvalues[A]}$$

```
{-3, -1}
```

The phase portrait shows that all orbits approach the origin as $t \to \infty$. The origin is an *asymptotically stable node*.

```
PhasePlot[Evaluate[A.{x, y}], {t, -2, 3},
  {x, -2, 2}, {y, -2, 2}, InitialPoints → {{-1, 0}, {1, 0},
    {0, -1}, {0, 1}, {1, 1}, {-1, -1}, {-1, 1}, {1, -1}}];
```

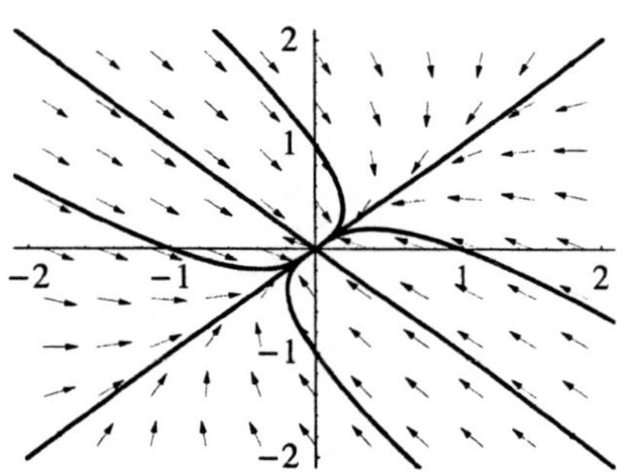

A zero eigenvalue

This matrix has a zero eigenvalue; the other is negative.

```
A := (-1   1
       1  -1); Eigenvalues[A]
```

```
{-2, 0}
```

There is an entire line of equilibrium points, and the phase portrait shows that all orbits flow move along straight lines toward an equilibrium point as $t \to \infty$.

```
PhasePlot[Evaluate[A.{x, y}], {t, -1, 3}, {x, -2, 2}, {y, -2, 2}];
```

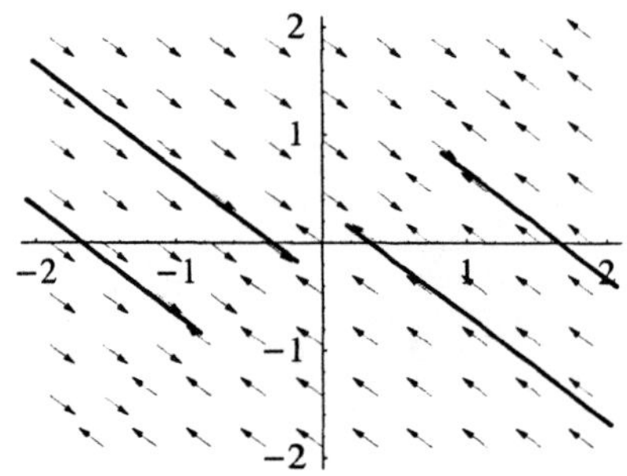

This matrix has a zero eigenvalue; the other is positive.

```
A := (1  1
      1  1); Eigenvalues[A]
```

```
{0, 2}
```

Again, there is an entire line of equilibrium points. But the phase portrait shows that all orbits flow along straight lines *away* from the line of equilibrium points.

```
PhasePlot[Evaluate[A.{x, y}], {t, -1, 2}, {x, -2, 2}, {y, -2, 2}];
```

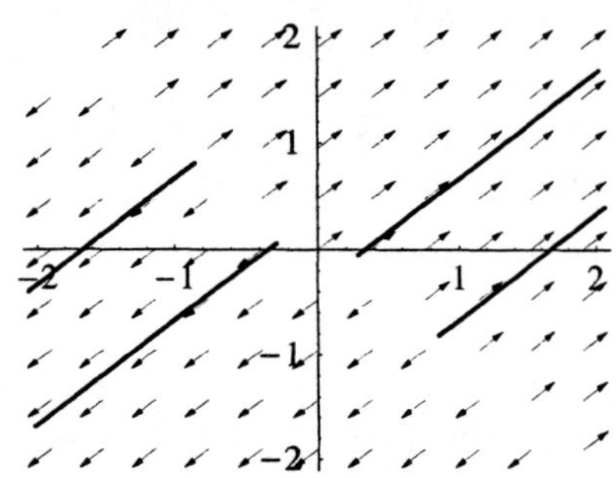

■ Dependence on a Parameter

Consider the following matrix, in which α is a parameter.

$$A := \begin{pmatrix} -1 & -1 \\ 2 & \alpha \end{pmatrix}$$

A plot of the eigenvalues' real parts reveals much about the way solutions of $\mathbf{y}' = A\mathbf{y}$ change as α varies.

```
evals = Eigenvalues[A]
```

$$\left\{\frac{1}{2}\left(-1+\alpha-\sqrt{-7+2\alpha+\alpha^2}\right),\ \frac{1}{2}\left(-1+\alpha+\sqrt{-7+2\alpha+\alpha^2}\right)\right\}$$

```
Plot[Evaluate[Re[evals]], {α, -6, 5}, AxesLabel → {α, Re[λ]}];
```

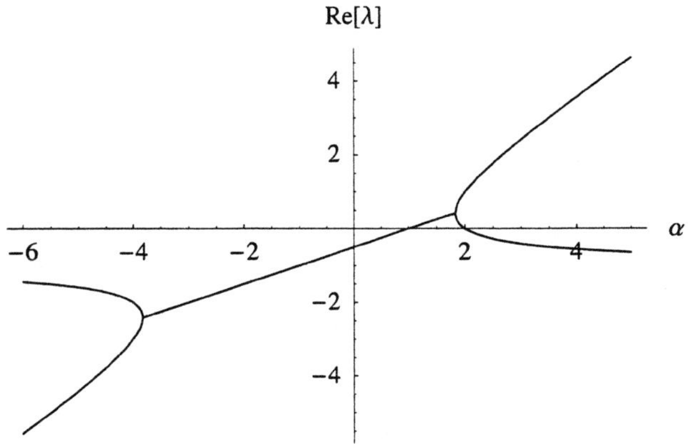

The two graphs coincide when the eigenvalues are complex. (Why?) The ends of that straight portion of the graphs correspond to the repeated real case, which happens when α satisfies the quadratic equation $-7 + 2\alpha + \alpha^2 = 0$.

```
Solve[-7 + 2 α + α^2 == 0, α]
```

$$\{\{\alpha \to -1 - 2\sqrt{2}\},\ \{\alpha \to -1 + 2\sqrt{2}\}\}$$

So we see that A has negative real eigenvalues when $\alpha \le -1 - 2\sqrt{2}$, complex eigenvalues when $-1 - 2\sqrt{2} < \alpha < -1 + 2\sqrt{2}$, and a positive real eigenvalue when $\alpha \le -1 + 2\sqrt{2}$. Of particular importance is the value of α where the eigenvalues are imaginary. This clearly occurs at $\alpha = 1$. Moreover, as α increases through 1, the real parts of the eigenvalues change from negative to positive. A fourth α-value of interest is $\alpha = 2$, where one of the real eigenvalues changes sign.

At each of these critical values of α the phase portrait of the system undergoes an abrupt structural transition from one fundamental type to another—in other words, a *bifurcation*. Each critical value of α is said to be a *bifurcation point*.

The following command creates an animation in which α varies from 0.25 to 2.5 in steps of 0.125. Eight of the resulting frames are shown after.

```
Needs["DiffEqs`DEGraphics`"]
```

```
Do[PhasePlot[Evaluate[A.{x, y}], {t, -20, 20}, {x, -2, 2}, {y, -2, 2},
  InitialPoints → {{.5, .5}, {-.5, -.5}}, ShowNullclines → True,
  Ticks → False, PlotLabel → "α = " <> ToString[α]], {α, .25, 2.5, .125}]
```

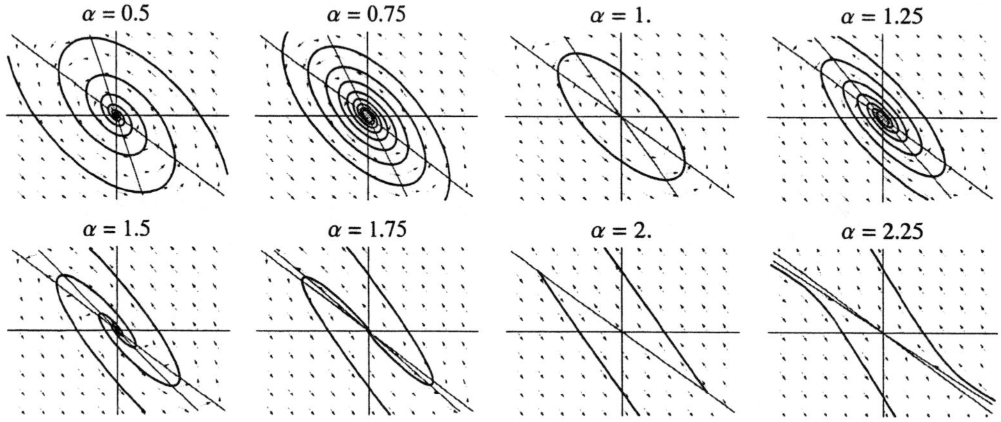

■ Exercises

For each matrix A in Exercises 1–8, do the following:

(a) Compute the eigenvalues and decide whether the origin will be (i) a stable spiral point, (ii) a center, (iii) an unstable spiral point, (iv) an stable node, (v) an unstable node, or (vi) a saddle point for the system $\mathbf{y}' = A\mathbf{y}$.

(b) Use `PhasePlot` to verify your answer to (a).

1. $\begin{pmatrix} -2 & 1 \\ 1 & -2 \end{pmatrix}$
2. $\begin{pmatrix} 0 & -1 \\ 1 & 0 \end{pmatrix}$
3. $\begin{pmatrix} 1 & -1 \\ 1 & 1 \end{pmatrix}$
4. $\begin{pmatrix} 1 & 2 \\ 3 & -4 \end{pmatrix}$
5. $\begin{pmatrix} 3 & -1 \\ 1 & -1 \end{pmatrix}$
6. $\begin{pmatrix} -1 & 3 \\ -1 & 1 \end{pmatrix}$
7. $\begin{pmatrix} -3 & -1 \\ 2 & -1 \end{pmatrix}$
8. $\begin{pmatrix} 1 & -1 \\ 1 & 4 \end{pmatrix}$

In each matrix A in Exercises 9–15, α is a parameter. Compute the eigenvalues in terms of α, and create a plot of their real parts. Identify the bifurcation points, and describe how the structure of the eigenvalues and the phase portrait of $\mathbf{y}' = A\mathbf{y}$ change at each one.

9. $\begin{pmatrix} 1 & 1 \\ -1 & \alpha \end{pmatrix}$
10. $\begin{pmatrix} -1 & 2 \\ \alpha & -1 \end{pmatrix}$
11. $\begin{pmatrix} -1 & \alpha \\ 1 & \alpha \end{pmatrix}$
12. $\begin{pmatrix} -1 & \alpha \\ \alpha & -1 \end{pmatrix}$
13. $\begin{pmatrix} -1 + \frac{1}{2}\alpha & -1 \\ 1 & -1 + 2\alpha \end{pmatrix}$
14. $\begin{pmatrix} \alpha & -1 \\ 1 & -1 \end{pmatrix}$
15. $\begin{pmatrix} \alpha & -1 \\ 1 & -2 \end{pmatrix}$

Chapter 8

Nonlinear Autonomous Systems in the Plane

8.1 Equilibrium Points and Stability

We begin here an exploration of two-dimensional, autonomous, nonlinear systems

$$\frac{dx}{dt} = f(x, y), \quad \frac{dy}{dt} = g(x, y),$$

where f and g are continuous functions of two variables. A parametric curve $(x(t), y(t))$ corresponding to a solution of the system is called an *orbit.* A *phase portrait* of the system is a picture of a number of orbits sufficient to represent the qualitative behaviors of all orbits.

The function $F(x, y) = (f(x, y), g(x, y))$ is the system's *vector field.* Scaling the vector field to have constant length produces a *direction field* of the system. A curve described by $f(x, y) = 0$ is an *x-nullcline*, and a curve described by $g(x, y) = 0$ is a *y-nullcline*. The *equilibrium points* (a.k.a. critical points) of such a system are the solutions of the algebraic system

$$f(x, y) = 0, \quad g(x, y) = 0.$$

Thus an equilibrium point is a point where an x-nullcline intersects a y-nullcline.

Notions of stability/instability of equilibrium points may be defined informally as follows. An equilibrium point E^* is:

- *asymptotically stable* if every orbit that comes sufficiently close to E^* approaches E^* as $t \to \infty$;
- *neutrally stable* if every orbit that comes sufficiently close to E^* stays close to E^* as $t \to \infty$;
- *unstable* if it is neither neutrally stable nor asymptotically stable.

Example 8.1.1

Consider the system

$$x' = y - x, \quad y' = 1 - x^2 - y^2.$$

The isoclines are the line $y = x$ and the circle $x^2 + y^2 = 1$. The intersection of these comprises two equilibrium points.

```
Solve[{y - x == 0, 1 - x² - y² == 0}, {x, y}]
```

$$\left\{\left\{x \to -\frac{1}{\sqrt{2}}, y \to -\frac{1}{\sqrt{2}}\right\}, \left\{x \to \frac{1}{\sqrt{2}}, y \to \frac{1}{\sqrt{2}}\right\}\right\}$$

A phase portrait indicates that $\left(-\frac{1}{\sqrt{2}}, -\frac{1}{\sqrt{2}}\right)$ is unstable and $\left(\frac{1}{\sqrt{2}}, \frac{1}{\sqrt{2}}\right)$ is asymptotically stable.

```
<< DiffEqs`DEGraphics`
```

```
PhasePlot[{y - x, 1 - x² - y²}, {t, -2, 4}, {x, -2, 2}, {y, -1.5, 1.5},
  InitialPoints → {{-1, 1}, {-1, -.25}, {-1, -1}, {1, 0}, {0, 0}, {1, -.85}},
  ShowNullclines → True, ShowEquilibriumPoints → True,
  AspectRatio → Automatic];
```

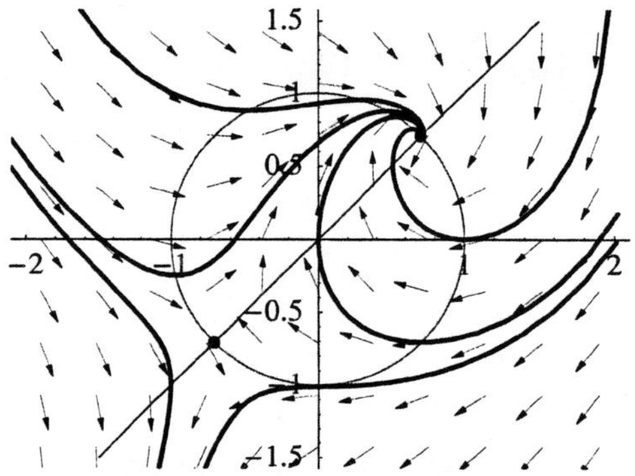

■ The Linearized Stability Principle

The stability/instability of an *isolated* equilibrium point can often be determined by *linearization.* The linearization of the vector field (f, g) at an equilibrium point (x^*, y^*) is

$$L(x, y) = \mathcal{J}(x^*, y^*)\begin{pmatrix} x - x^* \\ y - y^* \end{pmatrix}, \quad \text{where} \quad \mathcal{J}(x^*, y^*) = \begin{pmatrix} f_x(x^*, y^*) & f_y(x^*, y^*) \\ g_x(x^*, y^*) & g_y(x^*, y^*) \end{pmatrix}.$$

The matrix $\mathcal{J}(x^*, y^*)$ is the *Jacobian* of the vector field at (x^*, y^*). The *principle of linearized stability* says that the behavior of orbits of the nonlinear system near (x^*, y^*) is qualitatively the same as the behavior of orbits near $(0, 0)$ of the linear system

$$\begin{pmatrix} x \\ y \end{pmatrix}' = \mathcal{J}(x^*, y^*)\begin{pmatrix} x \\ y \end{pmatrix},$$

provided that the eigenvalues of $\mathcal{J}(x^*, y^*)$ are distinct and neither has zero real part.

Example 8.1.2

A model of competition between two species leads to the equations

$$\frac{dx}{dt} = x(1 - x - ay), \quad \frac{dy}{dt} = y(1 - y - bx),$$

where a and b are positive parameters. So let us begin by defining

```
f[x_, y_] := x (1 - x - a y); g[x_, y_] := y (1 - y - b x)
```

and solving for the system's equilibrium points:

```
eqpts = Solve[{f[x, y] == 0, g[x, y] == 0}, {x, y}] // Simplify
```

$$\left\{\{x \to 0,\ y \to 0\},\ \{x \to 1,\ y \to 0\},\ \left\{x \to \frac{-1+a}{-1+a\,b},\ y \to \frac{-1+b}{-1+a\,b}\right\},\ \{y \to 1,\ x \to 0\}\right\}$$

The third of these has positive coordinates only when either $a, b < 1$ or $a, b > 1$. The Jacobian matrix (as a function of x and y) is

$$\texttt{jac[x\_, y\_]} = \begin{pmatrix} \partial_x f[x, y] & \partial_y f[x, y] \\ \partial_x g[x, y] & \partial_y g[x, y] \end{pmatrix};\ \texttt{jac[x, y] // MatrixForm}$$

$$\begin{pmatrix} 1 - 2x - a y & -a x \\ -b y & 1 - b x - 2 y \end{pmatrix}$$

Now, at each equilibrium point, we evaluate the Jacobian and compute its eigenvalues.

```
J1 = jac[x, y] /. eqpts[[1]] // Simplify
```

{{1, 0}, {0, 1}}

```
evals1 = Eigenvalues[J1]
```

{1, 1}

```
J2 = jac[x, y] /. eqpts[[2]] // Simplify
```

{{-1, -a}, {0, 1 - b}}

```
evals2 = Eigenvalues[J2]
```

{-1, 1 - b}

```
J3 = jac[x, y] /. eqpts[[3]] // Simplify
```

$$\left\{\left\{\frac{1-a}{-1+a\,b},\ -\frac{(-1+a)\,a}{-1+a\,b}\right\},\ \left\{-\frac{(-1+b)\,b}{-1+a\,b},\ \frac{1-b}{-1+a\,b}\right\}\right\}$$

```
evals3 = Eigenvalues[J3] // Simplify
```

$$\left\{-1,\ \frac{(-1+a)\,(-1+b)}{-1+a\,b}\right\}$$

```
J4 = jac[x, y] /. eqpts[[4]] // Simplify
```

{{1 - a, 0}, {-b, -1}}

```
evals4 = Eigenvalues[J4]
```

{-1, 1 - a}

From these calculations, one can readily see that:

- (0, 0) is an unstable node;
- (1, 0) is an unstable node if $0 < b < 1$ and a stable node if $b > 1$;
- $(\frac{1-a}{1-ab}, \frac{1-b}{1-ab})$ is a saddle point (and therefore unstable) if $a, b > 1$ and a stable node if $a, b < 1$;
- (0, 1) is an unstable node if $0 < a < 1$ and a stable node if $a > 1$.

The following phase portraits illustrate the four cases:

$$\text{(i) } a < 1 < b, \quad \text{(ii) } b < 1 < a, \quad \text{(iii) } a,\, b < 1, \text{ and } \quad \text{(iv) } a,\, b > 1.$$

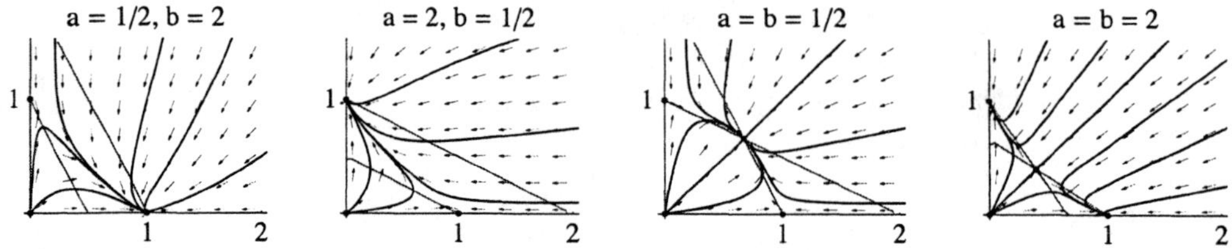

■ Exercises

1. Consider the system

$$x' = y - 4x\,(x^2 - 1), \quad y' = 1 - ax - y,$$

where a is a positive parameter. Use the nullclines of the system to show that the system has exactly one equilibrium point. Then for each of the parameter values $a = 2, 4, 6$, do the following:

(a) Use `NSolve` to find the equilibrium point.

(b) Compute the eigenvalues of the Jacobian at the equilibrium point.

(c) Infer from the eigenvalues the nature of the equilibrium point. (Is it stable or unstable? Is it a spiral point, a node, or a saddle point?)

(d) Create a phase portrait on the rectangle $-2 \le x \le 2$, $-4 \le y \le 4$, including the nullclines.

2. For the same system as in Exercise 1, solve for the equilibrium point in terms of a. (You'll get one real solution and two complex ones. Ignore the complex ones.) Then compute the eigenvalues of the Jacobian at the equilibrium point (also in terms of a), and plot the real parts of the eigenvalues versus a for $0 \le a \le 10$. Use this graph to describe how the nature of the equilibrium point changes as a varies over $[0, \infty)$.

3. Consider the system

$$x' = -x\,y, \quad y' = \tfrac{1}{2}\,(x^2 - y^2).$$

(a) Verify that (0, 0) is the only equilibrium point of the system and that the Jacobian matrix there is the zero matrix. Conclude that linearization provides no information about the phase portrait.

(b) Create a phase portrait on the rectangle $-4 \le x \le 4$, $-2 \le y \le 2$. Include the nullclines in the plot.

(c) Show that all (nontrivial) orbits trace out circles.

4. With specific values for all parameters except ρ, the spruce budworm model from Section 5.3 becomes

$$\frac{dy}{dt} = \frac{1}{4}\,y(1 - y) - \frac{1}{10}\,w, \qquad \frac{dw}{dt} = 2w\left(1 - \frac{w}{6\,y}\right) - \rho\,\frac{8\,w^2}{y^2 + 8\,w^2}\,.$$

(a) Show that with $\rho = 1.12$ this system has four equilibrium points in the first quadrant, three of which have positive coordinates. Show further that one of them is asymptotically stable and the other three are unstable. Create a phase portrait to illustrate. (Include the nullclines and the equilibrium points in the plot.) What does a typical orbit do as $t \to \infty$?

(b) Show that with $\rho = 1.11$ the system still has four equilibrium points in the first quadrant, three of which have positive coordinates. Show further that *all four are unstable*. Create a phase portrait to illustrate. (Include the nullclines and the equilibrium points in the plot.) What does a typical orbit do as $t \to \infty$?

8.2 Limit Cycles

In this section we continue our study of autonomous systems of the form

$$\frac{dx}{dt} = f(x, y), \quad \frac{dy}{dt} = g(x, y),$$

where as always we assume that f and g are continuous. We say that a region $\mathcal{M}$ in the plane is *forward-invariant* if no orbit can escape $\mathcal{M}$; in other words, any orbit that is in $\mathcal{M}$ at some time T remains in $\mathcal{M}$ for all $t > T$. The basic question we're interested in is this: *If $\mathcal{M}$ is a* bounded *forward-invariant region and $(x(t), y(t))$ is an orbit that is in $\mathcal{M}$ for $t \ge T$, what can happen to $(x(t), y(t))$ as $t \to \infty$?* One possibility is very simple: If $\mathcal{M}$ contains a stable equilibrium point E^*, then $(x(t), y(t))$ can approach E^* as $t \to \infty$. A second possibility is that $(x(t), y(t))$ may be periodic, retracing forever some closed curve in $\mathcal{M}$ as $t \to \infty$. A third possibility is that $(x(t), y(t))$ may spiral *toward* (and approach) some periodic orbit as $t \to \infty$. In this case, the periodic orbit approached by $(x(t), y(t))$ is called a (stable) *limit cycle*. There is a fourth possibility that we will discuss later; however, the following theorem can be stated now.

Theorem (Poincaré-Bendixson) If $\mathcal{M}$ is a closed, bounded, forward-invariant region, and no orbit in $\mathcal{M}$ approaches an equilibrium point as $t \to \infty$, then *every* orbit in $\mathcal{M}$ is either periodic or approaches a limit cycle as $t \to \infty$.

A closed, bounded, forward-invariant region can often be observed by looking at the system's direction field. One only needs to draw the boundary in such a way that the direction field is always either tangent to the boundary or pointing into the region. However, we will not be overly concerned here with verifying the hypotheses of the theorem in specific cases. The theorem mainly serves to inform us that *an orbit that is bounded for $t \ge 0$ is* likely *to approach a limit cycle if it does not approach an equilibrium point*.

Example 8.2.1

Consider the system

$$x' = -y(4 - x^2 - y^2) + x(2 - 5x^2 - y^2),$$
$$y' = x(4 - x^2 - y^2) + y(3 - 2x^2 - y^2).$$

in which the components of the vector field are

```
f[x_, y_] := -y (4 - x² - y²) + x (2 - 5 x² - y²);
g[x_, y_] := x (4 - x² - y²) + y (3 - 2 x² - y²);
```

The only (real) equilibrium point is the origin.

```
NSolve[{f[x, y] == 0, g[x, y] == 0}, {x, y}] // TableForm
x → -0.210188 - 0.754007 i      y → -2.01349 + 0.10946 i
x → -0.210188 + 0.754007 i      y → -2.01349 - 0.10946 i
x → 0.210188 - 0.754007 i       y → 2.01349 + 0.10946 i
x → 0.210188 + 0.754007 i       y → 2.01349 - 0.10946 i
x → 1.06838 + 0.342175 i        y → -1.56067 + 0.988858 i
x → 1.06838 - 0.342175 i        y → -1.56067 - 0.988858 i
x → -1.06838 + 0.342175 i       y → 1.56067 + 0.988858 i
x → -1.06838 - 0.342175 i       y → 1.56067 - 0.988858 i
x → 0.                          y → 0.
```

The Jacobian matrix and its eigenvalues at (0, 0) are

$$\texttt{jac[x\_, y\_]} = \begin{pmatrix} \partial_x \texttt{f[x, y]} & \partial_y \texttt{f[x, y]} \\ \partial_x \texttt{g[x, y]} & \partial_y \texttt{g[x, y]} \end{pmatrix}\texttt{; jac[x, y] // MatrixForm}$$

$$\begin{pmatrix} 2 - 15x^2 + 2xy - y^2 & -4 + x^2 - 2xy + 3y^2 \\ 4 - 3x^2 - 4xy - y^2 & 3 - 2x^2 - 2xy - 3y^2 \end{pmatrix}$$

```
Eigenvalues[jac[0, 0]]
```

$$\left\{\frac{1}{2}\left(5 - 3 i \sqrt{7}\right), \frac{1}{2}\left(5 + 3 i \sqrt{7}\right)\right\}$$

Since these eigenvalues are complex with positive real parts, (0, 0) is an unstable spiral point. The phase portrait of the system reveals a football-shaped limit cycle.

```
Needs["DiffEqs`DEGraphics`"];
PhasePlot[{f[x, y], g[x, y]}, {t, 0, 2.5}, {x, -3, 3}, {y, -3, 3},
 InitialPoints → {{-3, -3}, {-3, 3}, {3, -3}, {3, 3}, {.05, .05}, {-.05, -.05}},
 AspectRatio → Automatic, ShowNullclines → True, StepSize → .01];
```

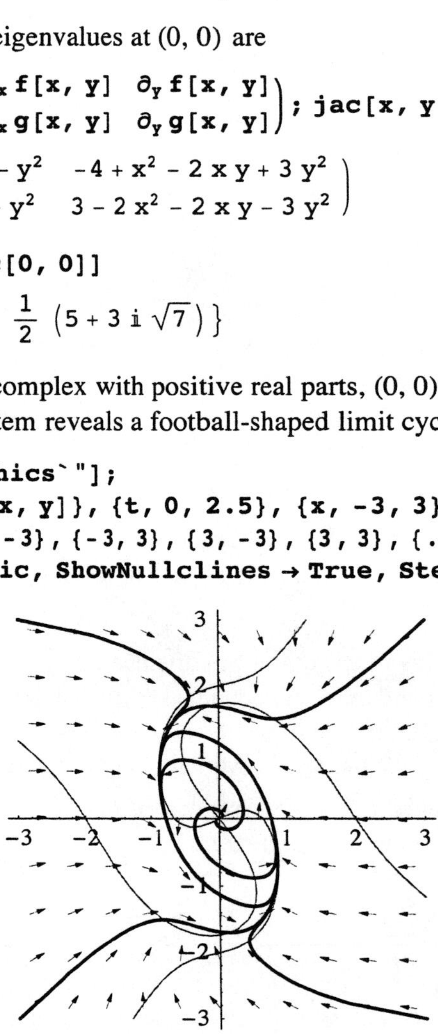

It is easy to see that the direction field points into the circle $x^2 + y^2 = 4$. (This can also be verified algebraically.) Thus $\mathcal{M} = \{(x, y) \mid x^2 + y^2 \le 4\}$ is a closed, bounded, forward-invariant region. Since the only equilibrium point in $\mathcal{M}$ is an unstable spiral point, every orbit in $\mathcal{M}$ must either be periodic or approach a limit cycle in $\mathcal{M}$. In this particular case, the phase portrait makes it clear that *every* nontrivial orbit approaches the same limit cycle.

Example 8.2.2

Consider the system

$$x' = ax - y + x^2, \quad y' = x + ay + x^2.$$

This system has two equilibrium points (unless $a = -1$), one of which is the origin.

```
f[x_, y_] := a x - y + x²; g[x_, y_] := x + a y + x²;

eqpts = Solve[{f[x, y] == 0, g[x, y] == 0}, {x, y}]
```

$$\left\{\{y \to 0,\ x \to 0\},\ \left\{y \to -\frac{(-1+a)\,(1+a^2)}{(1+a)^2},\ x \to \frac{-1-a^2}{1+a}\right\}\right\}$$

The Jacobian is

```
jac[x_, y_] = (∂x f[x, y]  ∂y f[x, y]
               ∂x g[x, y]  ∂y g[x, y]); jac[x, y] // MatrixForm
```

$$\begin{pmatrix} a+2x & -1 \\ 1+2x & a \end{pmatrix}$$

At (0, 0) the Jacobian has the following eigenvalues:

```
evals1 = Eigenvalues[jac[x, y] /. eqpts[[1]]]
```

$$\{-i + a,\ i + a\}$$

Thus (0, 0) is a stable spiral point when $a < 0$ and an unstable spiral point when $a > 0$. At the other equilibrium point, the eigenvalues are

```
evals2 = Eigenvalues[jac[x, y] /. eqpts[[2]]] // FullSimplify
```

$$\left\{\frac{-1+a-\sqrt{2+a^2\,(3+a\,(2+a))}}{1+a},\ \frac{-1+a+\sqrt{2+a^2\,(3+a\,(2+a))}}{1+a}\right\}$$

These eigenvalues are always real with opposite signs, as indicated by the following plot. Therefore, the second equilibrium point is a saddle point, provided that $a \neq -1$.

```
Plot[Evaluate[evals2], {a, -3, 6}];
```

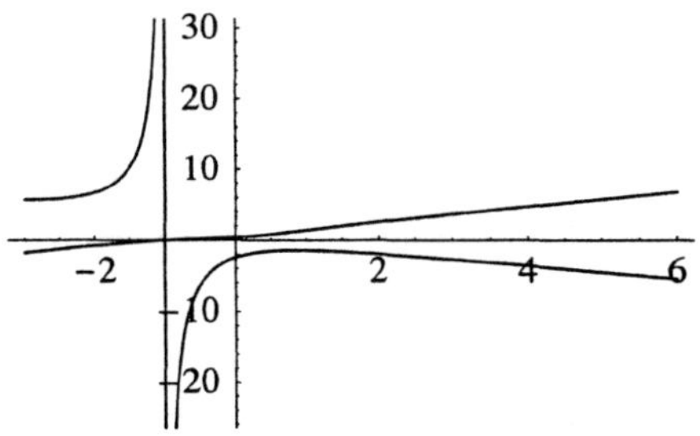

Now let's investigate the behavior of orbits near (0, 0) for small $|a|$. When $a < 0$, (0, 0) is a stable spiral point, and so orbits near (0, 0) will spiral in toward (0, 0) as $t \to \infty$, so there will be no surprises there. When $a > 0$, (0, 0) is an unstable spiral point, and so orbits near (0, 0) will spiral away from (0, 0) as t increases; however what is not clear is what those orbits actually do as $t \to \infty$, especially in light of the fact that the other equilibrium point is also unstable.

So let's create some pictures. The following plots show the orbit starting at $(-\frac{1}{2}, \frac{1}{2})$ for each of $a = -0.05,\ 0,\ 0.05$, and 0.1.

```
Needs["DiffEqs`DEGraphics`"];
Do[PhasePlot[{f[x, y], g[x, y]} /. {a → aa}, {t, 0, If[aa ≤ 0, 50, 20]},
  {x, -1.5, 1}, {y, -.5, 1.25}, InitialPoints → {{-.5, .5}},
  Arrows → {9, 7}, ScaleArrows → .15, Ticks → {{-1, 1}, {1}},
  ShowNullclines → True, ShowEquilibriumPoints → True,
  PlotLabel → "a = " <> ToString[aa]], {aa, -.05, .1, .05}]
```

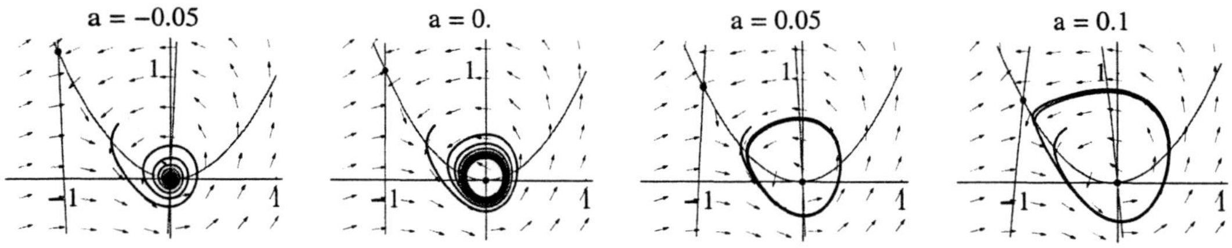

The first of these plots shows just what we expected. The second *suggests* that (0, 0) is a stable spiral point, something we could not learn from linearization. The third plot shows the orbit spiraling in toward (0, 0) but not being able to approach it as $t \to \infty$, because (0, 0) is unstable! Thus the orbit approaches a limit cycle as $t \to \infty$. The fourth plot shows the orbit spiraling outward, again approaching a limit cycle as $t \to \infty$. Can you see a closed, bounded, forward-invariant region in each of these pictures?

The Fourth Alternative

There is a fourth possibility for the asymptotic behavior of an orbit in a bounded, forward-invariant region. Such an orbit may spiral toward a closed graph that consists of a collection of (not necessarily distinct) equilibrium points $E_1, \ldots, E_k$, $k \geq 1$, and a chain of "connecting" orbits $(x_1(t), y_1(t)), \ldots, (x_k(t), y_k(t))$ such that

$$\lim_{t\to\infty} (x_i(t), y_i(t)) = E_i, \text{ for } i = 1, \ldots, k-1, \text{ and } \lim_{t\to\infty} (x_k(t), y_k(t)) = E_1;$$

$$\lim_{t\to-\infty} (x_1(t), y_1(t)) = E_k \text{ and } \lim_{t\to-\infty} (x_i(t), y_i(t)) = E_{i-1}, \text{ for } i = 2, \ldots, k.$$

The simplest situation of this type involves a single saddle point E and an orbit $(x(t), y(t))$ that approaches E both as $t \to -\infty$ and as $t \to \infty$. Such an orbit is called a *homoclinic* orbit.

The following example illustrates the case of a "double saddle connection," i.e., a close graph consisting of one saddle point ($E_1 = E_2$) and two distinct homoclinic orbits approaching it as $t \to -\infty$ and as $t \to \infty$.

Example 8.2.3

Consider the system

$$\begin{aligned} x' &= y + x(1-x^2)\,p(x, y), \\ y' &= x(1-x^2) - y\,p(x, y), \end{aligned} \qquad \text{where } p(x, y) = y^2 - x^2 + \tfrac{1}{2}x^4 .$$

Let's first define

```
p[x_, y_] := y^2 - x^2 + x^4 / 2;
f[x_, y_] := y + x (1 - x^2) p[x, y];
g[x_, y_] := x (1 - x^2) - y p[x, y];
```

There are three (real) equilibrium points: $(-1, 0)$, $(0, 0)$, and $(1, 0)$.

```
NSolve[{f[x, y] == 0, g[x, y] == 0}, {x, y}] // TableForm
y → -0.996423 - 0.713353 i      x → -1.31176 + 0.24284 i
y → -0.996423 + 0.713353 i      x → -1.31176 - 0.24284 i
y → 0.996423 - 0.713353 i       x → 1.31176 + 0.24284 i
y → 0.996423 + 0.713353 i       x → 1.31176 - 0.24284 i
y → 1.09868 - 0.45509 i         x → -1.09868 + 0.45509 i
y → 1.09868 + 0.45509 i         x → -1.09868 - 0.45509 i
y → -1.09868 - 0.45509 i        x → 1.09868 + 0.45509 i
y → -1.09868 + 0.45509 i        x → 1.09868 - 0.45509 i
y → 0.                          x → -1.
y → 0.                          x → 1.
y → 0.489688 + 0.590578 i       x → -0.343194 + 0.528712 i
y → 0.489688 - 0.590578 i       x → -0.343194 - 0.528712 i
y → -0.489688 + 0.590578 i      x → 0.343194 + 0.528712 i
y → -0.489688 - 0.590578 i      x → 0.343194 - 0.528712 i
y → 0.                          x → 0.
```

The Jacobian is

$$\texttt{jac[x\_, y\_]} = \begin{pmatrix} \partial_x f[x, y] & \partial_y f[x, y] \\ \partial_x g[x, y] & \partial_y g[x, y] \end{pmatrix} \texttt{ // Simplify;}$$

```
jac[x, y] // MatrixForm
```

$$\begin{pmatrix} \frac{15 x^4}{2} - \frac{7 x^6}{2} + y^2 - 3 x^2 (1 + y^2) & 1 + 2 x y - 2 x^3 y \\ 1 - 3 x^2 + 2 x y - 2 x^3 y & x^2 - \frac{x^4}{2} - 3 y^2 \end{pmatrix}$$

The eigenvalues of the Jacobian at the three equilibrium points are as follows:

```
Eigenvalues[jac[-1, 0]]
```

$$\left\{\frac{1}{4}\left(3 - i\sqrt{31}\right), \frac{1}{4}\left(3 + i\sqrt{31}\right)\right\}$$

```
Eigenvalues[jac[1, 0]]
```

$$\left\{\frac{1}{4}\left(3 - i\sqrt{31}\right), \frac{1}{4}\left(3 + i\sqrt{31}\right)\right\}$$

```
Eigenvalues[jac[0, 0]]
{-1, 1}
```

Thus all three equilibrium points are unstable. Note that $(\pm 1, 0)$ are each unstable spiral points, while $(0, 0)$ is a saddle point. A phase portrait reveals that there are two orbits that approach $(0, 0)$ as t approaches both $\pm\infty$. Together these orbits form a *Lissajous curve*, which can be parametrized by $x = \sqrt{2}\sin\theta$, $y = \sqrt{1/2}\sin 2\theta$.

```
Needs["DiffEqs`DEGraphics`"];
PhasePlot[{f[x, y], g[x, y]}, {t, 0, 12},
  {x, -2, 2}, {y, -1.5, 1.5}, InitialPoints →
   {{-2, -1}, {-2, 1}, {2, 1}, {2, -1}, {-1.05, 0}, {1.05, 0}},
  AspectRatio → Automatic, Method → Automatic, StepSize → .025,
  ShowEquilibriumPoints → True, ShowNullclines → True];
```

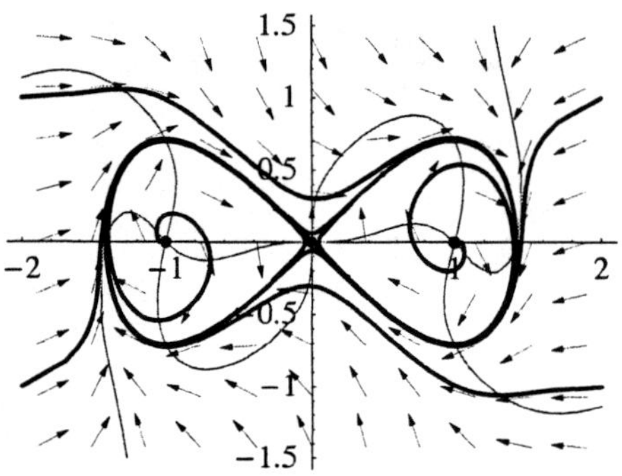

Notice that orbits inside each lobe of the Lissajous figure spiral outward toward the lobe. Orbits outside the Lissajous figure "spiral" around the entire figure.

■ Exercises

1. The system

$$x' = -3y + y^2, \quad y' = -2(x+y)^2 + 3xy + 5x + \tfrac{1}{2}y$$

has a circular, periodic orbit. Use `PhasePlot` (with `AspectRatio→Automatic` and `StepSize →.02`) to locate it experimentally. Then add two orbits to the plot—one inside and one outside the circular orbit—to show that the circular orbit is a stable limit cycle.

2. Repeat Exercise 1 for

$$x' = 3y - y^2, \quad y' = 2(x+y)^2 - 3xy - 5x - \tfrac{1}{2}y,$$

only this time the extra orbits should show that the circular orbit is an *unstable* limit cycle.

3. Consider the following system in which a is a nonnegative parameter:

$$x' = -xy^2 + ay, \quad y' = xy^2 - (a+1)y + 2.$$

(a) Verify that the system's only equilibrium point is $(a/2, 2)$, and compute the eigenvalues of the Jacobian there. Then plot the real parts of the eigenvalues versus a for $0 \le a \le 10$, and use the plot to describe the nature of the equilibrium point as a varies over $[0, \infty)$.

(b) Create a phase portrait for each of $a = 0, 2, 4, 5, 6, 8, 10$. Check that these phase portraits are consistent with your response to part (a).

4. Let $h(x, y) = 1 - x^2 - y^2$ and consider the system

$$x' = x\,h(x, y)^3 - y\,h(x, y)^2 - x^2 y, \quad y' = y\,h(x, y)^3 + x\,h(x, y)^2 + xy^2.$$

Use *Mathematica* to fill in the details of the following argument and to create an illustrative phase portrait.

The equilibrium points are (0, 0), (± 1, 0), (0, ± 1). The eigenvalues of the Jacobian at (0, 0) show that (0, 0) is an unstable spiral point. At the other equilibrium points, the Jacobian is the zero matrix. However, on the unit circle the slope equation becomes $\frac{dy}{dx} = -\frac{y}{x}$, which shows that four distinct orbits "connect" (± 1, 0), (0, ± 1), tracing out the four quarters of the unit circle. Along with the equilibrium points (± 1, 0), (0, ± 1), these orbits comprise a closed graph, which forms the boundary of a closed, bounded, forward-invariant region. Orbits inside spiral out toward this closed graph.

[This example is adapted from one in Borrelli, Coleman, and Boyce, *Differential Equations Laboratory Workbook* (Wiley and Sons: New York, 1992.]

5. For the same system as in Problem 4, use `NDSolve` to find the solution with initial values $x(0) = 0.5$, $y(0) = 0$, for $0 \le t \le 500$. Then plot $x(t)$ and $y(t)$ versus t. Explain why the graphs look the way they do.

8.3 Hopf Bifurcation

The term *bifurcation* is used to describe an abrupt structural change in the phase portrait of a system as a parameter in the system passes through some critical value. Consider, for example, the linear system $\mathbf{y}' = A\mathbf{y}$, where

$$A := \begin{pmatrix} a & -1 \\ 1 & a \end{pmatrix}$$

The eigenvalues of A are complex, and their real part is a.

```
Eigenvalues[A]
{-i + a, i + a}
```

Therefore, when $a < 0$, the origin is a stable spiral point, and when $a > 0$, the origin is an unstable spiral point. Thus the system experiences a bifurcation as a increases through 0, which corresponds to the eigenvalues crossing the imaginary axis in the complex plane.

For nonlinear systems, when the eigenvalues cross the imaginary axis as a parameter increases, a certain type of bifurcation can occur in which a stable spiral point "gives birth" to a stable limit cycle. This is known as the *Hopf bifurcation*, so named because the work of Eberhard Hopf in the 1940s. A version of the theorem is as follows:

Consider a nonlinear system

$$x' = f(a; x, y), \quad y' = g(a; x, y),$$

in which a is a parameter. Suppose that for each a in some interval J,

(i) there is an isolated equilibrium point $E(a)$ that depends continuously on a;

(ii) each of f and g have Taylor series expansions about $E(a)$ that converge in an open disk centered at $E(a)$;

(iii) the eigenvalues of the Jacobian $\mathcal{J}(E(a))$ are complex numbers $\gamma(a) \pm \beta(a)i$, with $\beta(a) \neq 0$.

Suppose further that

(iv) there is an a_0 in J where $\gamma(a_0) = 0$ and $\gamma(a)$ is increasing.

Under these assumptions, *if* $E(a_0)$ *is asymptotically stable*, then there is a stable limit cycle surrounding $E(a)$ for each a in some interval $(a_0, a_0 + \varepsilon)$.

Note carefully the most crucial and subtle part of the hypotheses: $E(a_0)$ *is asymptotically stable, while* $\gamma(a_0) = 0$. This is impossible in the linear case, since imaginary eigenvalues cause the origin to be a center point.

The system from Example 8.2.2,

$$x' = ax - y + x^2, \quad y' = x + ay + x^2,$$

provides what may be the simplest possible example of a Hopf bifurcation.

Example 8.3.1

Consider the system

$$x' = -xy^2 + 1, \quad y' = xy^2 - ay,$$

where a is a positive parameter. Let's first find the equilibrium points in terms of a.

```
f[x_, y_] := -x y^2 + 1; g[x_, y_] := x y^2 - a y

eqpts = Simplify[Solve[{f[x, y] == 0, g[x, y] == 0}, {x, y}], a > 0]
```

$$\left\{\left\{x \to a^2,\ y \to \frac{1}{a}\right\}\right\}$$

Next we compute the Jacobian and its eigenvalues at the equilibrium point. Note that the eigenvalues are imaginary when $a = 1$.

```
jac[x_, y_] = ( ∂x f[x, y]  ∂y f[x, y]
                ∂x g[x, y]  ∂y g[x, y] ); jac[x, y] // MatrixForm
```

$$\begin{pmatrix} -y^2 & -2xy \\ y^2 & -a + 2xy \end{pmatrix}$$

```
evals = Eigenvalues[jac[x, y]] /. eqpts[[1]] // Simplify
```

$$\left\{\frac{1}{2}\left(-\frac{1}{a^2} + a - \sqrt{\frac{1}{a^4} - \frac{6}{a} + a^2}\right),\ \frac{1}{2}\left(-\frac{1}{a^2} + a + \sqrt{\frac{1}{a^4} - \frac{6}{a} + a^2}\right)\right\}$$

A plot of the real parts of these eigenvalues is illuminating. Recall that the two graphs coincide where the eigenvalues are not real.

```
Plot[Evaluate[Re[evals]], {a, .1, 2.5},
  PlotRange → {-3, 2.2}, AxesLabel → {a, Re λ}];
```

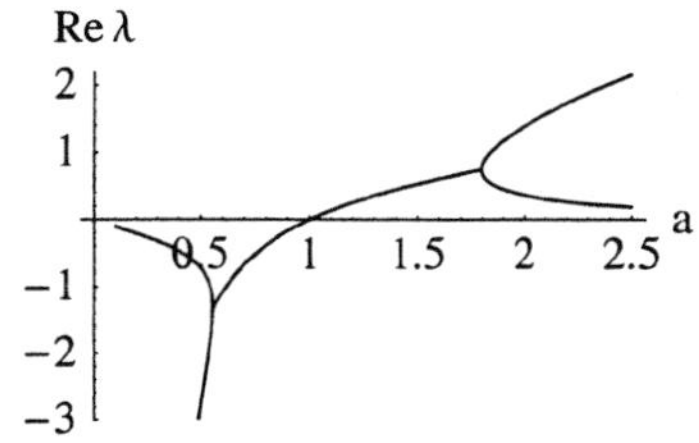

This plot tells us that as a increases through 1, the eigenvalues change from complex with negative real part to complex with positive real part. This means that we have a potential Hopf-bifurcation situation. A few phase portraits should now make things clear.

```
Needs["DiffEqs`DEGraphics`"];
Do[PhasePlot[{f[x, y], g[x, y]} /. a → aa,
  {t, 0, If[aa ≤ 1, 400, 75]}, {x, 0, 3}, {y, 0, 2.5},
  InitialPoints → {{0, 2}}, ShowNullclines → True,
  ShowEquilibriumPoints → True, Arrows → 7, Ticks → {{1, 2, 3}, {1, 2}},
  PlotLabel → "a = " <> ToString[aa]], {aa, .99, 1.02, .01}]
```

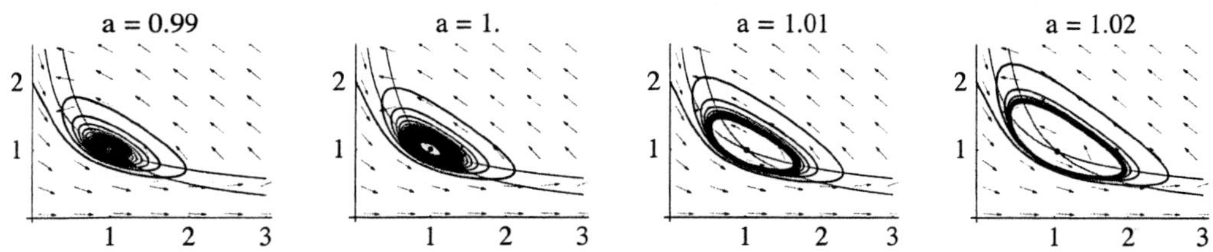

In the first of these plots, the equilibrium point is a stable spiral point, as we already knew. The second plot seems to indicate that the equilibrium point remains asymptotically stable when $a = 1$. If that is indeed true, a Hopf bifurcation must occur as a increases through 1. The third and fourth plots confirm this fact. (Remember that when $a > 1$ the equilibrium point is an *unstable* spiral point.)

■ Exercises

1. Consider the following system in which a is a parameter:

$$x' = ax + y, \quad y' = -x + ay - x^2 y.$$

(a) Find the equilibrium points in terms of a.

(b) One of the equilibrium points is $(0, 0)$. Check that there is a potential Hopf bifurcation there as a increases through 0.

(c) Find the eigenvalues of the Jacobian at the other equilibrium points. Using a plot of their real parts (versus a), show that those equilibrium points are always saddle points.

(d) Create phase portraits corresponding to $a = -0.1,\ 0,$ and 0.1 to demonstrate that a Hopf bifurcation does occur.

(e) Experimentally estimate the value of a at which the limit cycle disappears.

2. With specific values for all parameters except μ, the spruce budworm model from Section 5.3 becomes

$$\frac{dy}{dt} = \tfrac{1}{4}\, y\,(1 - y) - \mu\, w, \qquad \frac{dw}{dt} = 2\, w\left(1 - \frac{w}{6\, y}\right) - \frac{8\, w^2}{y^2 + 8\, w^2}\,.$$

   (a) Show that when $\mu = 0.16$ the equilibrium point with positive coordinates is a stable spiral point.

   (b) Show that when $\mu = 0.15$ the equilibrium point with positive coordinates is an unstable spiral point.

   (c) Create a pair of phase portraits to show that somewhere between $\mu = 0.16$ and $\mu = 0.15$, the equilibrium point with positive coordinates gives birth to a stable limit cycle. Include the nullclines in the plot, and give a possible "explanation" of the bifurcation in terms of where the intersection of the nullclines occurs.

3. Show that the system

$$x' = y\,(y - 1), \quad y' = -a - x\, y,$$

has an equilibrium point at $(-a,\, 1)$ that is a stable spiral point when $a < 0$ and an unstable spiral point when $0 < a < 2$. Use phase portraits to show that no limit cycle appears as a increases through 0. Why doesn't the Hopf bifurcation theorem apply here?

8.4 Chapter Projects

1. Consider the system

$$x' = y\left(\tfrac{12}{5} - \tfrac{3}{4}\,(x^2 + y^2)\right), \quad y' = 1 - x\left(\tfrac{12}{5} - \tfrac{3}{4}\,(x^2 + y^2)\right).$$

   (a) Find the equilibrium points and compute the eigenvalues of the Jacobian at each one. Infer from the eigenvalues the nature of each equilibrium point.

   (b) Create a phase portrait for the system on the square $-4 \le x \le 4$, $-4 \le y \le 4$. Identify and describe all of the *nonperiodic* orbits.

   (c) You should observe that the system *appears* to have circular orbits centered at $(x_0,\, 0)$ for some small negative x_0. Are those orbits really circular?

   [This system is borrowed from R. Grimshaw, *Nonlinear Ordinary Differential Equations*, (Boston: Blackwell Scientific, 1990) Grimshaw used its phase portrait for the cover of his book.]

2. Consider the system

$$x' = (x + 1)\,(a - x) - (x + 2)\, y^2\,, \quad y' = x\, y.$$

   (a) Find the equilibrium points in terms of a. Then, for each equilibrium point, compute the eigenvalues of the Jacobian there and infer from the eigenvalues the nature of the equilibrium point as a varies over $\mathbb{R}$. Whenever it is helpful, plot the real parts of the eigenvalues versus a. Identify any values of a at which there is a potential Hopf bifurcation.

(b) Create several plots—an animation if you like—showing the phase portrait of the system as a increases from -1 to 3. (Plot over the square $-2 \le x \le 2$, $-2 \le y \le 2$.) Are the phase portraits consistent with your analysis of the equilibrium points? Can you tell whether a Hopf bifurcation actually occurs? (In other words, are there any observable limit cycles?)

3. The *Brusselator*, a model of a hypothetical chemical reaction, gives rise to the system

$$x' = x^2 y - (a+1)x + b, \quad y' = -x^2 y + ax,$$

where a and b are positive parameters, and where x and y represent nonnegative quantities. For our purposes here, we'll set $b = 1$ to obtain the following system with one parameter:

$$x' = x^2 y - (a+1)x + 1, \quad y' = -x^2 y + ax.$$

(a) The system has one equilibrium point. Find it, and determine its nature as a varies over $(0, \infty)$. Identify any potential Hopf bifurcations.

(b) Create a few phase portraits that describe how solutions of the system change with a.

(c) Use the Internet to find the connection between the Brusselator and the 1977 Nobel Prize in chemistry.

4. Consider the system

$$x' = ax - y + x^2, \quad y' = x + ay + x^2$$

from Example 8.2.2, where we demonstrated that a Hopf bifurcation occurs at $a = 0$. It turns out that this same system also has a homoclinic orbit for a certain value (or values?) of a, as shown below.

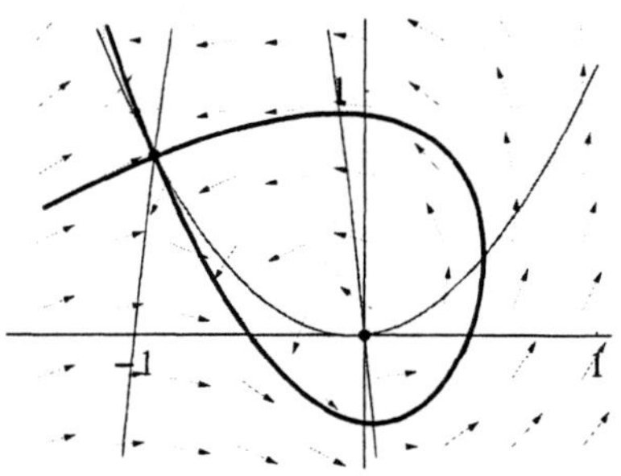

By experimenting with various initial points and values of a, reproduce this picture with `PhasePlot`. Then insert an initial point near the origin to show that orbits bounded by the homoclinic orbit spiral out toward it.

Chapter 9

A Gallery of Nonlinear Systems

In this brief chapter we display, without discussion, thirteen well-known nonlinear systems that are interesting because of either the behavior they exhibit or the applications from which they arise—or both. (Several of these have already appeared in this manual.)

The graphics commands here require functions in the DEGraphics package.

```
<< DiffEqs`DEGraphics`
```

In the interest of brevity, we set the following options for PhasePlot:

```
SetOptions[PhasePlot, ShowNullclines → True,
  ShowEquilibriumPoints → True, Ticks → None];
```

9.1 Competition

```
f[x_, y_] := k1 x (1 - x - a y);  g[x_, y_] := k2 y (1 - y - b x);

PhasePlot[{f[x, y], g[x, y]} /. {k1 → .5, k2 → .7, a → 3, b → 2},
  {t, -5, 15}, {x, 0, 1.3}, {y, 0, 1.2}, InitialPoints →
   {{1, .3}, {1, .625}, {1, 1}, {.5, 1}, {.15, .1}, {.3, .1}, {.2, .082}},
  Arrows → {9, 7}, ScaleArrows → .33, Ticks → {{.5, 1}, {.5, 1}}];
```

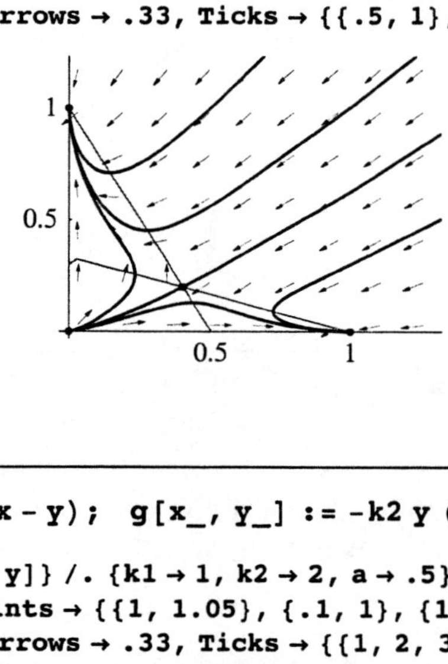

9.2 Predator-Prey

```
f[x_, y_] := k1 x (1 - a x - y);  g[x_, y_] := -k2 y (1 - x);

PhasePlot[{f[x, y], g[x, y]} /. {k1 → 1, k2 → 2, a → .5}, {t, -5, 15}, {x, 0, 2.5},
  {y, 0, 1.2}, InitialPoints → {{1, 1.05}, {.1, 1}, {1, 1.2}, {.4, 1}},
  Arrows → {9, 7}, ScaleArrows → .33, Ticks → {{1, 2, 3}, {.5, 1}}];
```

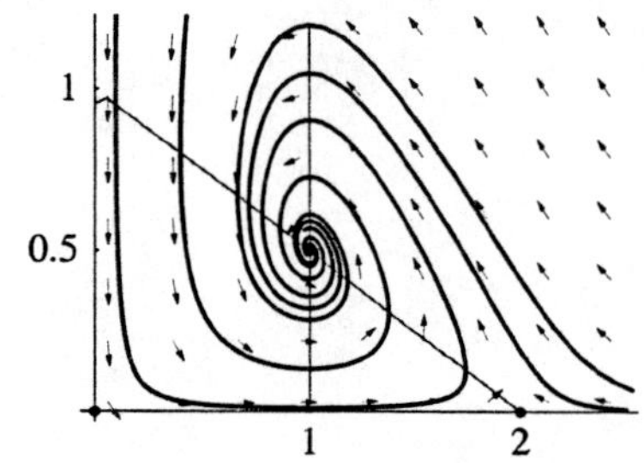

9.3 The Chemostat

```
f[x_, y_] := k (x y)/(1 + y) - x;  g[x_, y_] := - (x y)/(1 + y) - y + q;

PhasePlot[{f[x, y], g[x, y]} /. {k → 2, q → 2},
  {t, -5, 15}, {x, 0, 4}, {y, 0, 3}, InitialPoints →
   {{.1, 3}, {.75, 3}, {2, 2}, {3, 2}, {.2, .1}, {1, .1}, {2, .1}, {3, .1}, {4, .1}},
  Arrows → {9, 7}, Ticks → {{1, 2, 3, 4}, {1, 2, 3}}];
```

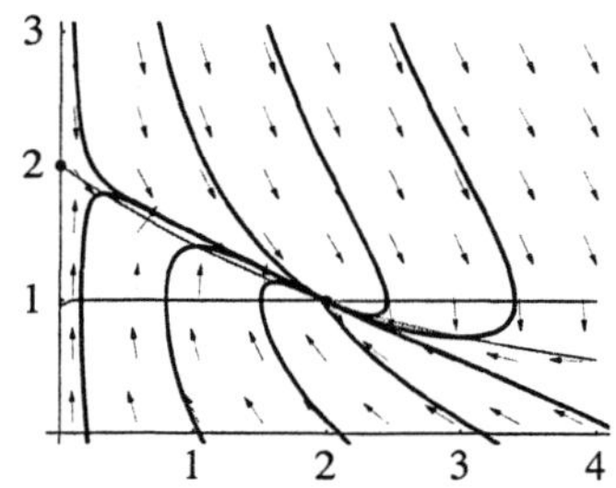

9.4 van der Pol's Equation

```
f[x_, y_] := y;  g[x_, y_] := -x + μ (1 - x²) y;

PhasePlot[{f[x, y], g[x, y]} /. μ → 1, {t, -5, 15}, {x, -4, 4}, {y, -3, 3},
  InitialPoints → {{.2, 0}, {3, 2}, {-2.5, 2}, {-3, -2}, {2.5, -2}},
  Arrows → {9, 7}, ScaleArrows → 2];
```

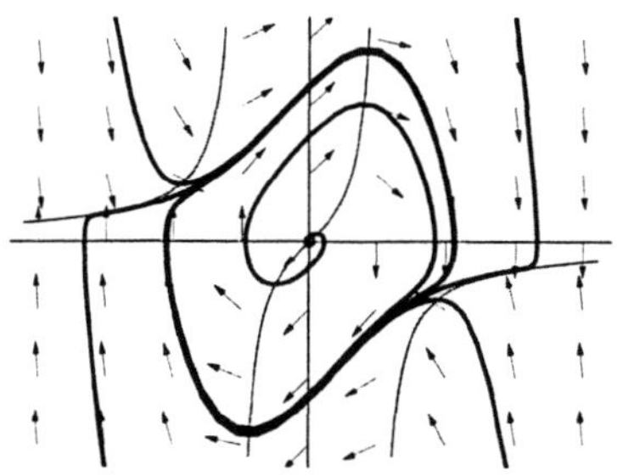

9.5 Duffing's Equation

```
f[x_, y_] := y;  g[t_, x_, y_] := a x - b x³ - c y + A Cos[ω t];

NDPlot[{x'[t] == f[x[t], y[t]], y'[t] == g[t, x[t], y[t]], x[0] == -1, y[0] == 1} /.
   {a → 1, b → 1, c → .2, A → .3, ω → 1}, {x[t], y[t]},
  {t, 0, 300}, MaxSteps → ∞, PlotPoints → 600, Ticks → None];
```

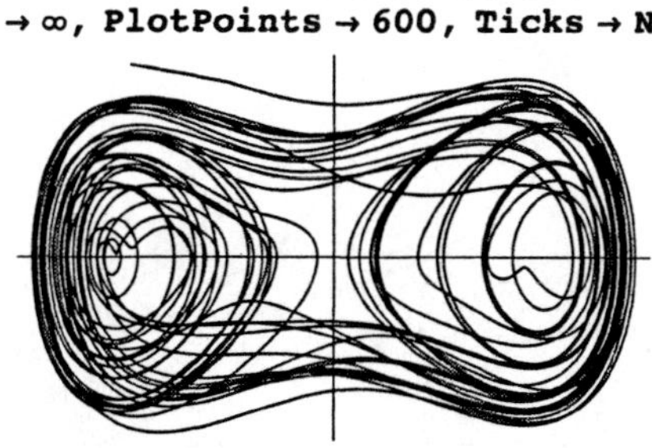

```
PoincareTimeSection[{f[x, y], g[t, x, y]} /. {a → 1, b → 1, c → .2, A → .3, ω → 1},
  {t, 0, 5000, 2 π}, {x, -1}, {y, 1}, Ticks → None, MaxSteps → ∞,
  AspectRatio → Automatic, PlotStyle → AbsolutePointSize[.5]];
```

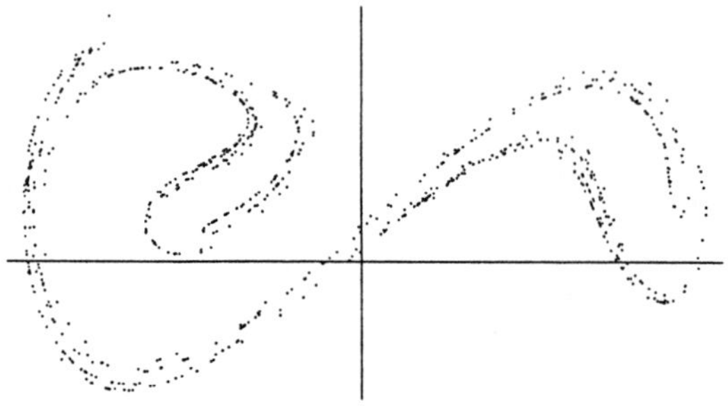

9.6 The Fitzhugh-Nagumo Equations

```
f[x_, y_] := 3 y - x (x^2 - 3) + σ;
g[x_, y_] := (7 - 10 x - 8 y) / 30

PhasePlot[{f[x, y], g[x, y]} /. σ → -1.010556585, {t, 0, 100},
  {x, -2.5, 2.5}, {y, -.6, 1.5}, InitialPoints → {{1, -.5}}, Arrows → {9, 7}];
```

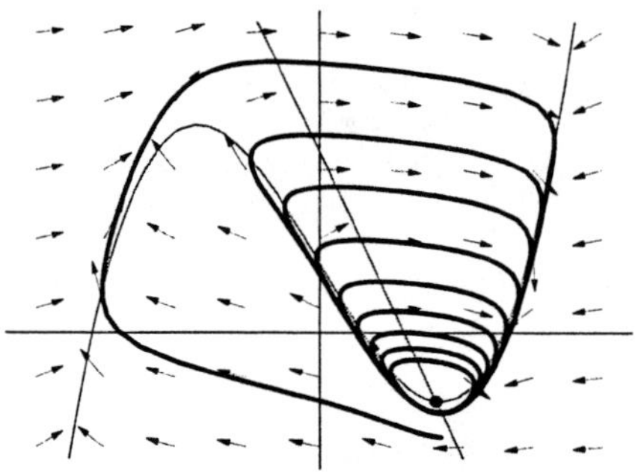

9.7 The Brusselator

```
f[x_, y_] := x^2 y - (a + 1) x + b;
g[x_, y_] := -x^2 y + a x;

PhasePlot[{f[x, y], g[x, y]} /. {a → 2.5, b → 1}, {t, 0, 40}, {x, 0, 3},
  {y, 0, 4}, InitialPoints → {{0, 0}, {1.0, 2.6}}, Arrows → {9, 7}];
```

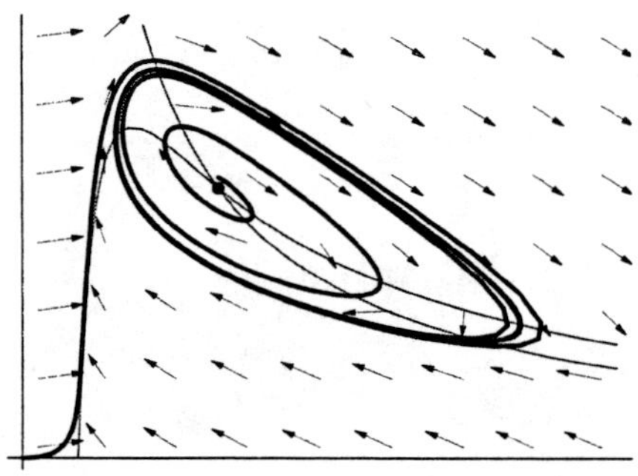

9.8 The Oregonator

```
oregonator = {x'[t] == a^-1 (q y[t] - x[t] y[t] + x[t] (1 - x[t])),
   y'[t] == b^-1 (-q y[t] - x[t] y[t] + z[t]), z'[t] == x[t] - z[t],
   x[0] == .065, y[0] == 1, z[0] == .05} /. {a → .01, b → 1 / 3, q → .005}
```

```
bz = NDPlot[oregonator, {x[t], y[t], z[t]},
   {t, 0, 30}, MaxSteps → ∞, PlotPoints → 200, BoxRatios → {1, 1, 1},
   WorkingPrecision → 20, Ticks → {{0, .1, .2}, {.8, 1, 1.2}, {0, .1, .2}}];
```

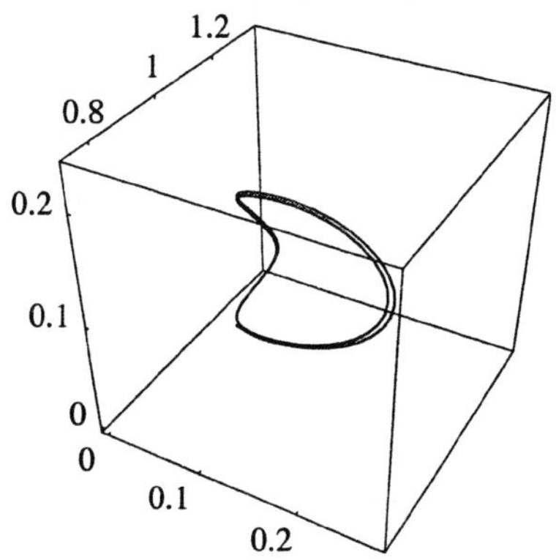


```
Needs["Graphics`Graphics3D`"]; Shadow[bz[[2]], BoxRatios → {1, 1, 1}];
```

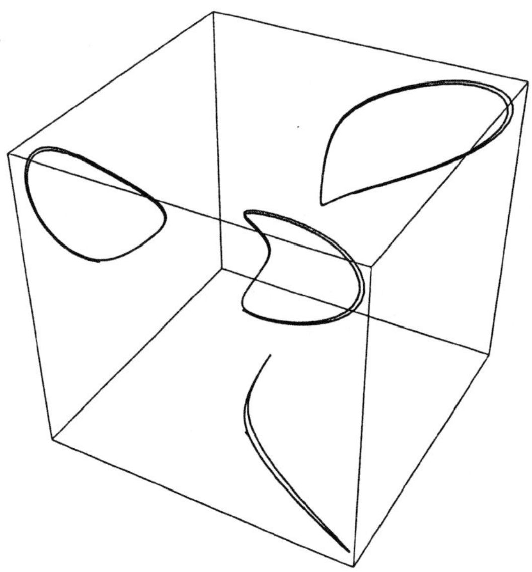

9.9 The Lorenz System

```
lorenz = {x'[t] == a (y[t] - x[t]),
   y'[t] == b x[t] - y[t] - x[t] z[t], z'[t] == x[t] y[t] - c z[t],
   x[0] == -8, y[0] == 8, z[0] == 27} /. {a → 10, b → 28, c → 8 / 3}
```

```
soln =
  NDPlot[lorenz, {x[t], y[t], z[t]}, {t, 0, 30}, MaxSteps → ∞, PlotPoints → 2000];
```

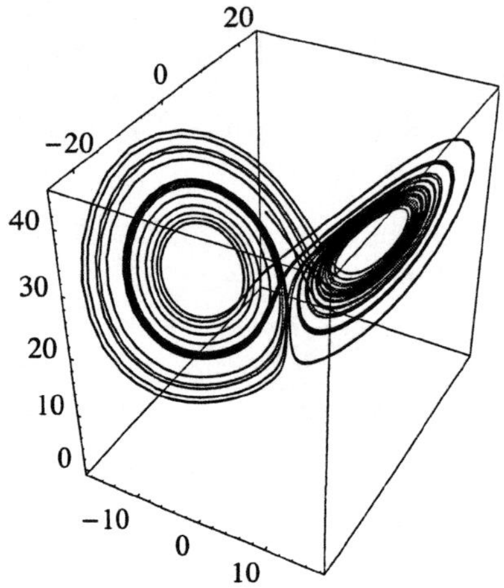


```
ViewProjections[soln[[1]], {t, 0, 30}, {x, y, z}, PlotPoints → 100, Ticks → None];
```

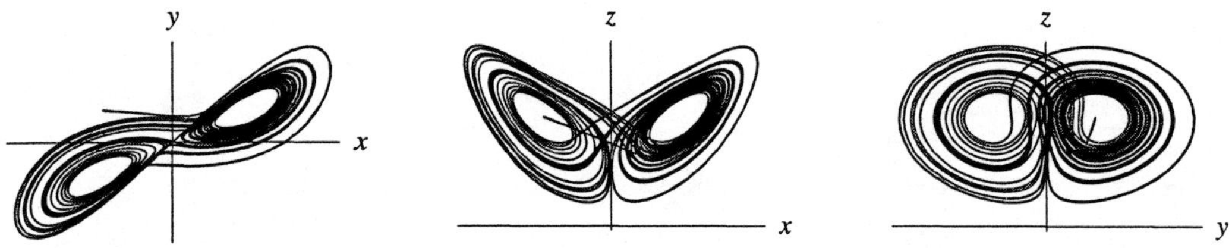

9.10 The Rössler System

```
roessler1 =
  {x'[t] == -y[t] - z[t], y'[t] == x[t] + a y[t], z'[t] == b - z[t] (c - x[t]),
    x[0] == 0, y[0] == 2, z[0] == 0} /. {a → .35, b → 2, c → 4}

curve1 = NDPlot[roessler1, {x[t], y[t], z[t]}, {t, 0, 100},
   MaxSteps → ∞, BoxRatios → {1, 1, 1}, PlotPoints → 1000, PlotRange → All];
```

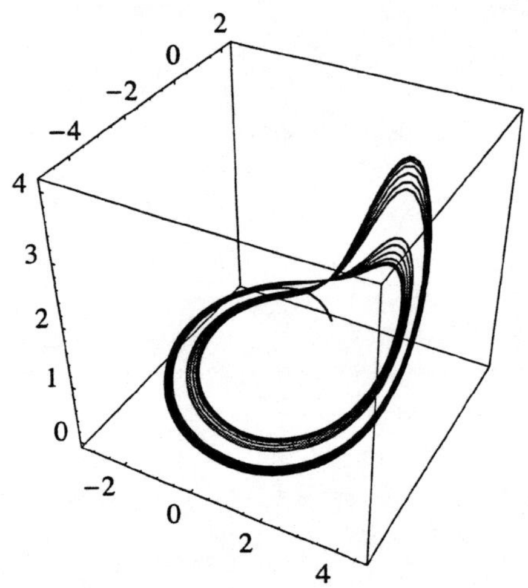

```
<< Graphics`Graphics3D`; Shadow[curve1[[2]], BoxRatios → {1, 1, 1}];
```

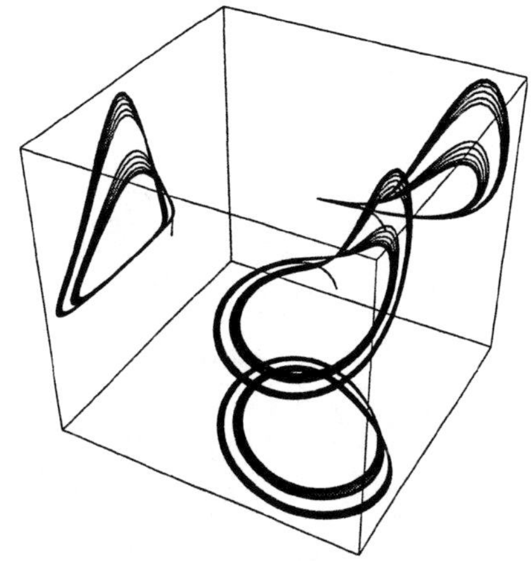

```
roessler2 =
 {x'[t] == -y[t] - z[t], y'[t] == x[t] + a y[t], z'[t] == b - z[t] (c - x[t]),
   x[0] == 0, y[0] == 2, z[0] == 0} /. {a → .46775565, b → 2, c → 4}
```

```
curve2 = NDPlot[roessler2, {x[t], y[t], z[t]},
   {t, 0, 100}, MaxSteps → ∞, WorkingPrecision → 20,
   BoxRatios → {1, 1, 1}, PlotPoints → 1500, PlotRange → All];
```

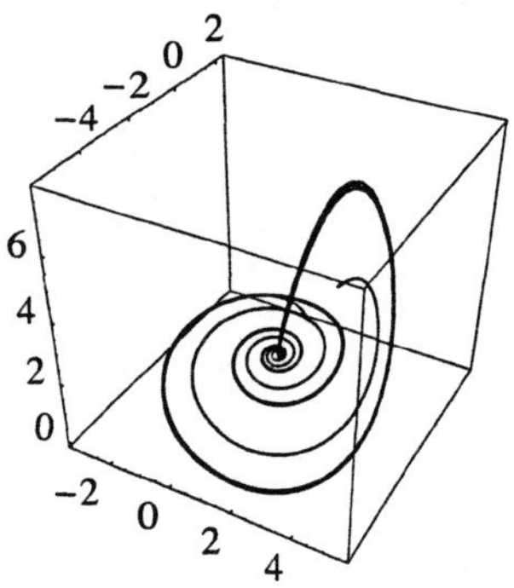


```
Shadow[curve2[[2]], BoxRatios → {1, 1, 1}];
```

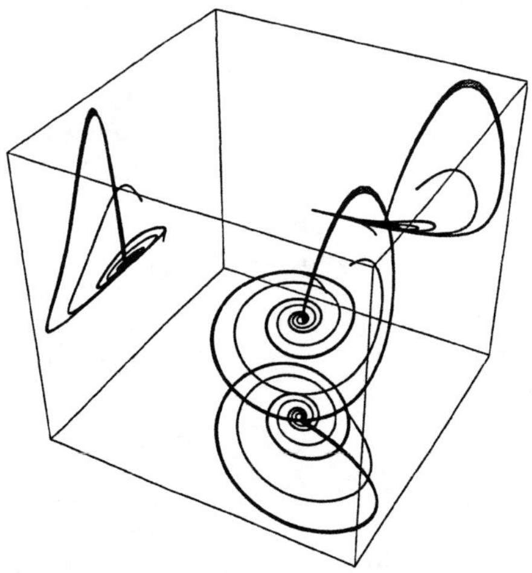

9.11 The Scroll Circuit

```
γ[v_] := -a v + 0.5 (a + b) (Abs[v + 1] - Abs[v - 1]) /. {a → .07, b → .1};
scroll = {x'[t] == -c γ[y[t] - x[t]], y'[t] == -z[t] - γ[y[t] - x[t]],
   z'[t] == k y[t], x[0] == 1, y[0] == 1, z[0] == 1} /. {c → 33, k → 1}
```

```
soln = NDPlot[scroll, {x[t], y[t], z[t]}, {t, 0, 250}, MaxSteps → ∞,
   BoxRatios → {1, 1, 1}, PlotPoints → 2000, PlotRange → All];
```

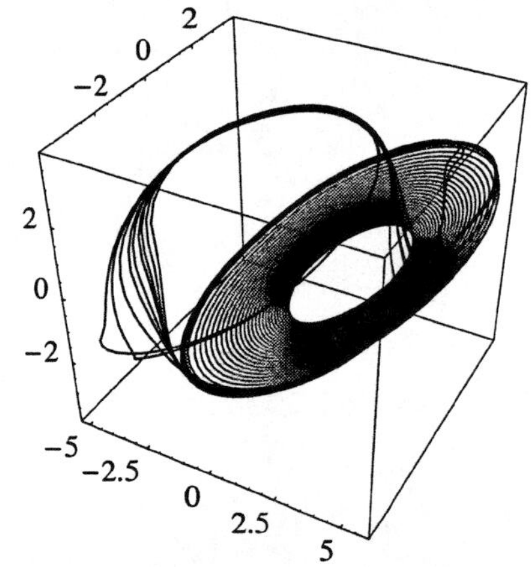


```
ViewProjections[soln[[1]], {t, 0, 250}, {x, y, z}, PlotPoints → 100, Ticks → None];
```

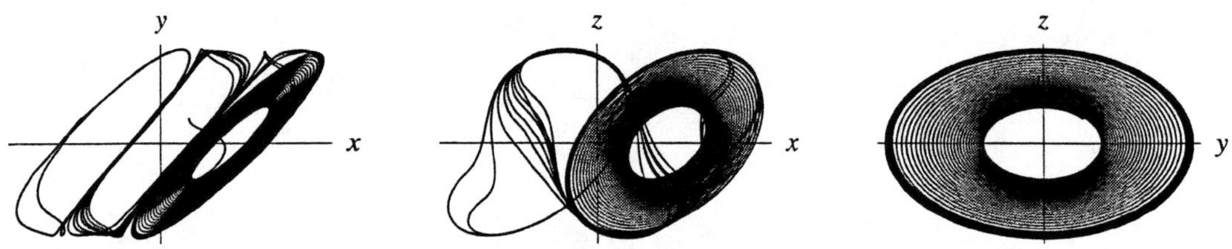

9.12 The Double Pendulum

```
dblPendsys =
  {α''[t] == -(9.8 m1 Sin[α[t]] + m2 Sin[α[t] - β[t]] (9.8 Cos[β[t]] + L1 Cos[α[t] - β[t]]
            α'[t]^2 + L2 β'[t]^2)) / (L1 (m1 + m2 Sin[α[t] - β[t]]^2)),
   β''[t] == Sin[α[t] - β[t]] ((m1 + m2) (9.8 Cos[α[t]] + L1 α'[t]^2) +
         m2 L2 Cos[α[t] - β[t]] β'[t]^2) / (L1 (m1 + m2 Sin[α[t] - β[t]]^2)),
   α[0] == 3.1, β[0] == 0, α'[0] == 0, β'[0] == 0} /. {m1 → 1, m2 → .25, L1 → 2, L2 → 1}
```

```
soln = NDPlot[dblPendsys, {α[t], β[t]},
   {t, 0, 100}, MaxSteps → ∞, PlotPoints → 2000, PlotRange → All];
```

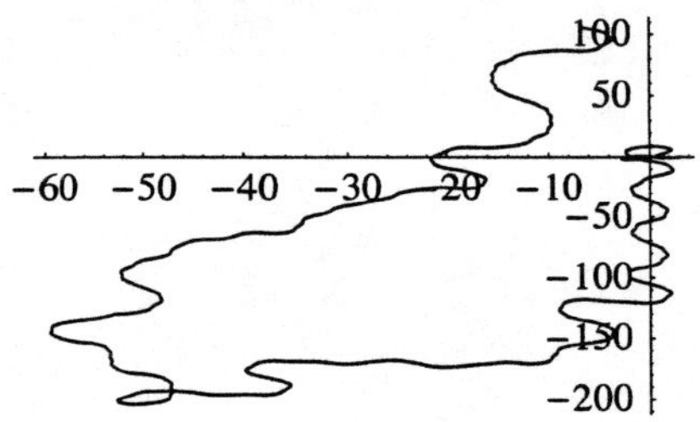

```
path2 = {L1 Sin[soln[[1, 1]]] + L2 Sin[soln[[1, 2]]],
    L1 Cos[soln[[1, 1]]] + L2 Cos[soln[[1, 2]]]} /. {L1 → 2, L2 → 1};
ParametricPlot[Evaluate[path2], {t, 0, 100}, PlotPoints → 1000,
  PlotRange → All, AspectRatio → Automatic];
```

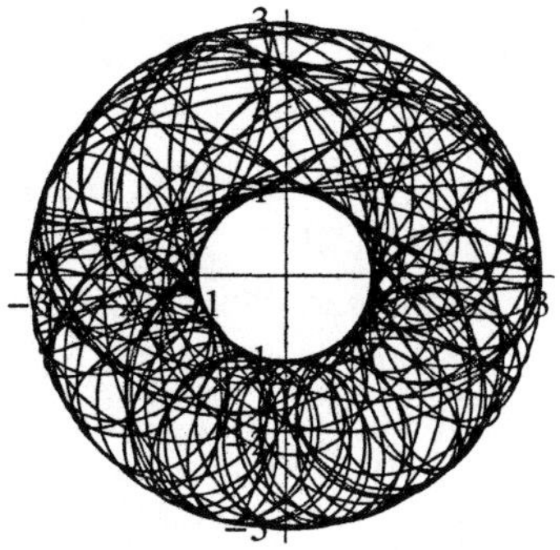

9.13 The Three-Body Problem

```
r1[t_] := (x1[t]^2 + y1[t]^2)^(1/2); r2[t_] := (x2[t]^2 + y2[t]^2)^(1/2);
r12[t_] := ((x1[t] - x2[t])^2 + (y1[t] - y2[t])^2)^(1/2);
```

```
ThreeBodySys =
  {x1''[t] == -((m0G + m1G) x1[t])/r1[t]^3 - (m2G x2[t])/r2[t]^3 - (m2G (x1[t] - x2[t]))/r12[t]^3,
   y1''[t] == -((m0G + m1G) y1[t])/r1[t]^3 - (m2G y2[t])/r2[t]^3 - (m2G (y1[t] - y2[t]))/r12[t]^3,
   x2''[t] == -((m0G + m2G) x2[t])/r2[t]^3 - (m1G x1[t])/r1[t]^3 + (m1G (x1[t] - x2[t]))/r12[t]^3,
   y2''[t] == -((m0G + m2G) y2[t])/r2[t]^3 - (m1G y1[t])/r1[t]^3 + (m1G (y1[t] - y2[t]))/r12[t]^3,
   x1[0] == 51, y1[0] == 16.5, x1'[0] == 0, y1'[0] == 0,
   x2[0] == 40, y2[0] == 0, x2'[0] == 0, y2'[0] == .5} /.
  {m0G → 13.4, m1G → .3335, m2G → 3.335};
```

```
NDPlot[ThreeBodySys, {{x1[t], y1[t]}, {x2[t], y2[t]}},
  {t, 0, 750}, MaxSteps → ∞, WorkingPrecision → 17, PlotRange → All,
  PlotPoints → 700, PlotStyle → {{Hue[0], Thickness[.005]}, {}}];
```

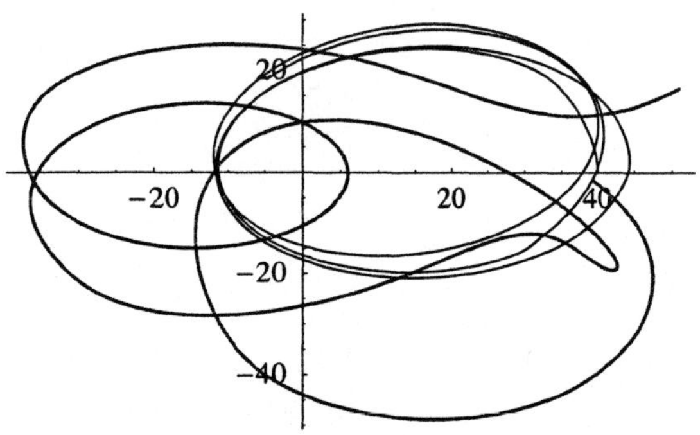

9.14 Chapter Projects

A. Chaos and Numerical Solutions

Chaotic behavior is characterized by extreme sensitivity to initial conditions. As a result, the accuracy of numerical solutions beyond a short time is always in question. One way of determining whether a system's behavior is chaotic is to generate multiple numerical solutions with the same initial conditions but different levels of accuracy. A convenient way of doing this in *Mathematica* is to use the `WorkingPrecision` option in `NDSolve` (or `NDPlot`).

For example, consider the Lorenz system from Section 9.9.

```
lorenz = {x'[t] == a (y[t] - x[t]),
    y'[t] == b x[t] - y[t] - x[t] z[t], z'[t] == x[t] y[t] - c z[t],
    x[0] == -8, y[0] == 8, z[0] == 27} /. {a → 10, b → 28, c → 8 / 3};
```

Let's compute the solution for $0 \le t \le 20$ with the default setting of `WorkingPrecision`, which is `$MachinePrecision`.

```
$MachinePrecision
    16
```

```
Needs["DiffEqs`DEGraphics`"];
soln = NDPlot[lorenz, {x[t], y[t], z[t]}, {t, 0, 20},
  PlotPoints → 2000, MaxSteps → ∞, Ticks → None]
```

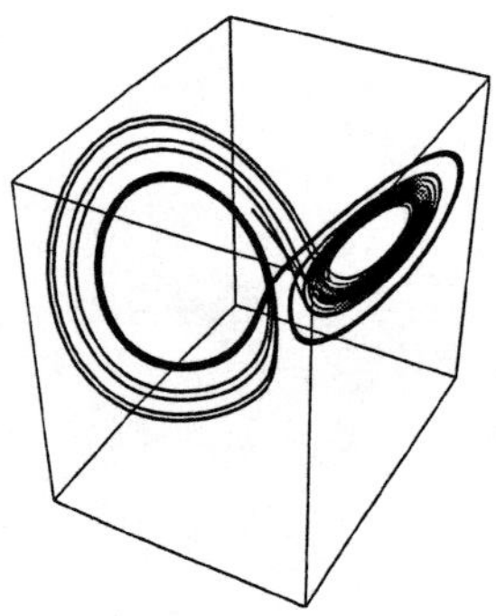

```
{{InterpolatingFunction[{{0., 20.}}, <>][t],
  InterpolatingFunction[{{0., 20.}}, <>][t],
  InterpolatingFunction[{{0., 20.}}, <>][t]}, - Graphics3D -}
```

The value of the approximate solution at $t = 20$ is

```
soln[[1]] /. t → 20
    {7.01312, 1.04721, 31.9967}
```

Now let's recompute the solution for $0 \le t \le 20$ with `WorkingPrecision` set to its default plus 1. (Notice the computation takes somewhat longer.)

```
soln = NDPlot[lorenz, {x[t], y[t], z[t]}, {t, 0, 20},
  PlotPoints → 1000, MaxSteps → ∞,
  Ticks → None, WorkingPrecision → $MachinePrecision + 1]
```

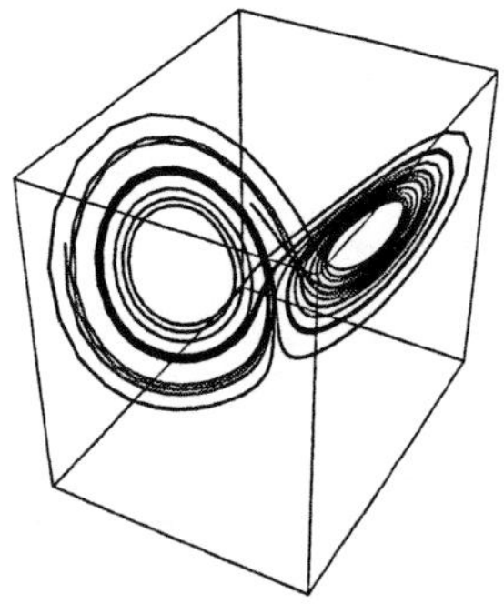

```
{{InterpolatingFunction[{{0, 20.000000000000000}}, <>][t],
  InterpolatingFunction[{{0, 20.000000000000000}}, <>][t],
  InterpolatingFunction[{{0, 20.000000000000000}}, <>][t]}, - Graphics3D -}
```

The value of the approximate solution at $t = 20$ is nowhere near what it was before.

```
soln[[1]] /. t → 20
```

```
{-16.63647591186122, -20.3127407722554, 34.6869177378848}
```

1. Demonstrate that similar sensitivity occurs with:

   (a) the Rössler system from Section 9.10;

   (b) the double pendulum from Section 9.12.

2. Demonstrate that the Oregonator from Section 9.8 does not exhibit this sensitivity.

3. Investigate the question for the three-body system from Section 9.13. (*Suggestion:* Using the same parameters and initial values as in Section 9.13, compute the solution for $0 \le t \le 6000$.)

B. Animation of the Double Pendulum

Use the pendulum animation from Section 6.2 as a starting point for creating an animation of the double pendulum. Once you're successful, experiment with different initial conditions.

C. Animation of the Three-Body Problem

Create an animation of the solution of the three-body problem plotted in Section 9.13. Include in it small disks representing the three bodies. (The largest of them is fixed at the origin.) Then experiment with different masses and initial conditions.

Chapter 10

Fourier Series and Sturm-Liouville Eigenvalue Problems

10.1 Fourier Series

Given a function f defined on an interval $[-L, L]$, the *Fourier coefficients* of f are

$$a_n = \frac{1}{L} \int_{-L}^{L} f(x) \cos(\tfrac{n\pi}{L} x)\, dx, \quad n = 0, 1, 2, \ldots,$$

$$b_n = \frac{1}{L} \int_{-L}^{L} f(x) \sin(\tfrac{n\pi}{L} x)\, dx, \quad n = 1, 2, 3, \ldots.$$

The *Fourier series* of f is the function $f_{\sim}$ defined by

$$f_{\sim}(x) = \frac{1}{2} a_0 + \sum_{n=1}^{\infty} (a_n \cos(\tfrac{n\pi}{L} x) + b_n \sin(\tfrac{n\pi}{L} x)).$$

Example 10.1.1

Let f be the function defined on $[-\pi, \pi]$ by

$$f(x) = \begin{cases} 0, & \text{if } x \le 0, \\ x, & \text{if } x \ge 0. \end{cases}$$

```
f[x_] := x UnitStep[x];
Plot[f[x], {x, -π, π}, Ticks → {{-π, -π / 2, π / 2, π}, {π / 2, π}},
  AspectRatio → Automatic, PlotStyle → Thickness[.008]];
```

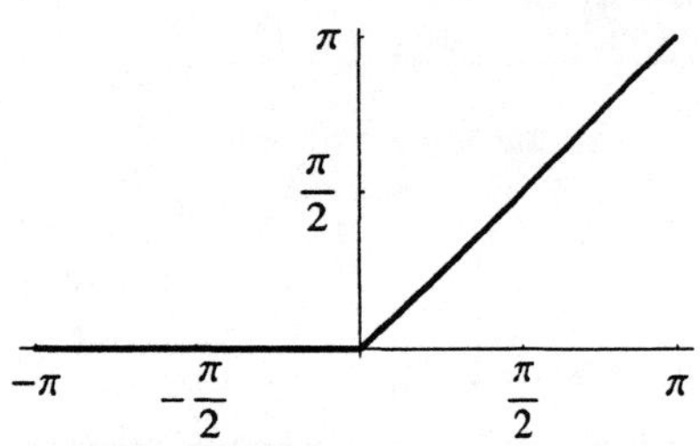

To compute the Fourier coefficients of f, we'll first observe that, since $f(x) = 0$ for $x < 0$, we need only integrate over $[0, \pi]$. Thus the Fourier coefficients are as follows:

```
a[0] = 1/π ∫_0^π x dx
```

$\frac{\pi}{2}$

```
a[n_] = Simplify[1/π ∫_0^π x Cos[n x] dx, n ∈ Integers]
```

$$\frac{-1+(-1)^n}{n^2\,\pi}$$

```
b[n_] = Simplify[1/π ∫_0^π x Sin[n x] dx, n ∈ Integers]
```

$$-\frac{(-1)^n}{n}$$

The tenth partial sum, for instance, of the Fourier series of f is

```
fourierf10[x_] = 1/2 a[0] + Sum[(a[n] Cos[n x] + b[n] Sin[n x]), {n, 1, 10}]
```

$$\frac{\pi}{4}-\frac{2\,\text{Cos}[x]}{\pi}-\frac{2\,\text{Cos}[3\,x]}{9\,\pi}-\frac{2\,\text{Cos}[5\,x]}{25\,\pi}-\frac{2\,\text{Cos}[7\,x]}{49\,\pi}-\frac{2\,\text{Cos}[9\,x]}{81\,\pi}+\text{Sin}[x]-\frac{1}{2}\,\text{Sin}[2\,x]+\frac{1}{3}\,\text{Sin}[3\,x]-\frac{1}{4}\,\text{Sin}[4\,x]+\frac{1}{5}\,\text{Sin}[5\,x]-\frac{1}{6}\,\text{Sin}[6\,x]+\frac{1}{7}\,\text{Sin}[7\,x]-\frac{1}{8}\,\text{Sin}[8\,x]+\frac{1}{9}\,\text{Sin}[9\,x]-\frac{1}{10}\,\text{Sin}[10\,x]$$

Here is a plot of this partial sum along with f on $[-\pi, \pi]$.

```
Plot[{fourierf10[x], f[x]}, {x, -π, π},
  AspectRatio → Automatic, Ticks → {{-π, -π / 2, π / 2, π}, {π / 2, π}},
  PlotStyle → {{Thickness[.007], Hue[0]}, {}}];
```

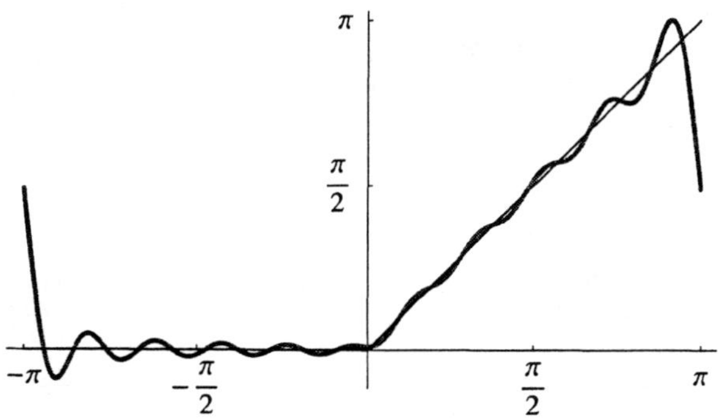

The Fourier series and any of its partial sums are periodic functions defined on all of $\mathbb{R}$ with period 2π. The following graph shows three cycles of the tenth partial sum.

```
Plot[fourierf10[x], {x, -3 π, 3 π},
  AspectRatio → Automatic, Ticks → {π Range[-3, 3], {π / 2, π}}];
```

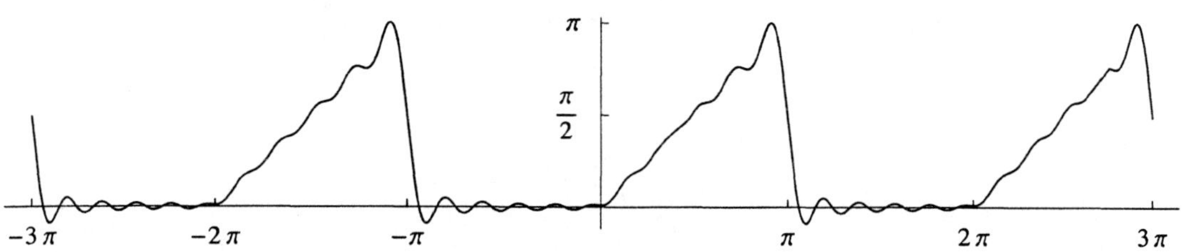

For comparison, let's also look at the 30th partial sum on the same interval.

$$\texttt{fourierf30[x\_]} = \frac{1}{2}\,\texttt{a[0]} + \sum_{n=1}^{30} (\texttt{a[n] Cos[n x] + b[n] Sin[n x]});$$

```
Plot[fourierf30[x], {x, -3 π, 3 π}, AspectRatio → Automatic,
  PlotRange → All, Ticks → {π Range[-3, 3], {π / 2, π}}];
```

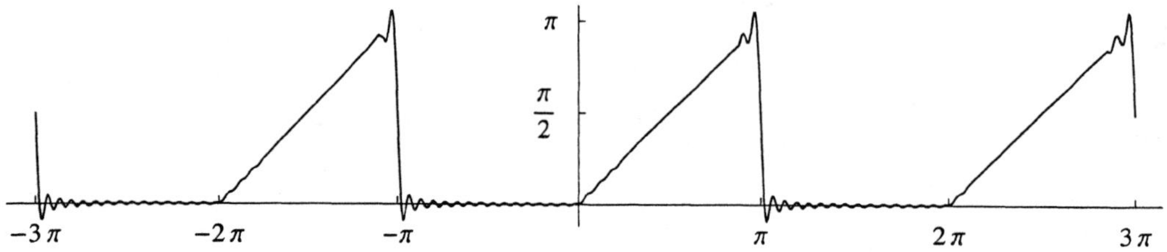

Notice that this partial sum agrees quite well with f on $(-\pi, \pi)$ except near the endpoints, where a kind of *ringing* occurs. This ringing is known as *Gibb's phenomenon*.

- For convenience, let's define a function that will compute a partial sum of a function's Fourier series. (This function, as well as the others in shaded cells that follow, is defined in the `FourierTools` package.)

$$\texttt{FourierSum[f\_, \{x\_, minusL\_, L\_\}, m\_Integer] :=}$$
$$\texttt{Module}\left[\left\{an = \int_{-L}^{L} f \cos\left[\frac{n\pi x}{L}\right] dx,\ bn = \int_{-L}^{L} f \sin\left[\frac{n\pi x}{L}\right] dx\right\},\right.$$
$$\left.\texttt{If}\left[minusL == -L,\ \frac{1}{2L}\int_{-L}^{L} f\, dx + \frac{1}{L}\sum_{n=1}^{m}\left(an \cos\left[\frac{n\pi x}{L}\right] + bn \sin\left[\frac{n\pi x}{L}\right]\right),\ \{\}\right]\right]$$

Example 10.1.2

Suppose that we want to find the Fourier series of the unit step function on the interval $[-2, 2]$. The tenth partial sum of the Fourier series is

```
f10[x_] = FourierSum[UnitStep[x], {x, -2, 2}, 10] // Expand
```

$$\frac{1}{2} + \frac{2\,\text{Sin}[\frac{\pi x}{2}]}{\pi} + \frac{2\,\text{Sin}[\frac{3\pi x}{2}]}{3\pi} + \frac{2\,\text{Sin}[\frac{5\pi x}{2}]}{5\pi} + \frac{2\,\text{Sin}[\frac{7\pi x}{2}]}{7\pi} + \frac{2\,\text{Sin}[\frac{9\pi x}{2}]}{9\pi}$$

```
Plot[{f10[x], UnitStep[x]}, {x, -2, 2}, AspectRatio → Automatic];
```

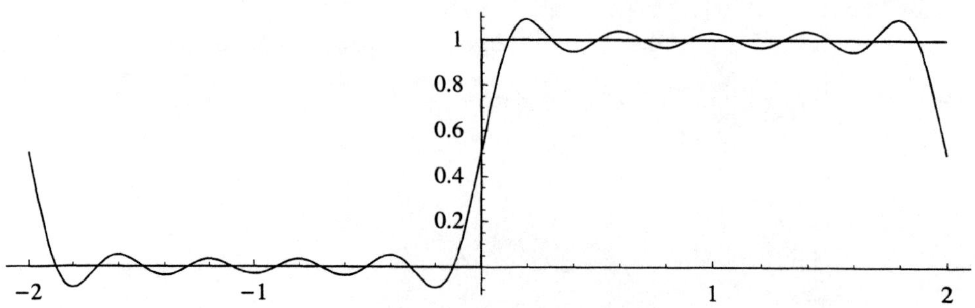

- In many situations it is not practical or desirable to find exact values of a function's Fourier coefficients. So it will be useful to have a numerical version of `FourierSum`.

```
NFourierSum[f_, {x_, minusL_, L_}, m_Integer] :=
 If[minusL == -L, Chop[1/(2 L) NIntegrate[f, {x, -L, L}] +
   1/L Sum[((NIntegrate[f Cos[n π x/L], {x, -L, L}]) Cos[n π x/L] +
        (NIntegrate[f Sin[n π x/L], {x, -L, L}]) Sin[n π x/L]), {n, 1, m}]], {}]
```

Example 10.1.3

Suppose that we want to find the Fourier series of the following function on the interval $[-1, 1]$.

```
f[x_] := UnitStep[x] 4 x^2 (1 - x^2);
Plot[f[x], {x, -1, 1}, AspectRatio → Automatic,
 Ticks → {Automatic, {.5, 1}}, PlotStyle → AbsoluteThickness[1]];
```

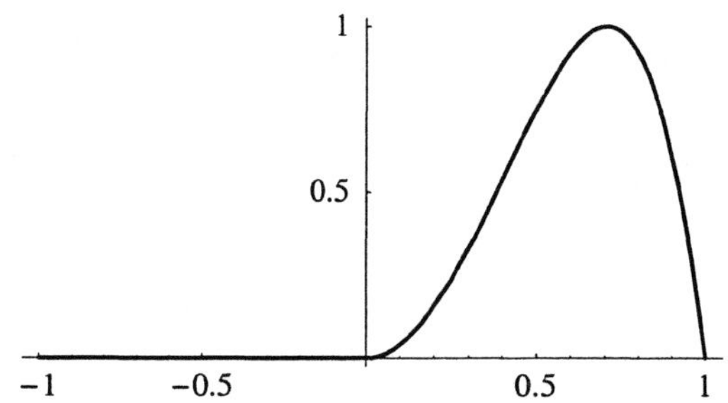

The following gives the eighth partial sum with numeric coefficients.

```
f8[x_] = NFourierSum[f[x], {x, -1, 1}, 8]
```

```
0.266667 - 0.174965 Cos[π x] - 0.141046 Cos[2 π x] +
 0.0778962 Cos[3 π x] - 0.0468108 Cos[4 π x] + 0.0308459 Cos[5 π x] -
 0.0217554 Cos[6 π x] + 0.0161318 Cos[7 π x] - 0.0124245 Cos[8 π x] +
 0.404638 Sin[π x] - 0.193509 Sin[2 π x] + 0.0356421 Sin[3 π x] -
 0.0241887 Sin[4 π x] + 0.00805562 Sin[5 π x] -
 0.00716701 Sin[6 π x] + 0.00297156 Sin[7 π x] - 0.00302358 Sin[8 π x]
```

```
Plot[f8[x], {x, -1, 1}, PlotRange → All, AspectRatio → Automatic];
```

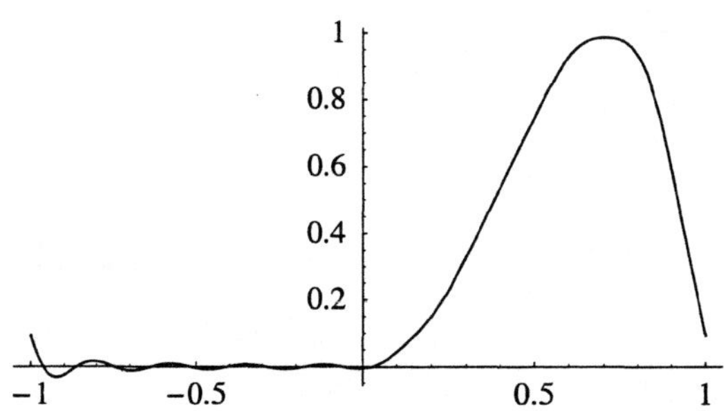

■ Fourier Sine and Cosine Series

Given a function f defined on an interval $[0, L]$, the *Fourier cosine coefficients* of f are

$$a_n = \frac{2}{L}\int_0^L f(x)\cos\left(\frac{n\pi}{L}x\right)dx, \quad n = 0, 1, 2, \ldots,$$

and the *Fourier cosine series* of f is the function $f_{\cos}$ defined by

$$f_{\cos}(x) = \frac{1}{2}a_0 + \sum_{n=1}^{\infty} a_n \cos\left(\frac{n\pi}{L}x\right).$$

The Fourier cosine series of f is the Fourier series on $[-L, L]$ of the *even extension* of f defined by

$$f_{\text{even}}(x) = \begin{cases} f(-x), & \text{if } x < 0, \\ f(x), & \text{if } x \geq 0. \end{cases}$$

The *Fourier sine coefficients* of f are

$$b_n = \frac{2}{L}\int_0^L f(x)\sin\left(\frac{n\pi}{L}x\right)dx, \quad n = 1, 2, 3, \ldots,$$

and the *Fourier sine series* of f is the function $f_{\sin}$ defined by

$$f_{\sin}(x) = \sum_{n=1}^{\infty} b_n \sin\left(\frac{n\pi}{L}x\right).$$

The Fourier sine series of f is the Fourier series on $[-L, L]$ of the *odd extension* of f defined by

$$f_{\text{odd}}(x) = \begin{cases} -f(-x), & \text{if } x < 0, \\ 0, & \text{if } x = 0, \\ f(x), & \text{if } x > 0. \end{cases}$$

The following define convenient functions for computing partial sums of sine and cosine series. (These are in the `FourierTools` package.)

```
FourierCosSum[f_, {x_, 0, L_}, m_Integer] :=
  Module[{an = 2 ∫_0^L f Cos[(n π x)/L] dx }, 1/L ∫_0^L f dx + 1/L Σ_{n=1}^m an Cos[(n π x)/L]]

FourierSinSum[f_, {x_, 0, L_}, m_Integer] :=
  Module[{bn = 2 ∫_0^L f Sin[(n π x)/L] dx}, 1/L Σ_{n=1}^m bn Sin[(n π x)/L]]

NFourierCosSum[f_, {x_, 0, L_}, m_Integer] :=
  Chop[1/L NIntegrate[f, {x, 0, L}] +
    2/L Σ_{n=1}^m ((NIntegrate[f Cos[(n π x)/L], {x, 0, L}]) Cos[(n π x)/L])]
```

```
NFourierSinSum[f_, {x_, 0, L_}, m_Integer] :=
  Chop[2/L Sum[((NIntegrate[f Sin[n π x/L], {x, 0, L}]) Sin[n π x/L]), {n, 1, m}]]
```

Example 10.1.4

Consider the function $f(x) = x^2 (3 - 2x)$ on the interval $[0, 1]$.

```
Plot[x^2 (3 - 2 x), {x, 0, 1}];
```

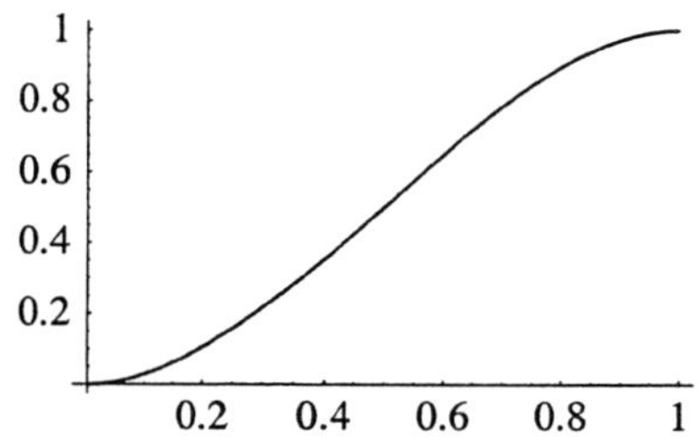

The first three (nonzero) terms of the Fourier cosine series do a good job of approximating f on $[0, 1]$.

```
fcos3[x_] = FourierCosSum[x^2 (3 - 2 x), {x, 0, 1}, 3]
```

$$\frac{1}{2} - \frac{48 \cos[\pi x]}{\pi^4} - \frac{16 \cos[3 \pi x]}{27 \pi^4}$$

```
Plot[{fcos3[x], x^2 (3 - 2 x)}, {x, 0, 1}];
```

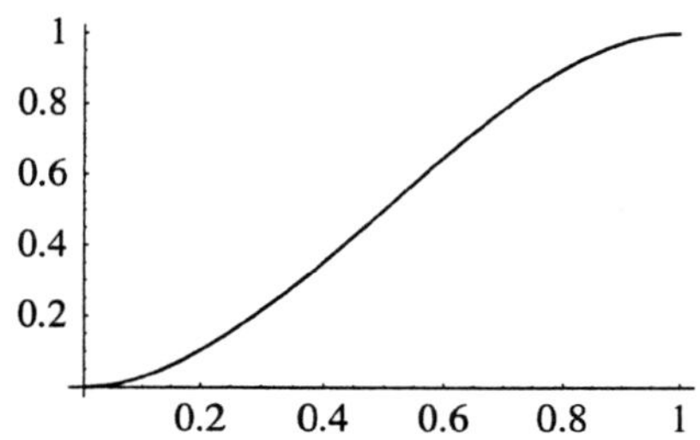

Notice how f compares to its Fourier cosine series on $[-1, 2]$.

```
Plot[{fcos3[x], x^2 (3 - 2 x)}, {x, -1, 2},
  PlotRange → {-.5, 1.5}, Ticks → {{-1, 1, 2}, {1}}];
```

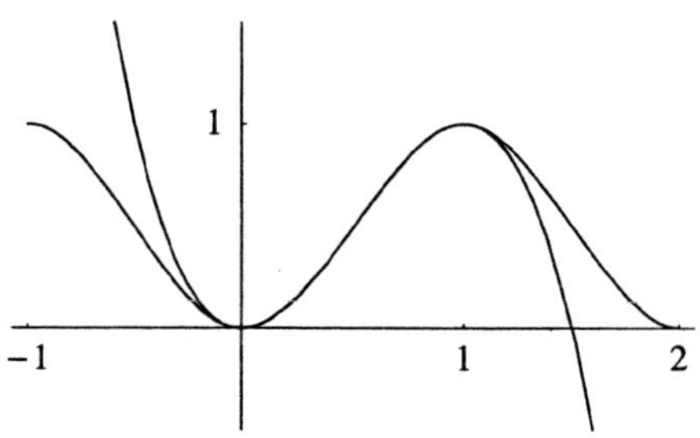

The first six terms of the Fourier sine series do a much poorer job of approximating f.

```
fsin6[x_] = FourierSinSum[x^2 (3 - 2 x), {x, 0, 1}, 6]
```

$$2\left(-\frac{6}{\pi^3}-\frac{-6-\pi^2}{\pi^3}\right)\text{Sin}[\pi\,x]+2\left(-\frac{3}{4\,\pi^3}+\frac{-6-4\,\pi^2}{8\,\pi^3}\right)\text{Sin}[2\,\pi\,x]+2\left(-\frac{2}{9\,\pi^3}-\frac{-6-9\,\pi^2}{27\,\pi^3}\right)\text{Sin}[3\,\pi\,x]+$$
$$2\left(-\frac{3}{32\,\pi^3}+\frac{-6-16\,\pi^2}{64\,\pi^3}\right)\text{Sin}[4\,\pi\,x]+2\left(-\frac{6}{125\,\pi^3}-\frac{-6-25\,\pi^2}{125\,\pi^3}\right)\text{Sin}[5\,\pi\,x]+2\left(-\frac{1}{36\,\pi^3}+\frac{-6-36\,\pi^2}{216\,\pi^3}\right)\text{Sin}[6\,\pi\,x]$$

```
Plot[{fsin6[x], x^2 (3 - 2 x)}, {x, 0, 1}, PlotRange → All];
```

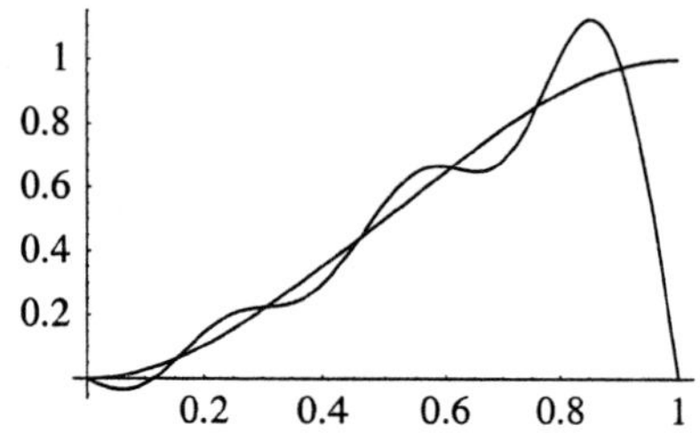


```
Plot[{fsin6[x], x^2 (3 - 2 x)}, {x, -1, 2},
  PlotRange → {-1.5, 1.5}, Ticks → {{-1, 1, 2}, {-1, 1}}];
```

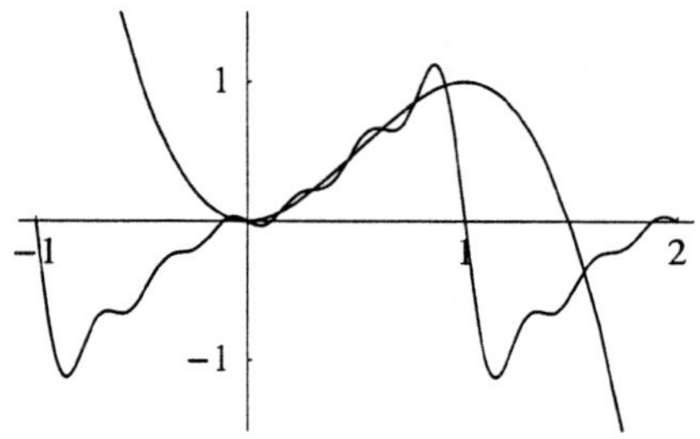

Example 10.1.5

Consider the function $f(x) = 6\,x^2(1 - x)$ on the interval [0, 1].

```
Plot[6 x^2 (1 - x), {x, 0, 1}];
```

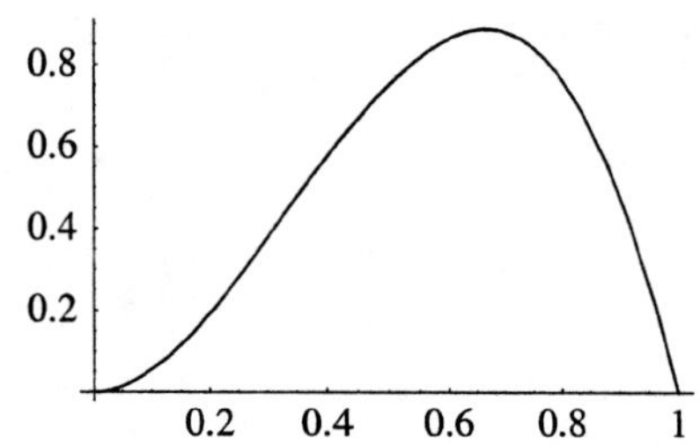

The first six terms of the Fourier cosine series are as follows:

```
fcos5[x_] = FourierCosSum[6 x^2 (1 - x), {x, 0, 1}, 5] // Simplify
```

$$\frac{1}{2}+\frac{12\,(-12+\pi^2)\,\text{Cos}[\pi\,x]}{\pi^4}-\frac{3\,\text{Cos}[2\,\pi\,x]}{\pi^2}-\frac{16\,\text{Cos}[3\,\pi\,x]}{9\,\pi^4}+$$
$$\frac{4\,\text{Cos}[3\,\pi\,x]}{3\,\pi^2}-\frac{3\,\text{Cos}[4\,\pi\,x]}{4\,\pi^2}-\frac{144\,\text{Cos}[5\,\pi\,x]}{625\,\pi^4}+\frac{12\,\text{Cos}[5\,\pi\,x]}{25\,\pi^2}$$

```
Plot[{fcos5[x], 6 x^2 (1 - x)}, {x, 0, 1}];
```

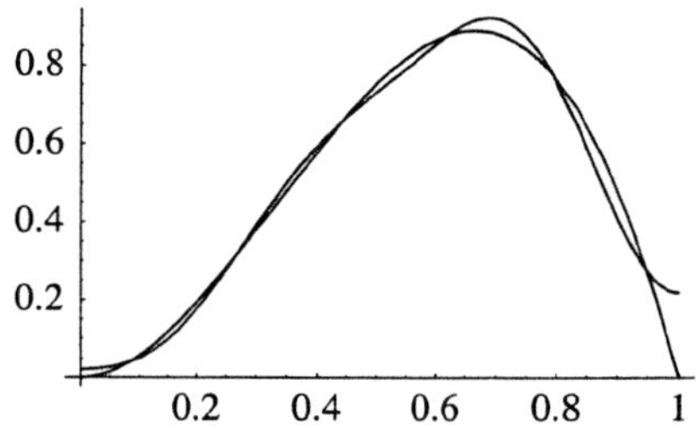


```
Plot[{fcos5[x], 6 x^2 (1 - x)}, {x, -1, 2},
  PlotRange → {-.25, 1.25}, Ticks → {{-1, 1, 2}, {1}}];
```

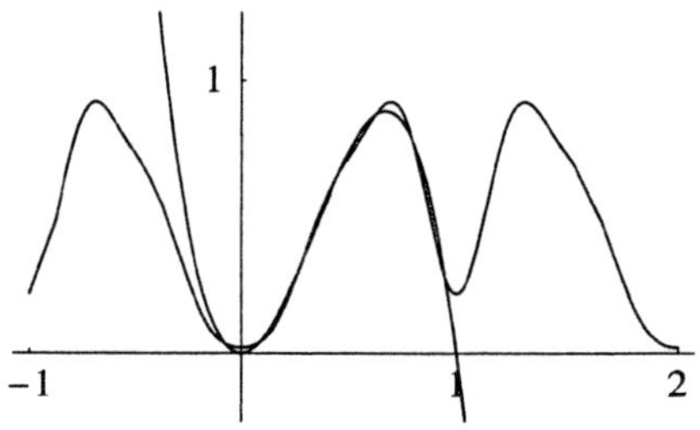

The first five terms of the Fourier sine series give a better approximation of f on [0, 1].

```
fsin5[x_] = FourierSinSum[6 x^2 (1 - x), {x, 0, 1}, 5]
```

$$\frac{24\,\text{Sin}[\pi x]}{\pi^3} - \frac{9\,\text{Sin}[2\pi x]}{\pi^3} + \frac{8\,\text{Sin}[3\pi x]}{9\pi^3} - \frac{9\,\text{Sin}[4\pi x]}{8\pi^3} + \frac{24\,\text{Sin}[5\pi x]}{125\pi^3}$$

```
Plot[{fsin5[x], 6 x^2 (1 - x)}, {x, 0, 1}, PlotRange → All];
```

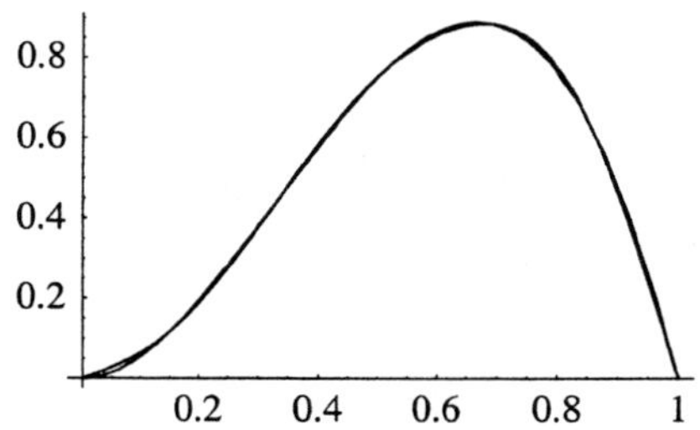


```
Plot[{fsin5[x], 6 x^2 (1 - x)}, {x, -1, 2},
  PlotRange → {-1.25, 1.25}, Ticks → {{-1, 1, 2}, {-1, 1}}];
```

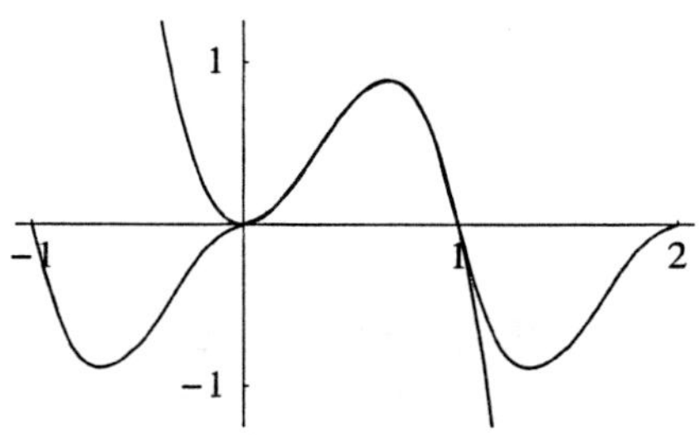

■ Exercises

For the given function f in Exercises 1–5,

(a) Compute the Fourier coefficients $a_0, a_1, \ldots, a_{10}$ and $b_1, b_2, \ldots, b_{10}$.

(b) Create a plot showing $f(x)$ together with the partial sum

$$\frac{1}{2} a_0 + \sum_{n=1}^{m} (a_n \cos(\tfrac{n\pi}{L} x) + b_n \sin(\tfrac{n\pi}{L} x))$$

for each of $m = 1, 3, 5,$ and 10.

1. $f(x) = \max\{\sin x, 0\}$ on $[-\pi, \pi]$
2. $f(x) = (\max\{\sin x, 0\})^2$ on $[-\pi, \pi]$
3. $f(x) = x - |x| - 1$ on $[-1, 1]$
4. $f(x) = x^2 + x$ on $[-1, 1]$
5. $f(x) = x(1 - x^4)$ on $[-1, 1]$
6. Find the Fourier cosine series of $\sin x$ on $[0, \pi]$, and and the following partial sum with 11 nonzero terms. Include the graph of $\sin x$ for comparison. Then plot the same functions on $[-\pi, \pi]$.
7. Find the Fourier sine series of $\cos x$ on $[0, \pi]$, and and the following partial sum with 10 nonzero terms. Include the graph of $\cos x$ for comparison. Then plot the same functions on $[-\pi, \pi]$.
8. Find the Fourier cosine series of $1 - x$ on $[0, 1]$, and and the following partial sum with 11 nonzero terms. Include the graph of $1 - x$ for comparison. Then plot the same functions on $[-1, 1]$.
9. Find the Fourier sine series of $1 - x$ on $[0, 1]$, and and the following partial sum with 10 nonzero terms. Include the graph of $1 - x$ for comparison. Then plot the same functions on $[-1, 1]$.
10. Find the Fourier series of each of the following on $[-\pi, \pi]$. Are the results surprising?

    (a) $f(x) = \sin(x - \frac{\pi}{4})$ (b) $f(x) = \sin(3x)\cos(5x)$
11. Create a function `datafn` that interpolates random data on the interval $[-1, 1]$ by entering

```
data = Table[{x, Random[Real, {-1, 1}]}, {x, -1, 1, .1}];
datafn = Interpolation[data];
Plot[datafn[x], {x, -1, 1}];
```

Then, for $m = 4, 8, 12$, use `NFourierSum` to compute the mth partial sum of the Fourier series of `datafn`, and plot the result along with `datafn`. Repeat the whole process several times with different data. You should observe that in nearly all cases, the 12th partial sum does a reasonably good job of approximating `datafn`, while the 8th partial sum does a poor job. Why do you think this is?

10.2 Sturm-Liouville Eigenvalue Problems

Suppose that p, k, and q are continuous functions on $[a, b]$ with p and k positive and k' continuous. An operator $\mathcal{S}$ defined for C^2 functions ψ on $[a, b]$ by

$$\mathcal{S}\psi = \frac{1}{p(x)}\left((k(x)\,\psi')' + q(x)\,\psi\right)$$

is called a (regular) *Sturm-Liouville operator*. The simplest example has $p(x) = k(x) = 1$ and $q(x) = 0$ and hence is simply the second-derivative operator $\mathcal{S}\psi = \psi''$.

An *eigenvalue problem* associated with a (regular) Sturm-Liouville operator $\mathcal{S}$ consists of the equation

$$\mathcal{S}\psi = \lambda\,\psi,$$

along with a pair of boundary conditions of the form

$$\alpha\,\psi(a) - (1-\alpha)\,\psi'(a) = 0,$$

$$\beta\,\psi(b) + (1-\beta)\,\psi'(b) = 0,$$

where α and β are numbers in $[0, 1]$. Any number λ for which there is a nonzero solution ψ is called an *eigenvalue* of $\mathcal{S}$. A corresponding nonzero solution ψ is called an *eigenfunction* of $\mathcal{S}$.

Example 10.2.1

Let $\mathcal{S}\psi = k\,\psi''$, where k is a positive constant, and consider the eigenvalue problem

$$\mathcal{S}\psi = \lambda\,\psi,$$

$$\psi(0) = \psi(L) = 0.$$

If we try to solve the problem all at once with `DSolve`, we find only the zero solution, which is of no interest.

```
ψ[x] /. DSolve[{k ψ''[x] == λ ψ[x], ψ[0] == 0, ψ[L] == 0}, ψ[x], x]
{0}
```

The key is to use only one boundary condition as a first step.

```
soln = ψ[x] /. DSolve[{k ψ''[x] == λ ψ[x], ψ[0] == 0}, ψ[x], x][[1]]
```

$$-e^{-\frac{x\sqrt{\lambda}}{\sqrt{k}}}\left(-1 + e^{\frac{2x\sqrt{\lambda}}{\sqrt{k}}}\right) C[2]$$

Then the value at $x = L$ is

```
ψatL = soln /. x → L
```

$$-e^{-\frac{L\sqrt{\lambda}}{\sqrt{k}}}\left(-1 + e^{\frac{2L\sqrt{\lambda}}{\sqrt{k}}}\right) C[2]$$

So we want determine all λ for which this quantity is zero. Clearly no positive λ will work. With $\lambda = 0$, the expression is zero, but this leads only to $\psi = 0$. So all the eigenvalues must be negative, in which case the expression above is complex. For convenience, let $\lambda = -\omega^2$, and replace $\sqrt{\lambda}$ with $i\,\omega$ in the expression above.

```
ψatL = ψatL /. √λ → i ω
```

$$-e^{-\frac{i L \omega}{\sqrt{k}}} \left(-1 + e^{\frac{2 i L \omega}{\sqrt{k}}}\right) C[2]$$

Next apply `ExpToTrig`.

```
ExpToTrig[ψatL] // Simplify
```

$$-2\, i\, C[2]\, \mathrm{Sin}\left[\frac{L\,\omega}{\sqrt{k}}\right]$$

This reveals that $\lambda = -\omega^2$ is an eigenvalue, provided that

$$\frac{L\,\omega}{\sqrt{k}} = n\,\pi \text{ for some positive integer } n.$$

This gives a sequence of values for ω,

$$\omega_n = n\,\pi\,\sqrt{k}\,/L,\ n = 1, 2, 3, \ldots,$$

which in turn gives the desired sequence of eigenvalues,

$$\lambda_n = -n^2\,\pi^2\,k/L^2,\ n = 1, 2, 3, \ldots .$$

Corresponding (imaginary) eigenfunctions are

```
ExpToTrig[soln /. {√λ → i n π √k / L}] // Simplify
```

$$-2\, i\, C[2]\, \mathrm{Sin}\left[\frac{n\,\pi\,x}{L}\right]$$

Finally, setting `C[1]` equal to $i/2$ gives real eigenfunctions:

```
% /. C[2] → i / 2
```

$$\mathrm{Sin}\left[\frac{n\,\pi\,x}{L}\right]$$

Example 10.2.2

Let $\mathcal{S}\,\psi = \psi''$, where k is a positive constant, and consider the eigenvalue problem

$$\mathcal{S}\,\psi = \lambda\,\psi,$$

$$\psi(0) - \psi'(0) = \psi(1) = 0.$$

As in the previous example, if we try to solve the problem all at once with `DSolve`, we find only the zero solution.

```
ψ[x] /. DSolve[{ ψ''[x] == λ ψ[x], ψ[0] - ψ'[0] == 0, ψ[1] == 0}, ψ[x], x]
{0}
```

So we use only one boundary condition, the one at 0, as a first step.

```
soln = ψ[x] /.
  DSolve[{ψ''[x] == λ ψ[x], ψ[0] - ψ'[0] == 0}, ψ[x], x][[1]] // Simplify
```

$$\frac{e^{-x\sqrt{\lambda}}\left(-1+e^{2x\sqrt{\lambda}}\left(1+\sqrt{\lambda}\right)+\sqrt{\lambda}\right)C[2]}{-1+\sqrt{\lambda}}$$

Then the value at $x = 1$ is

```
ψat0 = soln /. x → 1
```

$$\frac{e^{-\sqrt{\lambda}}\left(-1+e^{2\sqrt{\lambda}}\left(1+\sqrt{\lambda}\right)+\sqrt{\lambda}\right)C[2]}{-1+\sqrt{\lambda}}$$

So we want determine all λ for which the quantity

$$z := -1 + e^{2\sqrt{\lambda}}\left(1+\sqrt{\lambda}\right)+\sqrt{\lambda}$$

is zero. A plot of the left side versus λ indicates that there are no positive solutions.

```
Plot[Evaluate[z], {λ, 0, 2}];
```

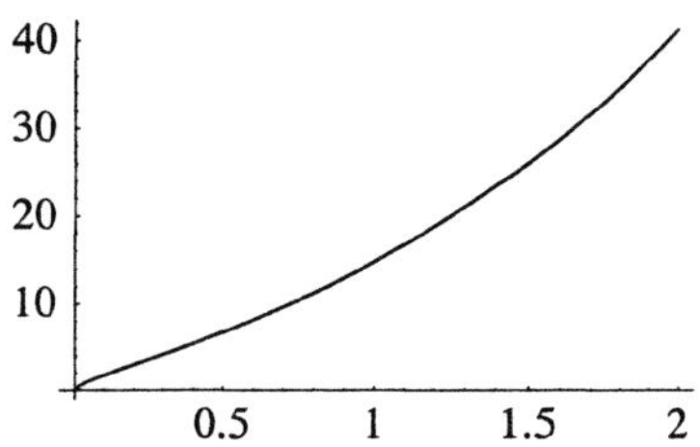

For further confirmation, note that the derivative with respect to λ is always positive when $\lambda > 0$.

```
∂λ z // Simplify
```

$$\frac{1+e^{2\sqrt{\lambda}}\left(3+2\sqrt{\lambda}\right)}{2\sqrt{\lambda}}$$

With $\lambda = 0$, the equation above is true, but this leads only to $\psi = 0$. So all the eigenvalues must be negative. For convenience, let $\lambda = -\omega^2$, and replace $\sqrt{\lambda}$ with $i\,\omega$.

```
z = z /. √λ → i ω
```

$$-1 + e^{2 i \omega}(1 + i\,\omega) + i\,\omega$$

Next apply `ExpToTrig`.

```
z = ExpToTrig[z] // Simplify
```

$$2\,i\,(\mathrm{Cos}[\omega] + i\,\mathrm{Sin}[\omega])\,(\omega\,\mathrm{Cos}[\omega] + \mathrm{Sin}[\omega])$$

This reveals that $\lambda = -\omega^2$ is an eigenvalue, provided that ω satisfies

$$\omega\cos\omega + \sin\omega = 0,$$

which for convenience we rewrite as

$$\omega = -\tan\omega.$$

This transcendental equation must be solved numerically. However, if we graph each side, we can see that there is an increasing sequence of positive solutions.

```
Plot[{ω, -Tan[ω]}, {ω, 0, 20}, PlotRange → {-10, 20}];
```

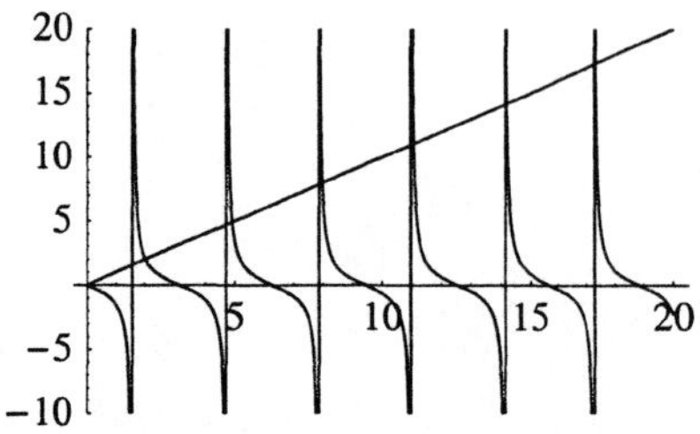

Given this sequence ω_1, ω_2, ω_3, ..., the resulting eigenvalues will be

$$\lambda_n = -\omega_n^2, \quad n = 1, 2, 3, \ldots,$$

and corresponding eigenfunctions will be

```
soln = ExpToTrig[soln /. {√λ → ⅈ ω}] // Simplify
```

$$\frac{2\,C[2]\,(\omega\,\text{Cos}[x\,\omega] + \text{Sin}[x\,\omega])}{ⅈ + \omega}$$

A judicious choice of `C[2]` makes these eigenfunctions real.

```
soln = soln /. C[2] → (ⅈ + ω)/2
```

```
ω Cos[x ω] + Sin[x ω]
```

Finally, let's compute the first several eigenvalues and eigenfunctions.

```
guesses = {2, 5, 8, 11.1, 14.3};
ωlist = Map[FindRoot[ω + Tan[ω] == 0, {ω, #}] &, guesses]
```

```
{{ω → 2.02876}, {ω → 4.91318}, {ω → 7.97867}, {ω → 11.0855}, {ω → 14.2074}}
```

```
λlist = -(ω /. ωlist)^2
```

```
{-4.11586, -24.1393, -63.6591, -122.889, -201.851}
```

```
ψfns = soln /. ωlist
```

```
{2.02876 Cos[2.02876 x] + Sin[2.02876 x],
 4.91318 Cos[4.91318 x] + Sin[4.91318 x], 7.97867 Cos[7.97867 x] + Sin[7.97867 x],
 11.0855 Cos[11.0855 x] + Sin[11.0855 x], 14.2074 Cos[14.2074 x] + Sin[14.2074 x]}
```

```
Plot[Evaluate[ψfns], {x, 0, 1}];
```

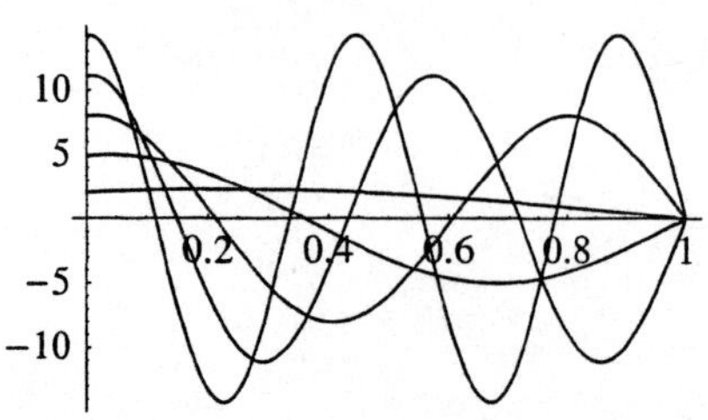

Example 10.2.3

Consider the eigenvalue problem

$$e^{-x}\left((e^{3x}\psi')' + 2e^{3x}\psi\right) = \lambda\psi,$$

$$\psi(0) = \psi(1) = 0.$$

First we find a solution that satisfies $\psi(0) = 0$.

```
de := e^-x (∂x (e^(3 x) ψ'[x]) + 2 e^(3 x) ψ[x]) == λ ψ[x];
soln = ψ[x] /. DSolve[{de, ψ[0] == 0}, ψ[x], x][[1]] // Simplify
```

$$-\frac{1}{4}\sqrt{e^{-2x}}\,\sqrt{\lambda}\,C[2]\,\mathrm{Sech}\left[\sqrt{\lambda}\right]\mathrm{Sinh}\left[\left(-1+\sqrt{e^{-2x}}\right)\sqrt{\lambda}\right]$$

The value of this solution at $x = 1$ is

```
ψat1 = soln /. x → 1
```

$$-\frac{\sqrt{\lambda}\,C[2]\,\mathrm{Sech}\left[\sqrt{\lambda}\right]\mathrm{Sinh}\left[\left(-1+\frac{1}{e}\right)\sqrt{\lambda}\right]}{4\,e}$$

This is nonzero if $\lambda > 0$; hence there are no positive eigenvalues. Also, $\lambda = 0$ leads only to the zero solution. Since all eigenvalues are therefore negative, let $\lambda = -\omega^2$.

```
soln = soln /. √λ → i ω
```

$$\frac{1}{4}\sqrt{e^{-2x}}\,\omega\,C[2]\,\mathrm{Sec}[\omega]\,\mathrm{Sin}\left[\left(-1+\sqrt{e^{-2x}}\right)\omega\right]$$

```
ψat1 = soln /. x → 1
```

$$\frac{\omega\,C[2]\,\mathrm{Sec}[\omega]\,\mathrm{Sin}\left[\left(-1+\frac{1}{e}\right)\omega\right]}{4\,e}$$

From this we observe that the boundary condition at 1 is satisfied if

$$\left(1-\frac{1}{e}\right)\omega = n\pi \quad \text{for some integer } n \geq 1.$$

Therefore we have the sequence of eigenvalues

$$\lambda_n = -\omega_n^2 = -\frac{n^2\pi^2 e^2}{(e-1)^2}, \quad n = 1, 2, 3, \ldots$$

with corresponding eigenfunctions

```
soln /. ω → (n π e)/(e - 1) // Simplify
```

$$\frac{e\sqrt{e^{-2x}}\,n\pi\,C[2]\,\mathrm{Sec}\left[\frac{e\,n\pi}{-1+e}\right]\mathrm{Sin}\left[\frac{e\left(-1+\sqrt{e^{-2x}}\right)n\pi}{-1+e}\right]}{4\,(-1+e)}$$

A judicious choice of `C[2]` yields the eigenfunctions

```
ψ[n_][x_] := e^-x Sin[(n π e)/(e - 1) (1 - e^-x)]
```

The first five of these are plotted as follows.

```
Plot[Evaluate[Table[ψ[n][x], {n, 1, 5}]], {x, 0, 1}];
```

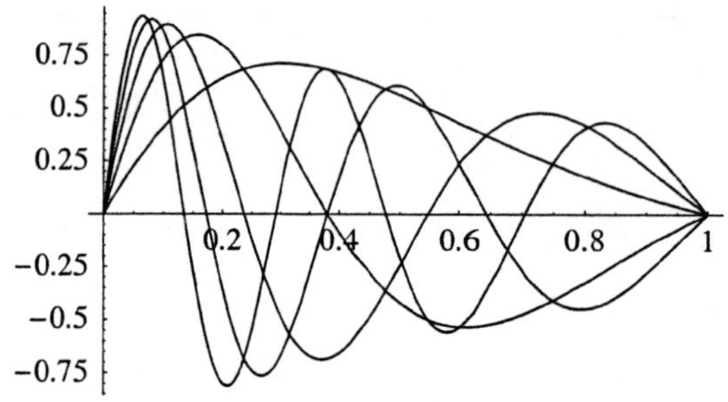

Exercises

Find the eigenvalues and eigenfunctions of the given Sturm-Liouville operator, and plot the first several eigenfunctions.

1. $\mathcal{S}\psi = \psi''$, $\psi(0) = 0$, $\psi'(1) = 0$
2. $\mathcal{S}\psi = \psi''$, $\psi(0) - \psi'(0) = 0$, $\psi'(\pi) = 0$
3. $\mathcal{S}\psi = \psi''$, $2\,\psi(0) - \psi'(0) = 0$, $\psi(0) + 3\,\psi'(\pi) = 0$
4. $\mathcal{S}\psi = e^{-3x}\,(e^{-x}\,\psi')'$, $\psi(0) = \psi(1) = 0$
5. $\mathcal{S}\psi = (x+1)^2\,((x+1)^2\,\psi')'$, $\psi(0) = \psi(1) = 0$
6. $\mathcal{S}\psi = \sqrt{2-x^2}\,\left(\sqrt{2-x^2}\,\psi'\right)'$, $\psi(-1) = \psi(1) = 0$
7. $\mathcal{S}\psi = (2-x^2)\,((2-x^2)\,\psi')'$, $\psi(0) = \psi(1) = 0$
8. $\mathcal{S}\psi = \frac{1}{1+x^2}\left(\frac{1}{1+x^2}\,\psi'\right)'$, $\psi(0) = \psi(1) = 0$

10.3 Eigenfunction Expansions

Eigenfunction expansions are generalized Fourier series in which the role of sines and cosines is played by the eigenfunctions of a Sturm-Liouville operator. Suppose that a regular Sturm-Liouville operator $\mathcal{S}$ is defined for C^2 functions on $[a, b]$ by

$$\mathcal{S}\psi = \frac{1}{p(x)}\,((k(x)\,\psi')' + q(x)\,\psi).$$

Given a sequence of eigenfunctions $\psi_0, \psi_1, \psi_2, \ldots$ corresponding to distinct eigenvalues $\lambda_0, \lambda_1, \lambda_2, \ldots$, any piecewise-continuous function f on $[a, b]$ has the eigenfunction expansion

$$\tilde{f}(x) = \sum_{n=0}^{\infty} c_n\,\psi_n(x), \text{ where } c_n = \frac{\int_a^b f(x)\,\psi_n(x)\,p(x)\,dx}{\int_a^b \psi_n^2(x)\,p(x)\,dx}, \; n = 0, 1, 2, \ldots .$$

The function p is called the *weight function.* Also, if the ψ_n are scaled so that the denominator in the coefficient formula is 1, then the ψ_n are said to be *normalized.*

Example 10.3.1

Consider the Sturm-Liouville eigenvalue problem

$$\frac{1}{1+x^2}\left(\frac{1}{1+x^2}\,\psi'\right)' = \lambda\,\psi,$$
$$\psi(0) = 0,\ \ \psi'(2) = 0.$$

and suppose that we want the associated eigenfunction expansion of

$$f(x) = (x-1)\,h(x-1),$$

where h is the Heaviside unit step function.

```
f[x_] := (x - 1) UnitStep[x - 1];
Plot[f[x], {x, 0, 2}, AspectRatio → Automatic, Ticks → {{1, 2}, {.5, 1}}];
```

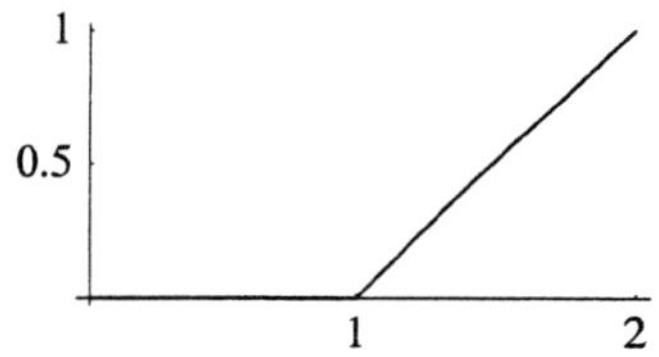

We first set about finding the eigenvalues and eigenfunctions. The boundary condition at 0 is satisfied by any of the following solutions.

```
de := 1/(1 + x²) ∂ₓ (1/(1 + x²) ψ'[x]) == λ ψ[x];
soln = ψ[x] /. DSolve[{de, ψ[0] == 0}, ψ[x], x][[1]] // Simplify
```

$i\,C[2]\,\text{Sinh}\left[\frac{1}{3}\,x\,(3+x^2)\,\sqrt{\lambda}\right]$

Now we examine the derivative at $x = 2$.

```
∂ₓ soln /. x → 2 // Simplify
```

$5\,i\,\sqrt{\lambda}\,C[2]\,\text{Cosh}\left[\frac{14\sqrt{\lambda}}{3}\right]$

This cannot be zero if $\lambda > 0$. Also, $\lambda = 0$ leads only to the zero solution. So λ must be negative; hence we set $\lambda = -\omega^2$, $\omega > 0$.

```
soln = soln /. √λ → i ω
```

$-C[2]\,\text{Sin}\left[\frac{1}{3}\,x\,(3+x^2)\,\omega\right]$

```
∂ₓ soln /. x → 2 // Simplify
```

$-5\,\omega\,C[2]\,\text{Cos}\left[\frac{14\,\omega}{3}\right]$

Thus the boundary condition at $x = 2$ is satisfied, provided that

$$\frac{14\omega}{3} = \frac{2n+1}{2}\pi \text{ for some integer } n \geq 0.$$

This leads to the sequence of eigenvalues

$$\lambda_n = -9\pi^2 \frac{(2n+1)^2}{28^2}, \quad n = 0, 1, 2, \ldots .$$

Corresponding eigenfunctions are

```
ψ[n_][x_] = soln /. {ω → 3 (2 n + 1) π / 28, C[2] → -1}
```

$\text{Sin}\left[\frac{1}{28}(1+2n)\pi x(3+x^2)\right]$

```
Plot[Evaluate[Table[ψ[n][x], {n, 0, 4}]], {x, 0, 2}];
```

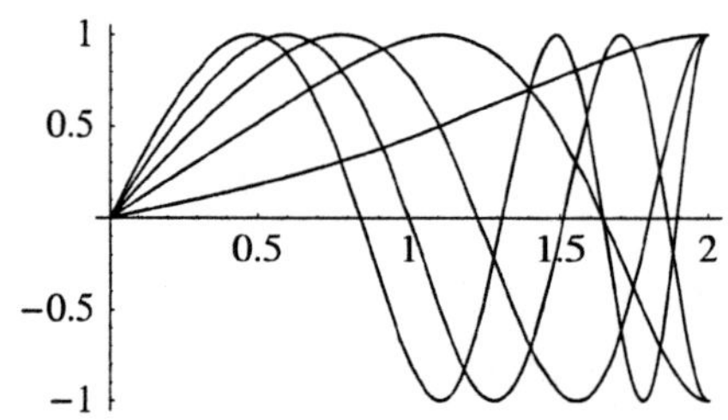

To normalize the eigenfunctions, we first compute

```
Simplify[∫_0^2 ψ[n][x]^2 (1 + x^2) dx, n ∈ Integers]
```

$\frac{7}{3}$

(Notice the presence of the weight function $p(x) = 1 + x^2$.) To make this be 1 instead, we multiply each eigenfunction by $\sqrt{3/7}$.

```
ψ[n_][x_] = Sqrt[3/7] ψ[n][x]
```

$\sqrt{\frac{3}{7}}\,\text{Sin}\left[\frac{1}{28}(1+2n)\pi x(3+x^2)\right]$

```
Simplify[∫_0^2 ψ[n][x]^2 (1 + x^2) dx, n ∈ Integers]
```

1

Now we're ready to compute the coefficients in the eigenfunction expansion of f. We need to use `NIntegrate`, since *Mathematica* won't be able to find an exact (i.e., symbolic) result. The following computes the first six coefficients and plots the resulting (partial) eigenfunction expansion. (Again note the presence of the weight function $p(x) = 1 + x^2$.)

```
coeffs = Table[NIntegrate[f[x] ψ[n][x] (1 + x^2), {x, 0, 2}], {n, 0, 5}]
{1.11222, -0.351822, -0.0507072, -0.0337232, 0.0375502, 0.0136755}
```

```
Plot[Evaluate[∑_{n=0}^{5} coeffs[[n + 1]] ψ[n][x]], {x, 0, 2},
  AspectRatio → Automatic, Ticks → {{1, 2}, {.5, 1}}];
```

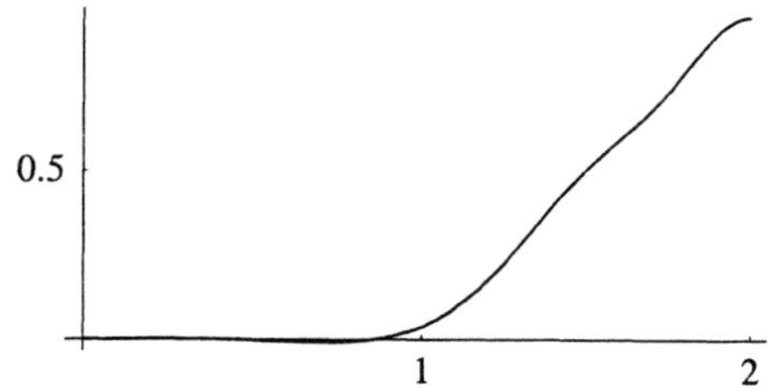

■ A Heat Conduction Problem

Let $y(x), 0 \leq x \leq 1$, be the steady-state temperature of a uniform, thin metal rod of length 1. The metal has thermal conductivity k, and one end is held at a fixed temperature $y(0) = 0$. The other end is perfectly insulated, giving rise to the boundary condition $y'(1) = 0$. Furthermore, heat is lost at a rate of $\rho y(x)$ (where $\rho > 0$) and generated at a rate of $f(x)$ along the entire length of the rod. Under these suppositions, y will satisfy the boundary-value problem

$$-k\,y'' + \rho y = f(x),$$

$$y(0) = 0, \;\; y'(1) = 0.$$

We assume further that f has the eigenfunction expansion

$$f(x) = \sum_{n=0}^{\infty} b_n\, \psi_n(x),$$

where the ψ_n are eigenfunctions associated with the eigenvalue problem

$$k\,\psi'' - \rho\,\psi = \lambda\,\psi,$$

$$\psi(0) = 0, \;\; \psi'(1) = 0.$$

Now suppose that the solution y has a similar eigenfunction expansion

$$y(x) = \sum_{n=0}^{\infty} c_n\, \psi_n(x).$$

Substituting this into the differential equation yields

$$-k\,y'' + \rho y = -\sum_{n=0}^{\infty} c_n (k\,\psi_n''(x) - \rho\,\psi_n(x)) = -\sum_{n=0}^{\infty} c_n\,\lambda_n\,\psi_n(x),$$

Therefore, y satisfies the differential equation, provided that

$$c_n = -\frac{b_n}{\lambda_n}, \;\; n = 0, 1, 2, \ldots .$$

Moreover, y *automatically satisfies the boundary conditions*, since all of the ψ_n do. Hence the solution is

$$y(x) = -\sum_{n=0}^{\infty} \frac{b_n}{\lambda_n}\, \psi_n(x).$$

Example 10.3.2

The boundary problem

$$-y'' + y = \sin \pi x,$$
$$y(0) = 0,\ y'(1) = 0,$$

is an example of the type of problems described above. To find a solution in the form of an eigenfunction expansion, we first need solve the eigenvalue problem

$$\psi'' - \psi = \lambda \psi,$$
$$\psi(0) = 0,\ \psi'(1) = 0.$$

Functions satisfying the differential equation and the boundary condition at 0 are as follows.

```
soln = ψ[x] /.
   DSolve[{ψ''[x] - ψ[x] == λ ψ[x], ψ[0] == 0}, ψ[x], x][[1]] // Simplify
```

$$-e^{-x\sqrt{1+\lambda}}\left(-1+e^{2x\sqrt{1+\lambda}}\right) C[2]$$

Next we examine the derivative at $x = 1$.

```
∂x soln /. x → 1 // Simplify
```

$$-e^{-\sqrt{1+\lambda}}\left(1+e^{2\sqrt{1+\lambda}}\right)\sqrt{1+\lambda}\ C[2]$$

This can be zero only if $1+\lambda \le 0$, and $\lambda = -1$ leads only to the zero solution. So let $1+\lambda = -\omega^2, \omega > 0$.

```
soln = soln /. √(1 + λ) → i ω // ExpToTrig // Simplify
```

$$-2\, i\, C[2]\, \mathrm{Sin}[x\,\omega]$$

Now we reexamine the derivative at $x = 1$.

```
∂x soln /. x → 1
```

$$-2\, i\, \omega\, C[2]\, \mathrm{Cos}[\omega]$$

This is zero if

$$\omega = \tfrac{2n+1}{2}\pi, \text{ for some integer } n \ge 0.$$

Thus we have eigenvalues given by

$$\lambda[n\_] := -\left(\frac{2\,n+1}{2}\pi\right)^2 + 1$$

Corresponding normalized eigenfunctions are

$$\psi[n\_][x\_] = \mathrm{Simplify}\left[\frac{\mathrm{Sin}[\frac{2n+1}{2}\pi x]}{\sqrt{\int_0^1 \mathrm{Sin}[\frac{2n+1}{2}\pi x]^2\, dx}},\ n \in \mathrm{Integers}\right]$$

$$\sqrt{2}\ \mathrm{Sin}\left[\frac{\pi x}{2} + n\,\pi\,x\right]$$

The coefficients in the eigenfunction expansion of $\sin \pi x$ are

```
b[n_] = ∫₀¹ Sin[π x] ψ[n][x] ⅆx // Simplify
```

$$\frac{4\sqrt{2}\,\mathrm{Cos}[n\,\pi]}{3\,\pi - 4\,n\,\pi - 4\,n^2\,\pi}$$

The following is the graph of the first 16 terms. Notice its horizontal tangent at the right endpoint.

```
Plot[Evaluate[∑_{n=0}^{15} b[n] ψ[n][x]], {x, 0, 1}, PlotRange → All];
```

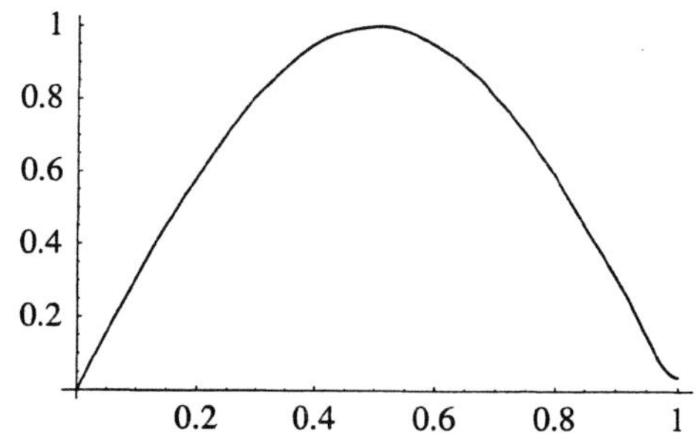

The first 16 terms in the expansion of the solution y are plotted as follows:

```
Plot[Evaluate[-∑_{n=0}^{15} b[n]/λ[n] ψ[n][x]], {x, 0, 1}, PlotRange → All];
```

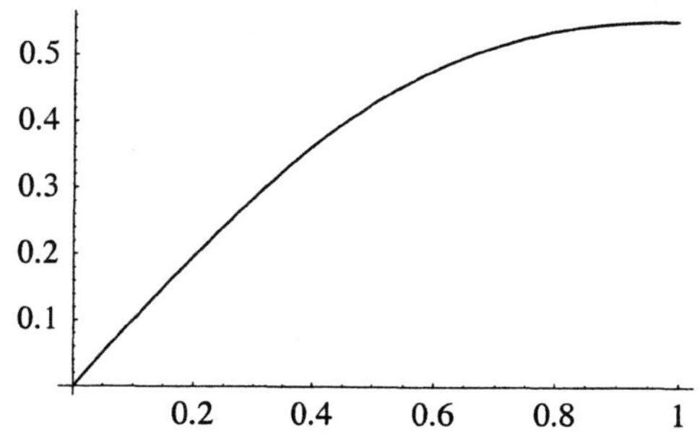

■ Exercises

The following are the same operators as in Exercises 1–8 in Section 10.2. In each case, compute, and plot, the 10th and 15th partial sums of the associated eigenfunction expansion of $f(x) = x$ on the appropriate interval. In most cases, you'll need to use `NIntegrate` to compute the coefficients.

1. $S\psi = \psi''$, $\psi(0) = 0$, $\psi'(1) = 0$

2. $S\psi = \psi''$, $\psi(0) - \psi'(0) = 0$, $\psi'(\pi) = 0$

3. $S\psi = \psi''$, $2\,\psi(0) - \psi'(0) = 0$, $\psi(0) + 3\,\psi'(\pi) = 0$

4. $S\psi = e^{-3x}\,(e^{-x}\,\psi')'$, $\psi(0) = \psi(1) = 0$

5. $\mathcal{S}\psi = (x+1)^2((x+1)^2\,\psi')'$, $\psi(0) = \psi(1) = 0$

6. $\mathcal{S}\psi = \sqrt{2-x^2}\left(\sqrt{2-x^2}\,\psi'\right)'$, $\psi(-1) = \psi(1) = 0$

7. $\mathcal{S}\,\psi = (2-x^2)\,((2-x^2)\,\psi')'$, $\psi(0) = \psi(1) = 0$

8. $\mathcal{S}\,\psi = \frac{1}{1+x^2}\left(\frac{1}{1+x^2}\,\psi'\right)'$, $\psi(0) = \psi(1) = 0$

In each of Exercises 9 and 10, find, and plot, the 12 th partial sum of an appropriate eigenfunction expansion of the solution.

9. $-y'' + y = x$, $y'(0) = 0$, $y'(1) = 0$

10. $-y'' + y = 1$, $y(0) - y'(0) = 0$, $y(1) = 0$

Chapter 11

Linear Partial Differential Equations

11.1 The 1-D Diffusion Problem

This section deals with the one-dimensional diffusion problem

$$\frac{\partial u}{\partial t} - k\,\frac{\partial^2 u}{\partial x^2} = 0 \quad \text{for } t > 0 \text{ and } 0 < x < L,$$

$$\left.\begin{aligned} \alpha\, u(t,0) - (1-\alpha)\,\frac{\partial u}{\partial x}(t,0) &= 0 \\ \beta\, u(t,L) + (1-\beta)\,\frac{\partial u}{\partial x}(t,L) &= 0 \end{aligned}\right\} \quad \text{for } t > 0,$$

$$u(0,x) = \varphi(x) \quad \text{for } 0 \le x \le L,$$

where the "diffusivity" k is a positive constant, and α, β are constants in $[0, 1]$. The associated Sturm-Liouville eigenvalue problem is

$$k\,\psi'' = \lambda\,\psi,$$
$$\alpha\,\psi(0) - (1-\alpha)\,\psi'(0) = 0$$
$$\beta\,\psi(L) + (1-\beta)\,\psi'(L) = 0.$$

Suppose that this solved by eigenvalues $\lambda_0, \lambda_1, \lambda_2, \ldots$ with corresponding eigenfunctions $\psi_0, \psi_1, \psi_2, \ldots$. Then, if the initial data function φ has the eigenfunction expansion

$$\varphi(x) = \sum_{n=0}^{\infty} b_n\,\psi_n(x),$$

it follows that the solution of the diffusion problem is

$$u(t,x) = \sum_{n=0}^{\infty} b_n\, e^{\lambda_n t}\,\psi_n(x).$$

Remark: It will simplify our discussions if we make note of the fact that the aforementioned eigenvalue problem has *only nonpositive eigenvalues*. Moreover, the eigenvalues are all *negative* except in the case where $\alpha = \beta = 0$ (i.e., zero-flux boundary conditions at both ends) in which the greatest eigenvalue $\lambda_0 = 0$ has the eigenfunction $\psi_0(x) = 1$.

Example 11.1.1

Let's solve the diffusion problem

$$\frac{\partial u}{\partial t} - \frac{\partial^2 u}{\partial x^2} = 0 \quad \text{for } t > 0 \text{ and } 0 < x < 1,$$
$$u(t,0) = \frac{\partial u}{\partial x}(t,1) = 0 \quad \text{for } t > 0,$$
$$u(0,x) = x^2 \quad \text{for } 0 \le x \le 1.$$

The first task is to solve the eigenvalue problem

$$\psi'' = \lambda \psi,$$
$$\psi(0) = \psi'(1) = 0.$$

Solutions satisfying $\psi'' = \lambda \psi$ and $\psi(0) = 0$ are

```
soln = ψ[x] /. DSolve[{ψ''[x] == λ ψ[x], ψ[0] == 0}, ψ[x], x][[1]]
```

$$-e^{-x\sqrt{\lambda}}\left(-1+e^{2x\sqrt{\lambda}}\right) C[2]$$

Since the eigenvalues will be negative, we set $\lambda = -\omega^2$ and use `ExpToTrig` to apply the Euler-DeMoivre formula.

```
soln = ExpToTrig[soln /. √λ → ⅈ ω] // Simplify
```

```
-2 ⅈ C[2] Sin[x ω]
```

The boundary condition $\psi'(1) = 0$ will be met if the following expression equals zero:

```
∂x soln /. x → 1
```

```
-2 ⅈ ω C[2] Cos[ω]
```

Thus we require that $\omega = \frac{2n+1}{2}\pi$ for some integer $n \geq 0$, and so the eigenvalues are given, for $n = 0, 1, 2, \ldots$, by

$$\lambda[n\_] := -\left(\frac{2n+1}{2}\pi\right)^2$$

and normalized eigenfunctions are given by

$$\psi[n\_][x\_] = \text{Simplify}\left[\frac{\text{Sin}\left[\frac{2n+1}{2}\pi x\right]}{\left(\int_0^1 \text{Sin}\left[\frac{2n+1}{2}\pi x\right]^2 dx\right)^{1/2}}, n \in \text{Integers}\right]$$

$$\sqrt{2}\,\text{Sin}\left[\frac{\pi x}{2} + n\pi x\right]$$

The coefficients in the eigenfunction expansion of x^2 are

$$b[n\_] = \text{Simplify}\left[\int_0^1 x^2\, \psi[n][x]\, dx, n \in \text{Integers}\right]$$

$$\frac{8\sqrt{2}\left(-2+(-1)^n(1+2n)\pi\right)}{(\pi+2n\pi)^3}$$

To check, let's plot the sum of the first six terms.

$$\varphi 5[x\_] = \sum_{n=0}^{5} b[n]\, \psi[n][x] \text{ // Simplify}$$

$$\frac{16(-2+\pi)\,\text{Sin}\left[\frac{\pi x}{2}\right]}{\pi^3} - \frac{16(2+3\pi)\,\text{Sin}\left[\frac{3\pi x}{2}\right]}{27\pi^3} + \frac{16(-2+5\pi)\,\text{Sin}\left[\frac{5\pi x}{2}\right]}{125\pi^3} - \frac{16(2+7\pi)\,\text{Sin}\left[\frac{7\pi x}{2}\right]}{343\pi^3} + \frac{16(-2+9\pi)\,\text{Sin}\left[\frac{9\pi x}{2}\right]}{729\pi^3} - \frac{16(2+11\pi)\,\text{Sin}\left[\frac{11\pi x}{2}\right]}{1331\pi^3}$$

```
Plot[{x², φ5[x]}, {x, 0, 1}];
```

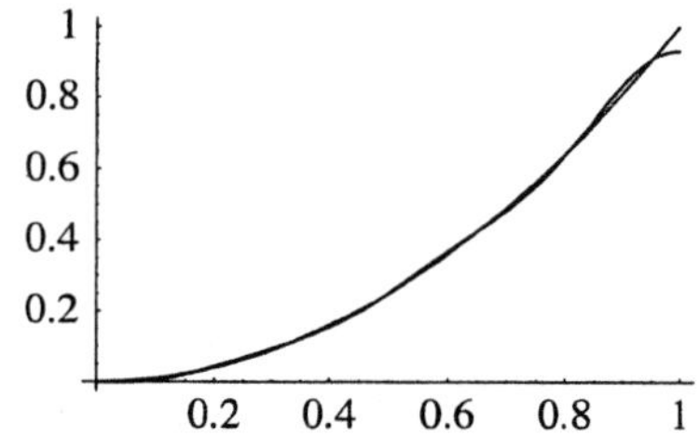

We next define the sum of the first $m + 1$ terms of the solution of the diffusion problem.

$$u[m\_][t\_, x\_] := \sum_{n=0}^{m} b[n]\, e^{\lambda[n]\, t}\, \psi[n][x]$$

Since we know the solution exactly at $t = 0$, let's also set

```
u[m_][0, x_] := x²
```

The following plot shows snapshots of an approximate solution at times $t = 0, 0.05, 0.10, \ldots, 0.25$.

```
Plot[Evaluate[Table[u[10][t, x], {t, 0, .25, .05}]],
  {x, 0, 1}, PlotRange → {0, 1}];
```

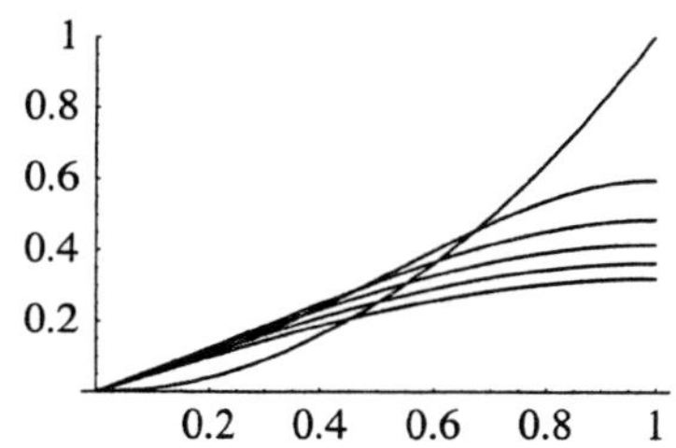

Here is a surface plot of the solution as a function of t and x.

```
Plot3D[Evaluate[u[10][t, x]], {t, 0, .5},
  {x, 0, 1}, BoxRatios → {2, 1.5, 1}, PlotRange → {0, 1.1},
  Ticks → {{0, .5}, {0, 1}, {1}}, AxesLabel → {t, x, u}];
```

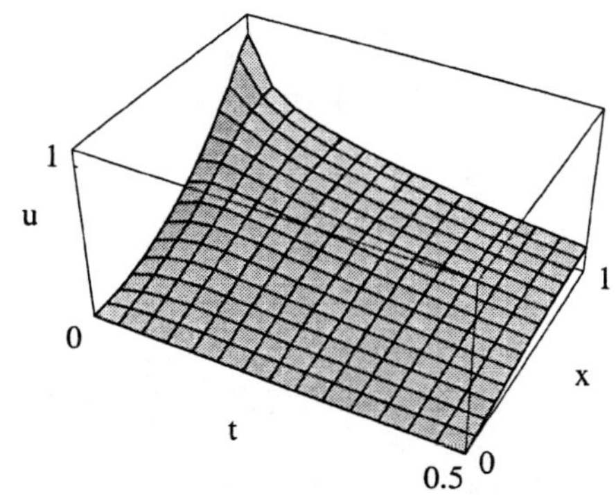

Example 11.1.2

Let's consider a variation on the problem in Example 11.1.1:

$$\frac{\partial u}{\partial t} - \frac{\partial^2 u}{\partial x^2} = 0 \quad \text{for } t > 0 \text{ and } 0 < x < 1,$$

$$u(t, 0) - \frac{\partial u}{\partial x}(t, 0) = \frac{\partial u}{\partial x}(t, 1) = 0 \quad \text{for } t > 0,$$

$$u(0, x) = x^2 \quad \text{for } 0 \le x \le 1.$$

The boundary conditions here are *Robin type* at $x = 0$ and *Neumann type* at $x = 1$. First we attack the associated eigenvalue problem

$$\psi'' = \lambda \psi,$$

$$\psi(0) - \psi'(0) = \psi'(1) = 0.$$

The differential equation and the boundary condition at $x = 0$ are satisfied by

```
soln = ψ[x] /. DSolve[{ψ''[x] == λ ψ[x], ψ[0] - ψ'[0] == 0}, ψ[x], x][[1]]
```

$$\frac{e^{-x\sqrt{\lambda}}\left(-1 + e^{2x\sqrt{\lambda}} + \sqrt{\lambda} + e^{2x\sqrt{\lambda}}\sqrt{\lambda}\right) C[2]}{-1 + \sqrt{\lambda}}$$

Since the eigenvalues will be negative, we set $\lambda = -\omega^2$.

```
soln = ExpToTrig[soln /. √λ → ⅈ ω] // Simplify
```

$$\frac{2\, C[2]\, (\omega \operatorname{Cos}[x\,\omega] + \operatorname{Sin}[x\,\omega])}{i + \omega}$$

For the boundary condition at $x = 1$ to be met, we demand that the following equal zero.

```
∂x soln /. x → 1
```

$$\frac{2\, C[2]\, (\omega \operatorname{Cos}[\omega] - \omega^2 \operatorname{Sin}[\omega])}{i + \omega}$$

Thus we find that ω must satisfy $\cos\omega = \omega \sin\omega$, i.e.,

$$\omega = \cot\omega.$$

A plot confirms that this equation has a sequence of positive solutions.

```
Plot[{ω, Cot[ω]}, {ω, 0, 20}, PlotRange → {-10, 20}];
```

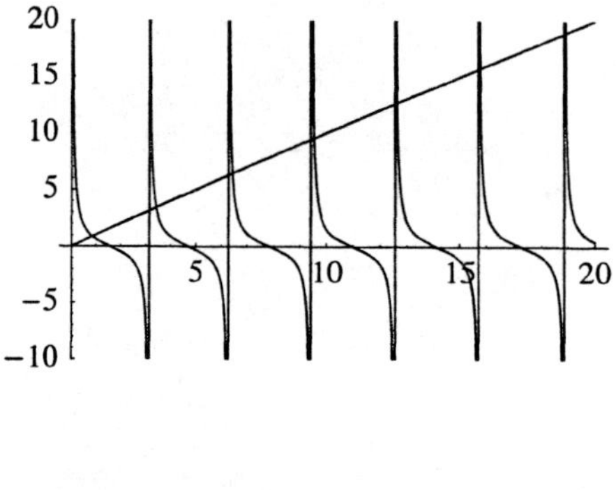

To find the first several values of ω, we'll need to use `FindRoot`, which requires an initial guess at each solution. The above plot indicates that each solution is slightly greater than a multiple of π. So let's define a list of ten guesses as follows:

```
guesses = π Range[0, 9] + .1
```

```
{0.1, 3.24159, 6.38319, 9.52478, 12.6664,
 15.808, 18.9496, 22.0911, 25.2327, 28.3743}
```

The corresponding list of solutions is found by *mapping* `FindRoot` through the list of guesses.

```
ωlist = ω /. Map[FindRoot[ω == Cot[ω], {ω, #}] &, guesses]
```

```
{0.860334, 3.42562, 6.4373, 9.52933, 12.6453,
 15.7713, 18.9024, 22.0365, 25.1724, 28.3096}
```

For convenience, let's now define ω and λ as follows:

```
ω[n_] := ωlist[[n + 1]]
```

```
λ[n_] := -ω[n]^2
```

Now we define normalized eigenfunctions.

$$\psi[n\_][x\_] := \frac{\omega[n]\ \mathrm{Cos}[\omega[n]\ x] + \mathrm{Sin}[\omega[n]\ x]}{\sqrt{\int_0^1 (\omega[n]\ \mathrm{Cos}[\omega[n]\ x] + \mathrm{Sin}[\omega[n]\ x])^2\, dx}}$$

```
Plot[Evaluate[Table[ψ[n][x], {n, 0, 3}]], {x, 0, 1}];
```

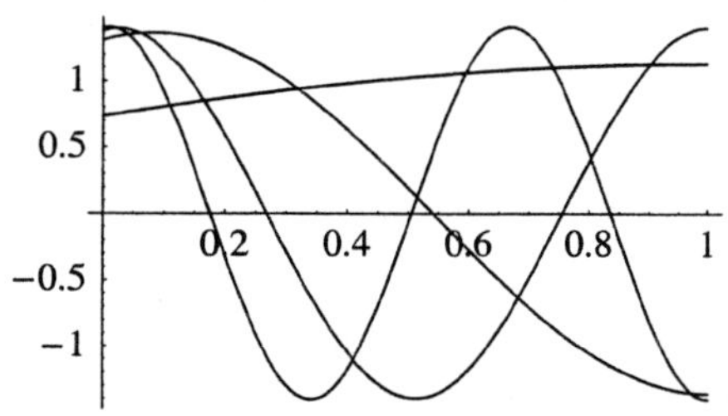

We next define the coefficients in the eigenfunction expansion of the initial data,

$$b[n\_] := \int_0^1 x^2\ \psi[n][x]\, dx$$

and then the sum of the first $m + 1$ terms of the solution of the diffusion problem:

$$u[m\_][t\_, x\_] := \sum_{n=0}^{m} b[n]\ e^{\lambda[n]\ t}\ \psi[n][x]$$

Since we know the solution exactly at $t = 0$, let's also set

$$u[m\_][0, x\_] := x^2$$

The following plot shows snapshots of an approximate solution at times $t = 0, 0.05, 0.10, \ldots, 0.25$.

```
Plot[Evaluate[Table[u[6][t, x], {t, 0, .25, .05}]],
  {x, 0, 1}, PlotRange → All];
```

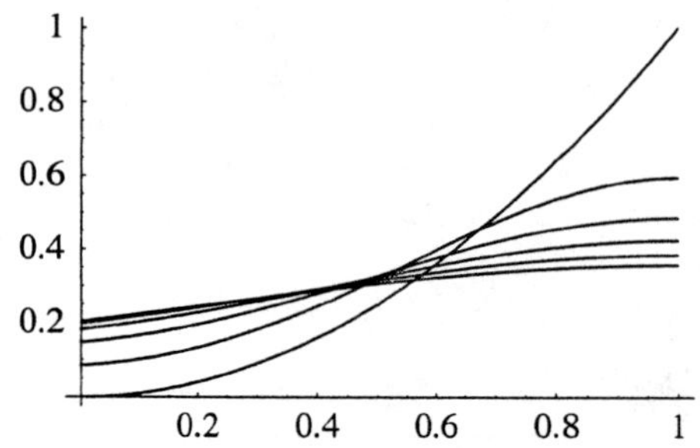

The following is a surface plot of the solution as a function of t and x.

```
Plot3D[Evaluate[u[6][t, x]], {t, 0, .25},
  {x, 0, 1}, PlotRange → {0, 1.05}, BoxRatios → {4, 3, 2},
  Ticks → {{0, .1, .2}, {0, 1}, {1}}, AxesLabel → {t, x, u}];
```

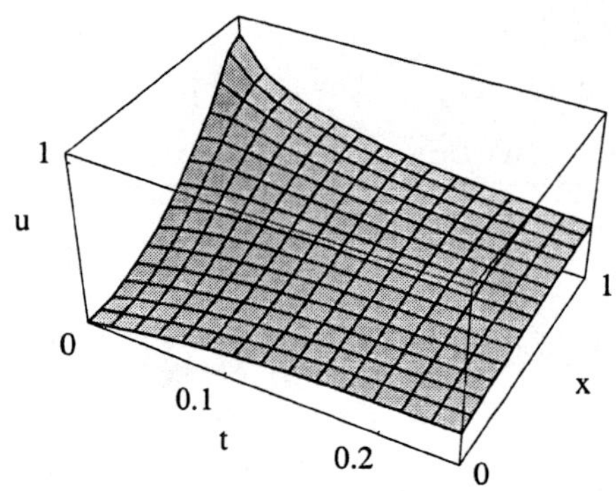

Example 11.1.3

In this example we'll consider the following diffusion problem with a nonhomogeneous differential equation (due to a *source term*), a nonhomogeneous boundary condition, and zero initial data.

$$\frac{\partial u}{\partial t} - \frac{\partial^2 u}{\partial x^2} = 10\cos\left(\frac{\pi}{2}\,x\right) \quad \text{for } t > 0 \text{ and } 0 < x < 1,$$

$$u(t, 0) = 0, \quad u(t, 1) = 1 \quad \text{for } t > 0,$$

$$u(0, x) = 0 \quad \text{for } 0 \le x \le 1.$$

The solution will be the superposition of two pieces,

$$u = v + w,$$

where w satisfies the time-independent problem

$$-w'' = 10\cos\left(\frac{\pi}{2}\,x\right) \quad \text{for } 0 < x < 1,$$

$$w(0) = 0, \quad w(1) = 1,$$

and v satisfies the following homogeneous problem with initial data given by $-w$:

$$\frac{\partial v}{\partial t} - \frac{\partial^2 v}{\partial x^2} = 0 \quad \text{for } t > 0 \text{ and } 0 < x < 1,$$
$$v(t, 0) = v(t, 1) = 0 \quad \text{for } t > 0,$$
$$v(0, x) = -w \quad \text{for } 0 \le x \le 1.$$

Be sure to convince yourself that $v + w$ will indeed satisfy the original diffusion problem.

Let's begin by using `DSolve` to find w.

```
Clear[w]; w[x_] = w[x] /.
  DSolve[{-w''[x] == 10 Cos[π x / 2], w[0] == 0, w[1] == 1}, w[x], x][[1]]
```

$$\frac{-40 + 40\,x + \pi^2\,x + 40\,\text{Cos}[\frac{\pi x}{2}]}{\pi^2}$$

```
Plot[w[x], {x, 0, 1}, PlotRange → All];
```

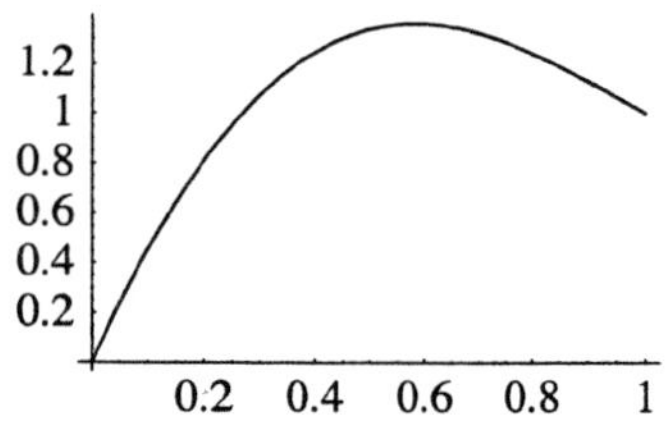

Now, to solve for v, we define eigenvalues and normalized eigenfunctions that solve the eigenvalue problem $\psi'' = \lambda\psi$, $\psi(0) = \psi(1) = 0$.

```
λ[n_] := -n² π²;
ψ[n_][x_] := √2 Sin[n π x]
```

Next we define the coefficients in the eigenfunction expansion of $-w$:

```
b[n_] := ∫₀¹ -w[x] ψ[n][x] dx
```

Now the first m terms of the eigenfunction expansion of v are defined by

```
v[m_][t_, x_] := Σ_{n=1}^{m} b[n] e^(λ[n] t) ψ[n][x]
```

and so the first m terms of the eigenfunction expansion of the solution u are defined by

```
u[m_][t_, x_] := v[m][t, x] + w[x]
```

The following plot shows snapshots of an approximate solution at $t = 0.003,\ 0.023,\ \ldots,\ 0.103$.

```
Plot[Evaluate[Table[u[10][t, x], {t, 0.003, .103, .02}]], {x, 0, 1}];
```

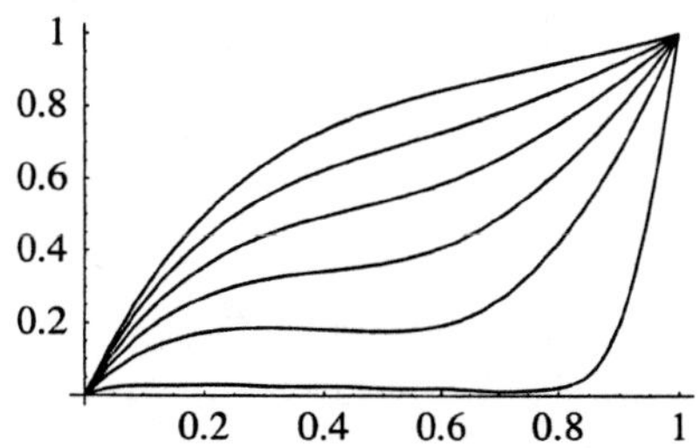

■ Exercises

Each of the following is a diffusion problem posed on the interval [0, 1]. In a manner similar to the examples in this section, create a plot showing the solution at an appropriately chosen sequence of times.

1. $\frac{\partial u}{\partial t} - \frac{\partial^2 u}{\partial x^2} = 0$
   $u(t, 0) = \frac{\partial u}{\partial x}(t, 1) = 0$
   $u(0, x) = x \sin^2\left(\frac{5\pi}{2} x\right)$

2. $\frac{\partial u}{\partial t} - \frac{\partial^2 u}{\partial x^2} = 0$
   $\frac{\partial u}{\partial x}(t, 0) = \frac{\partial u}{\partial x}(t, 1) = 0$
   $u(0, x) = x^2$

3. $\frac{\partial u}{\partial t} - \frac{\partial^2 u}{\partial x^2} = 0$
   $u(t, 0) = u(t, 1) = 0$
   $u(0, x) = x \sin^2(5\pi x)$

4. $\frac{\partial u}{\partial t} - \frac{\partial^2 u}{\partial x^2} = 0$
   $u(t, 0) - \frac{\partial u}{\partial x}(t, 0) = \frac{\partial u}{\partial x}(t, 1) = 0$
   $u(0, x) = x \sin^2(5\pi x)$

5. $\frac{\partial u}{\partial t} - \frac{\partial^2 u}{\partial x^2} + u = 0$
   $u(t, 0) = u(t, 1) = 1$
   $u(0, x) = 1 + \sin \pi x$

6. $\frac{\partial u}{\partial t} - \frac{\partial^2 u}{\partial x^2} = 2$
   $u(t, 0) = 0, \quad \frac{\partial u}{\partial x}(t, 1) = 0$
   $u(0, x) = 0$

11.2 The 2-D Laplace Equation

For convenience, we introduce the following standard notation for the two-dimensional *Laplace operator*:

$$\Delta u = \frac{\partial^2 u}{\partial x^2} + \frac{\partial^2 u}{\partial y^2}.$$

Given a two-dimensional domain Ω with boundary $\partial\Omega$, we wish to find a function u that satisfies the *Dirichlet problem for Laplace's equation*:

$$\Delta u = 0 \quad \text{for } (x, y) \in \Omega,$$

$$u(x, y) = g(x, y) \quad \text{for } (x, y) \in \partial\Omega.$$

One example of where such problems occur is the steady-state temperature distribution on a uniform, thin plate that is perfectly insulated on both faces and has a known temperature at each point along its edge.

■ Rectangular Domain

Consider the Dirichlet problem

$$\Delta u = 0 \quad \text{for } (x, y) \in \Omega,$$

$$u(x, y) = g(x, y) \text{ for } (x, y) \in \partial\Omega$$

where Ω is the rectangle where $0 < x < L,\ 0 < y < H$. Thus $\partial\Omega$ consists of the four edges of the rectangle, and the value of the solution is given along each of these edges as depicted below.

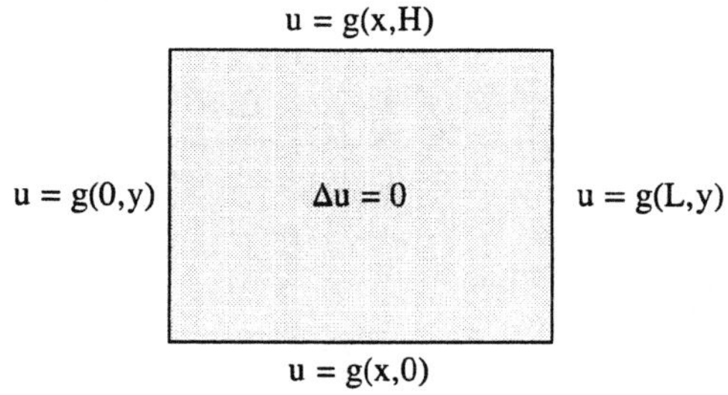

Let g have the following Fourier sine-series expansions along the respective edges of Ω :

$$g(x,0) = \sum_{n=1}^{\infty} a_n \sin(\tfrac{n\pi}{L}x),\quad g(x,H) = \sum_{n=1}^{\infty} b_n \sin(\tfrac{n\pi}{L}x),$$

$$g(0,y) = \sum_{n=1}^{\infty} c_n \sin(\tfrac{n\pi}{H}y),\quad g(L,y) = \sum_{n=1}^{\infty} d_n \sin(\tfrac{n\pi}{H}y).$$

The solution is then the superposition $u = u_a + u_b + u_c + u_d$ of the following pieces:

$$u_a(x,y) = \sum_{n=1}^{\infty} a_n \sin(\tfrac{n\pi}{L}x)\,\frac{\sinh(\frac{n\pi}{L}(H-y))}{\sinh(\frac{n\pi}{L}H)},\quad u_b(x,y) = \sum_{n=1}^{\infty} b_n \sin(\tfrac{n\pi}{L}x)\,\frac{\sinh(\frac{n\pi}{L}y)}{\sinh(\frac{n\pi}{L}H)},$$

$$u_c(x,y) = \sum_{n=1}^{\infty} c_n \sin(\tfrac{n\pi}{H}y)\,\frac{\sinh(\frac{n\pi}{H}(L-x))}{\sinh(\frac{n\pi}{H}L)},\quad u_d(x,y) = \sum_{n=1}^{\infty} d_n \sin(\tfrac{n\pi}{H}y)\,\frac{\sinh(\frac{n\pi}{H}x)}{\sinh(\frac{n\pi}{H}L)}.$$

Example 11.2.1

Let Ω be the rectangle in which $0 < x < 3$ and $0 < y < 3$, and suppose that boundary data are given by

$$g(x,0) = g(x,3) = \begin{cases} 4\,(x-1)\,(x-2), & \text{if } 1 \le x \le 2, \\ 0, & \text{otherwise,} \end{cases}$$

$$g(0,y) = g(3,y) = \begin{cases} 8\,(y-1)\,(2-y), & \text{if } 1 \le y \le 2 \\ 0, & \text{otherwise.} \end{cases}$$

We first compute Fourier sine coefficients for g along each edge.

```
L = 3; H = 3;
a[n_] = Simplify[2/L ∫_1^2 4 (x - 1) (x - 2) Sin[n π x/L] dx, n ∈ Integers];
b[n_] = Simplify[2/L ∫_1^2 4 (x - 1) (x - 2) Sin[n π x/L] dx, n ∈ Integers];
c[n_] = Simplify[2/H ∫_1^2 8 (y - 1) (2 - y) Sin[n π y/H] dy, n ∈ Integers];
d[n_] = Simplify[2/H ∫_1^2 8 (y - 1) (2 - y) Sin[n π y/H] dy, n ∈ Integers];
```

Next we define the pieces that we'll superimpose to form the solution.

```
ua[x_, y_] := Sum[a[n] Sin[n π x / L] Sinh[n π (H - y) / L] / Sinh[n π H / L], {n, 1, nterms}];

ub[x_, y_] := Sum[b[n] Sin[n π x / L] Sinh[n π y / L] / Sinh[n π H / L], {n, 1, nterms}];

uc[x_, y_] := Sum[c[n] Sin[n π y / H] Sinh[n π (L - x) / H] / Sinh[n π L / H], {n, 1, nterms}];

ud[x_, y_] := Sum[d[n] Sin[n π y / H] Sinh[n π x / H] / Sinh[n π L / H], {n, 1, nterms}];
```

Fifteen terms of each piece are superimposed and plotted as follows:

```
nterms := 15;
u[x_, y_] = ua[x, y] + ub[x, y] + uc[x, y] + ud[x, y];
Plot3D[u[x, y], {x, 0, L}, {y, 0, H},
  PlotRange → All, PlotPoints → 25, BoxRatios → Automatic];
```

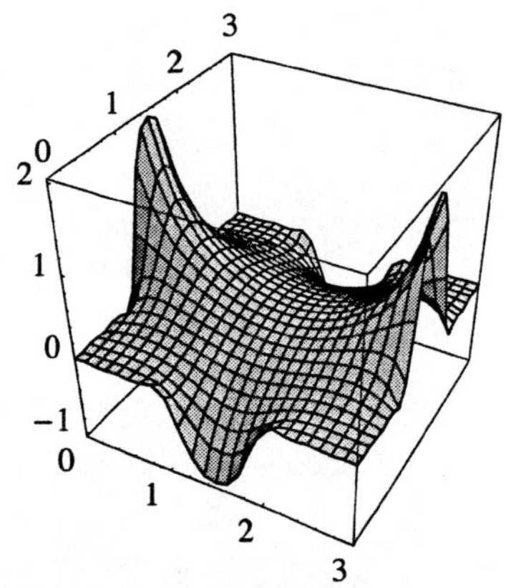

A contour plot also provides a good way to visualize the solution.

```
ContourPlot[Evaluate[u[x, y]], {x, 0, L}, {y, 0, H},
  ContourShading → False, AspectRatio → Automatic, PlotPoints → 30];
```

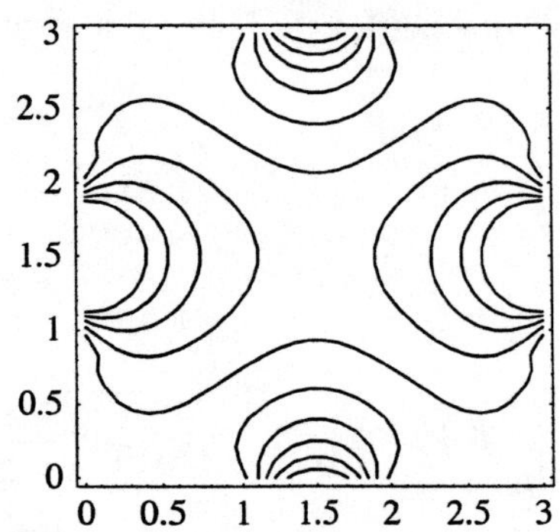

Other Boundary Conditions

Consider the problem indicated in the figure, where Neumann—or "zero flux"—boundary conditions are imposed on two adjacent edges.

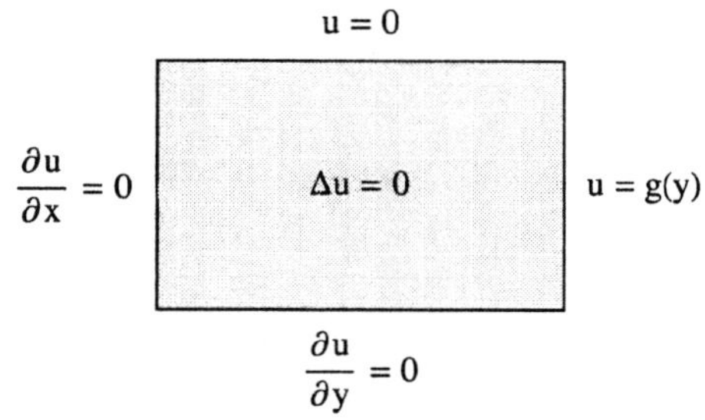

Separation of variables produces the (easily verified) solution

$$u(x, y) = \sum_{n=1}^{\infty} a_n \cos(\tfrac{(2n-1)\pi}{2H} y) \frac{\cosh(\frac{(2n-1)\pi}{2H} x)}{\cosh(\frac{(2n-1)\pi}{2H} L)},$$

where the a_n are given by

$$a_n = \tfrac{2}{H} \int_0^L \cos(\tfrac{(2n-1)\pi}{2H} y)\, g(y)\, dy.$$

Example 11.2.2

Let $g(y) = y(3 - y)$, and consider, on the rectangle $0 \le x \le 2, 0 \le y \le 3$, the problem just described. The necessary coefficients are computed as follows:

```
L = 2; H = 3;
a[n_] = Simplify[2/H ∫_0^H y (H - y) Cos[((2 n - 1) π y)/(2 H)] dy, n ∈ Integers]
```

$$-\frac{72\,(4\,(-1)^n + (-1 + 2\,n)\,\pi)}{(-1 + 2\,n)^3\,\pi^3}$$

Now let's compute 20 terms of the solution and plot the result.

```
u[x_, y_] := Sum_{n=1}^{20} a[n] Cos[((2 n - 1) π y)/(2 H)] Cosh[((2 n-1) π)/(2 H) x]/Cosh[((2 n-1) π)/(2 H) L];

Plot3D[Evaluate[u[x, y]], {x, 0, L}, {y, 0, H}, BoxRatios → Automatic];
```

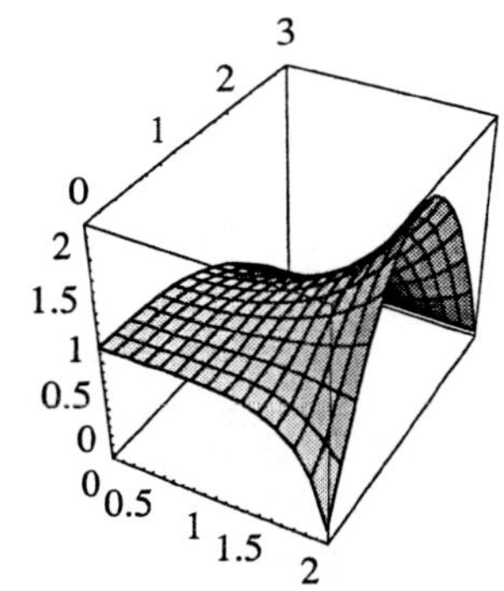

```
ContourPlot[Evaluate[u[x, y]], {x, 0, L}, {y, 0, H},
  ContourShading → False, AspectRatio → Automatic, PlotPoints → 30];
```

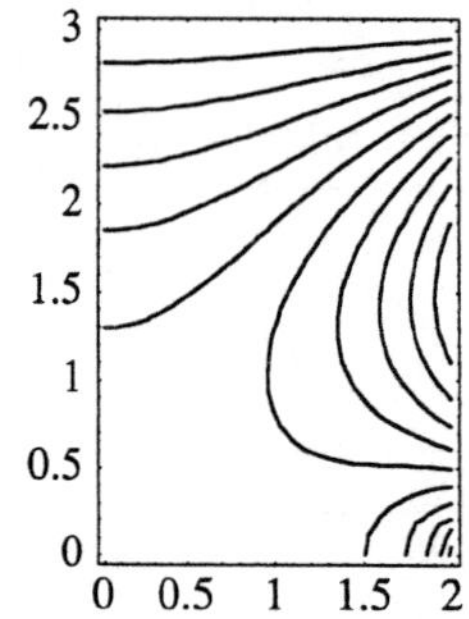

■ Circular Domain

The Laplace operator $\Delta u = \frac{\partial^2 u}{\partial x^2} + \frac{\partial^2 u}{\partial y^2}$ is given in polar coordinates by

$$\Delta u = \frac{\partial^2 u}{\partial r^2} + \frac{1}{r}\frac{\partial u}{\partial r} + \frac{1}{r^2}\frac{\partial^2 u}{\partial \theta^2}.$$

We consider here the Dirichlet problem on a disk of radius ρ:

$$\Delta u = 0 \quad \text{for } 0 < r < \rho,\ -\infty < \theta < \infty,$$

$$u(\rho, \theta) = g(\theta) \quad \text{for } -\infty < \theta < \infty,$$

$$u \text{ and } \frac{\partial u}{\partial r} \text{ bounded for } 0 \le r \le \rho,\ -\infty < \theta < \infty.$$

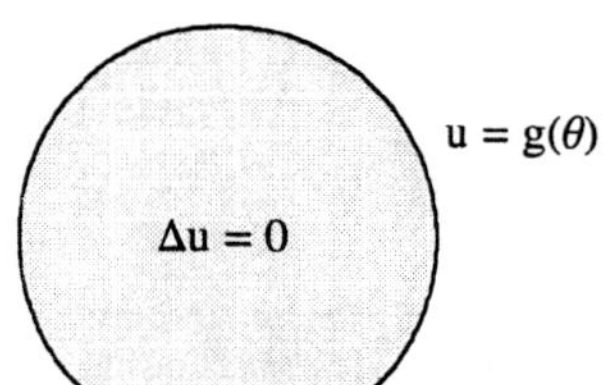

We assume that the boundary data $g(\theta)$ have a 2π-periodic Fourier series representation

$$g(\theta) = \frac{1}{2}a_0 + \sum_{n=1}^{\infty}(a_n \cos n\theta + b_n \sin n\theta).$$

Then the solution of the Dirichlet problem is

$$u(r, \theta) = \frac{1}{2}a_0 + \sum_{n=1}^{\infty}\left(\frac{r}{\rho}\right)^n (a_n \cos n\theta + b_n \sin n\theta).$$

Example 11.2.3

A simple case of the problem described above is one in which Ω is the unit disk, and the solution along its boundary is given by

$$g(\theta) = \sin 3\theta.$$

Since g is already in the desired Fourier-series form, the solution of the problem is simply

$$u(r, \theta) = r^3 \sin 3\theta.$$

To plot the solution, we use `ParametricPlot3D` as follows:

```
ParametricPlot3D[{r Cos[θ], r Sin[θ], r³ Sin[3 θ]}, {r, 0, 1},
  {θ, 0, 2 π}, PlotRange → All, PlotPoints → {15, 45}];
```

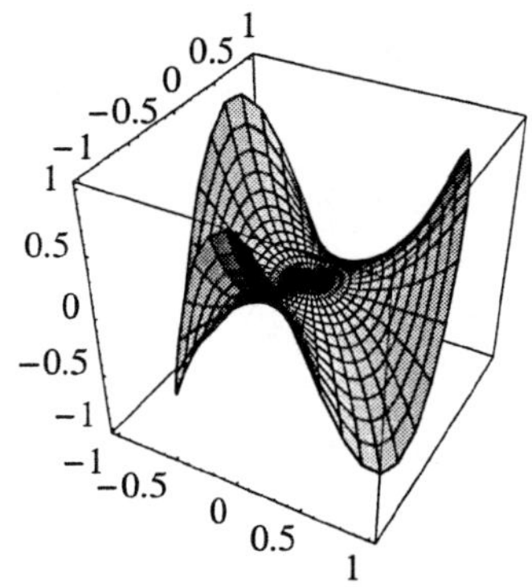

Example 11.2.4

Consider the Dirichlet problem on the unit disk with boundary data given by

$$g(\theta) = \begin{cases} \frac{2}{\pi}\,\theta, & \text{if } -\frac{\pi}{2} \le \theta \le \frac{\pi}{2}, \\ 0, & \text{elsewhere in } [-\pi, \pi]. \end{cases}$$

Our first task is to compute Fourier coefficients. Since g is an odd function, its Fourier series will be its Fourier sine series; so we define the coefficients by entering

```
b[n_] = Simplify[2/π (∫_0^(π/2) 2/π θ Sin[n θ] dθ), n ∈ Integers]
```

$$\frac{-2\, n\, \pi \,\text{Cos}[\frac{n\pi}{2}] + 4\, \text{Sin}[\frac{n\pi}{2}]}{n^2\, \pi^2}$$

Now let's compute and plot the sum of the first fifty terms of the solution's Fourier series.

```
u[r_, θ_] = Σ_(n=1)^50 b[n] r^n Sin[n θ];

ParametricPlot3D[{r Cos[θ], r Sin[θ], u[r, θ]}, {r, 0, 1},
  {θ, 0, 2 π}, PlotRange → All, PlotPoints → {10, 60}];
```

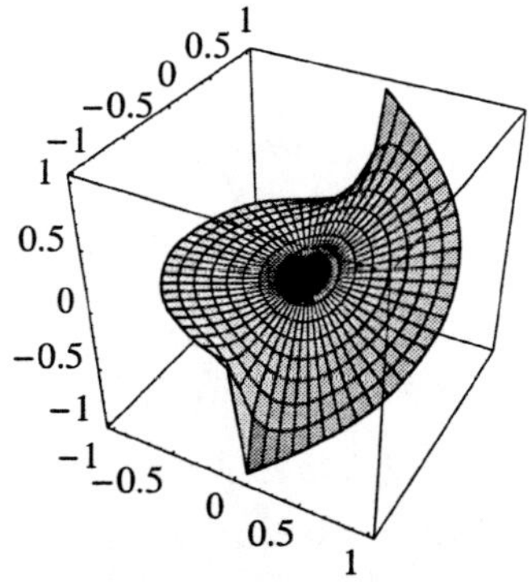

■ Exercises

Let Ω be the rectangle in which $0 < x < 1$ and $0 < y < 1$. For each set of boundary data in Exercises 1–3, create a surface plot and a contour plot of the solution of the Dirichlet problem for Laplace's equation.

1. $g(x, 0) = x, \ \ g(x, 1) = 1 - x, \ \ g(0, y) = y, \ \ g(1, y) = 1 - y$
2. $g(x, 0) = 0, \ \ g(x, 1) = 1 - x, \ \ g(0, y) = y^3, \ \ g(1, y) = 0$
3. $g(x, 0) = 0, \ \ g(x, 1) = 0, \ \ g(0, y) = 1 - |2y - 1|, \ \ g(1, y) = 0$

Let Ω be the unit disk. For each set of boundary data in Exercises 4–6, create a surface plot of the solution of the Dirichlet problem for Laplace's equation.

4. $g(\theta) = \begin{cases} 0, & \text{if } -\frac{\pi}{2} \le \theta < \frac{\pi}{2} \\ 1, & \text{otherwise} \end{cases}$

5. $g(\theta) = \begin{cases} 0, & \text{if } -\frac{\pi}{4} \le \theta < \frac{\pi}{4} \\ 1, & \text{otherwise} \end{cases}$

6. $g(\theta) = \begin{cases} 1, & \text{if } \frac{\pi}{3} \le \theta < \frac{2\pi}{3} \\ -1, & \text{if } -\frac{2\pi}{3} \le \theta < -\frac{\pi}{3} \\ 0, & \text{otherwise} \end{cases}$

7. Let Ω be the quarter-disk described by $0 < r < 1, 0 < \theta < \frac{\pi}{2}$, and consider the problem

$$\begin{aligned} \Delta u = 0 \ \ &\text{for } 0 < r < 1, \ 0 < \theta < \tfrac{\pi}{2}, \\ \tfrac{\partial u}{\partial \theta}(r, 0) = u(r, \tfrac{\pi}{2}) = 0 \ \ &\text{for } 0 \le r < 1, \\ u(1, \theta) = g(\theta) \ \ &\text{for } 0 \le \theta \le \tfrac{\pi}{2}. \end{aligned}$$

(a) Verify that the differential equation and the boundary conditions $\frac{\partial u}{\partial \theta}(r, 0) = u(r, \frac{\pi}{2}) = 0$ are satisfied by any convergent series of the form

$$u(r, \theta) = \sum_{n=1}^{\infty} b_n \, r^{2n-1} \cos((2n - 1)\theta).$$

(b) Suppose that $g(\theta) = \frac{1}{3}(\cos\theta - \cos 7\theta)$. Find and plot the solution of the Dirichlet problem.

8. Let Ω be the annular region described by $1 < r < \rho$, and consider the Dirichlet problem

$$\begin{aligned} \Delta u = 0 \ \ &\text{for } 1 < r < \rho, -\infty < \theta < \infty, \\ u(1, \theta) = 0 \ \ &\text{for } -\infty < \theta < \infty, \\ u(\rho, \theta) = g(\theta) \ \ &\text{for } -\infty < \theta < \infty. \end{aligned}$$

(a) Verify that the differential equation and the inner boundary condition $u(1, \theta) = 0$ are satisfied by any convergent series of the form

$$u(r, \theta) = \sum_{n=1}^{\infty} b_n (r^n - r^{-n}) \sin n\theta.$$

(b) Suppose that $\rho = 2$ and $g(\theta) = \sin 4\theta$. Find and plot the solution of the Dirichlet problem.

9. A nonhomogeneous Laplace equation is often called *Poisson's equation*. With a zero boundary condition, the boundary-value problem for Poisson's equation takes the form

$$\begin{aligned} -\Delta u = f(x, y) \ \ &\text{for } (x, y) \in \Omega, \\ u(x, y) = 0 \ \ &\text{for } (x, y) \in \partial\Omega. \end{aligned}$$

Suppose that Ω is the unit square ($0 < x < 1$, $0 < y < 1$) and that

$$f(x, y) = \sin \pi y \, (\sin \pi x - \sin 3\pi x + \tfrac{1}{5} \sin 5\pi x).$$

Find the solution in the form

$$u(x, y) = \sin \pi y \, (b_1 \sin \pi x + b_2 \sin 3\pi x + b_3 \sin 5\pi x),$$

and plot the result.

11.3 The 2-D Diffusion Problem

In this section we will take a look at the two-dimensional diffusion problem:

$$\begin{aligned}
\frac{\partial u}{\partial t} - k \, \Delta u &= 0 \quad \text{for } t > 0, \ (x, y) \in \Omega, \\
\alpha \, u + (1 - \alpha) \, \frac{\partial u}{\partial \mathbf{n}} &= 0 \quad \text{for } t > 0, \ (x, y) \in \partial\Omega, \\
u(0, x, y) &= \varphi(x, y) \quad \text{for } (x, y) \in \overline{\Omega} = \Omega \cup \partial\Omega.
\end{aligned}$$

As usual, Ω is a planar domain with boundary $\partial\Omega$. (Also recall that $\Delta u = \frac{\partial^2 u}{\partial x^2} + \frac{\partial^2 u}{\partial y^2}$.) The symbol $\frac{\partial u}{\partial \mathbf{n}}$ denotes the (outward) *normal derivative* of u on $\partial\Omega$, which is defined at each point along $\partial\Omega$ (where $\partial\Omega$ is smooth) as the directional derivative of u in the direction of the unit vector $\mathbf{n}$ that is perpendicular (normal) to $\partial\Omega$ and points toward the exterior of Ω.

■ Dirichlet Boundary Condition

When $\alpha = 1$, the problem described above becomes

$$\begin{aligned}
\frac{\partial u}{\partial t} - k \, \Delta u &= 0 \quad \text{for } t > 0, \ (x, y) \in \Omega, \\
u(t, x, y) &= 0 \quad \text{for } t > 0, \ (x, y) \in \partial\Omega, \\
u(0, x, y) &= \varphi(x, y) \quad \text{for } (x, y) \in \overline{\Omega}.
\end{aligned}$$

We will consider in the next two examples the cases in which Ω is a rectangle and a disk, respectively.

Example 11.3.1 (A Rectangular Domain)

Let Ω be the rectangle in which $0 < x < 2$ and $0 < y < 1$, and consider the problem described above with $k = 0.1$ and initial data given by

$$\varphi(x, y) = \begin{cases} x\,(1 - y), & \text{if } x \le 1; \\ 0, & \text{elsewhere in } \overline{\Omega}. \end{cases}$$

First we'll enter definitions for k and for the length and height of Ω.

```
k = 0.1; L = 2; H = 1;
```

Next we define the coefficients in the *double* Fourier sine series representation of φ:

$$\texttt{b[m\_, n\_] = Simplify}\left[\frac{4}{\texttt{L H}}\int_0^1\int_0^1 \texttt{x (1 - y) Sin}\left[\frac{\texttt{m}\,\pi}{\texttt{L}}\texttt{x}\right]\texttt{Sin}\left[\frac{\texttt{n}\,\pi}{\texttt{H}}\texttt{y}\right]\,d\texttt{x}\,d\texttt{y},\right.$$
$$\left.\{\texttt{n} \in \texttt{Integers},\ \texttt{m} \in \texttt{Integers}\}\right]$$

$$\frac{-4\,\texttt{m}\,\pi\,\texttt{Cos}\left[\frac{\texttt{m}\,\pi}{2}\right] + 8\,\texttt{Sin}\left[\frac{\texttt{m}\,\pi}{2}\right]}{\texttt{m}^2\,\texttt{n}\,\pi^3}$$

Now we compute the sum of 36 terms of the solution's double Fourier sine series representation at time t.

$$\texttt{u[t\_, x\_, y\_]} = \sum_{\texttt{n}=1}^{6}\sum_{\texttt{m}=1}^{6}\texttt{b[m, n] Exp}\left[-\texttt{k}\left(\frac{\texttt{m}^2}{\texttt{L}^2}+\frac{\texttt{n}^2}{\texttt{H}^2}\right)\pi^2\,\texttt{t}\right]\texttt{Sin}\left[\frac{\texttt{m}\,\pi}{\texttt{L}}\texttt{x}\right]\texttt{Sin}\left[\frac{\texttt{n}\,\pi}{\texttt{H}}\texttt{y}\right];$$

Since we know the exact value of the solution at $t = 0$, we also enter

```
u[0, x_, y_] := If[x ≤ 1, x (1 - y), 0];
```

The following command creates animation frames at each of $t = 0, 0.05, 0.10, \ldots, 0.5$. Six of the frames are shown below.

```
Do[Plot3D[u[t, x, y], {x, 0, L}, {y, 0, H},
   PlotRange → {0, 1}, BoxRatios → {2, 1, 1},
   PlotPoints → {30, 15}, PlotLabel → "t = " <> ToString[t],
   Ticks → {{0, 1, 2}, {0, 1}, {.5, 1}}], {t, 0, .5, .05}];
```

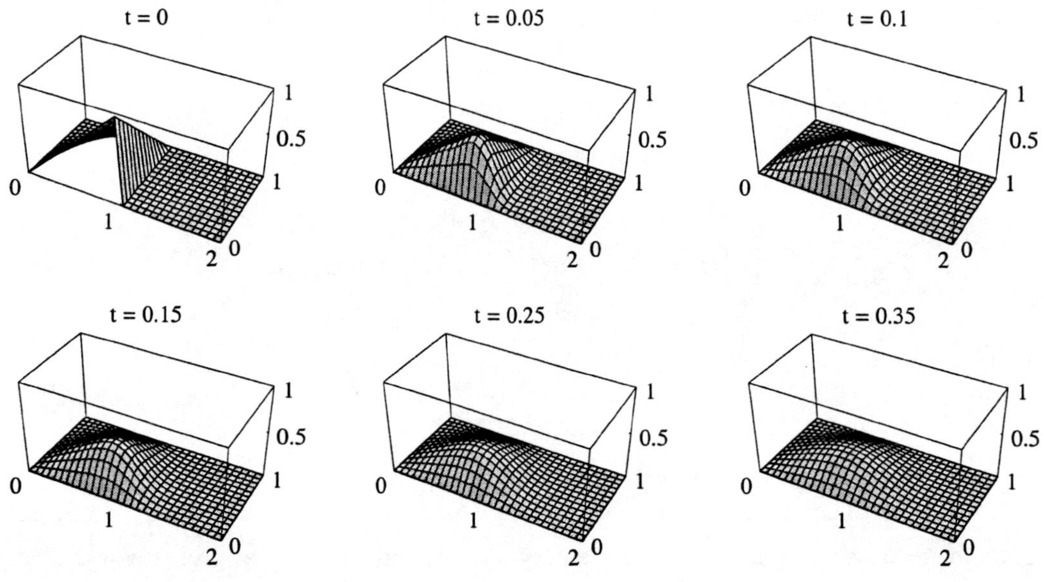

Example 11.3.2 (A Circular Domain)

Let Ω be the (open) disk $x^2 + y^2 < \rho^2 = 1$, and consider the diffusion problem described above with $k = 0.1$ and initial data given by

$$\rho := 1;\ \texttt{k} := 0.1;\ \varphi[\texttt{r\_}, \theta\texttt{\_}] := 15\,\texttt{Cos}[2\,\theta]^2\,(1-\texttt{r})^2\,\texttt{r}^2$$

```
ParametricPlot3D[{r Cos[θ], r Sin[θ], φ[r, θ]}, {r, 0, ρ}, {θ, -π, π},
  PlotPoints → {15, 45}, Ticks → {{-1, 0, 1}, {-1, 0, 1}, {.5}}];
```

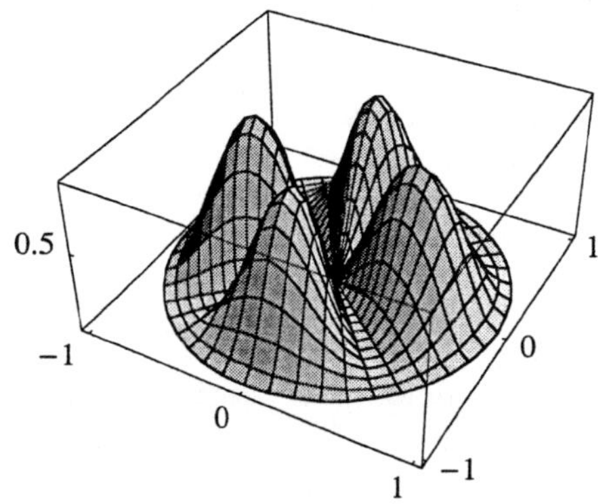

The solution of the diffusion problem will be in terms of the *Bessel functions* $J_m(r)$, $m = 0, 1, 2, \ldots$ and their positive zeros. *Mathematica*'s name for $J_m(r)$ is `BesselJ[m,r]`. The following plot shows the graphs of $J_0, \ldots, J_3$.

```
Plot[Evaluate[Table[BesselJ[m, r], {m, 0, 3}]], {r, 0, 10}];
```

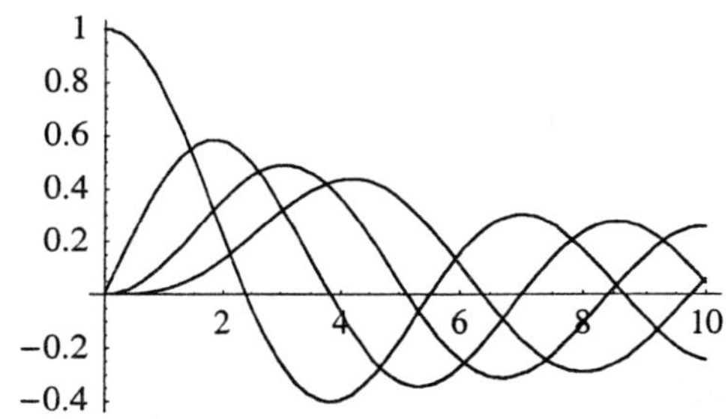

The zeros of $J_0, J_1, J_2, \ldots$ are easily found with the help of the `BesselZeros` package, which is loaded by entering

```
<< NumericalMath`BesselZeros`
```

The following builds a 6×5 table of these zeros:

```
Jzeros = Table[BesselJZeros[m, 5], {m, 0, 5}]

{{2.40483, 5.52008, 8.65373, 11.7915, 14.9309},
 {3.83171, 7.01559, 10.1735, 13.3237, 16.4706},
 {5.13562, 8.41724, 11.6198, 14.796, 17.9598},
 {6.38016, 9.76102, 13.0152, 16.2235, 19.4094},
 {7.58834, 11.0647, 14.3725, 17.616, 20.8269},
 {8.77148, 12.3386, 15.7002, 18.9801, 22.2178}}
```

For convenience, let's define $\sigma_{m,n}$ to be the nth positive zero of J_m, divided by ρ, for $m = 0, 1, 2, \ldots$.

```
σ[m_, n_] := Jzeros[[m + 1, n]] / ρ
```

The coefficients in the required *Fourier-Bessel* eigenfunction expansion of φ are:

$$\texttt{a[0, n\_] :=} \frac{\int_0^{\rho}\int_{-\pi}^{\pi}\varphi\texttt{[r, }\theta\texttt{] BesselJ[0, }\sigma\texttt{[0, n] r] r}\,d\theta\,dr}{\int_0^{\rho}\int_{-\pi}^{\pi}\texttt{BesselJ[0, }\sigma\texttt{[0, n] r]}^2\texttt{ r}\,d\theta\,dr}$$

$$\texttt{a[m\_, n\_] :=} \frac{\int_0^{\rho}\int_{-\pi}^{\pi}\varphi\texttt{[r, }\theta\texttt{] BesselJ[m, }\sigma\texttt{[m, n] r] Cos[m}\theta\texttt{] r}\,d\theta\,dr}{\int_0^{\rho}\int_{-\pi}^{\pi}\texttt{BesselJ[m, }\sigma\texttt{[m, n] r]}^2\texttt{ Cos[m}\theta\texttt{]}^2\texttt{ r}\,d\theta\,dr}\ \texttt{/; m > 0}$$

$$\texttt{b[m\_, n\_] :=} \frac{\int_0^{\rho}\int_{-\pi}^{\pi}\varphi\texttt{[r, }\theta\texttt{] BesselJ[m, }\sigma\texttt{[m, n] r] Sin[m}\theta\texttt{] r}\,d\theta\,dr}{\int_0^{\rho}\int_{-\pi}^{\pi}\texttt{BesselJ[m, }\sigma\texttt{[m, n] r]}^2\texttt{ Sin[m}\theta\texttt{]}^2\texttt{ r}\,d\theta\,dr}\ \texttt{/; m > 0}$$

Finally, the sum of 30 terms of the eigenfunction expansion of the solution at time t is

$$\texttt{u[t\_, r\_, }\theta\texttt{\_] =}$$
$$\sum_{m=0}^{5}\sum_{n=1}^{5} e^{-k\,\sigma[m,n]^2 t}\ \texttt{BesselJ[m, }\sigma\texttt{[m, n] r] (a[m, n] Cos[m}\theta\texttt{] + b[m, n] Sin[m}\theta\texttt{])}$$

$$0.52271\, e^{-0.578319\, t}\, \texttt{BesselJ[0, 2.40483 r]} - 0.330617\, e^{-3.04713\, t}\, \texttt{BesselJ[0, 5.52008 r]} - 0.193414\, e^{-7.4887\, t}\, \texttt{BesselJ[0, 8.65373 r]} + 0.0437413\, e^{-13.904\, t}\, \texttt{BesselJ[0, 11.7915 r]} - 0.0476553\, e^{-22.2932\, t}\, \texttt{BesselJ[0, 14.9309 r]} + 0.935129\, e^{-5.75829\, t}\, \texttt{BesselJ[4, 7.58834 r] Cos[4}\theta\texttt{]} + 0.645933\, e^{-12.2428\, t}\, \texttt{BesselJ[4, 11.0647 r] Cos[4}\theta\texttt{]} + 0.223963\, e^{-20.657\, t}\, \texttt{BesselJ[4, 14.3725 r] Cos[4}\theta\texttt{]} + 0.189817\, e^{-31.0322\, t}\, \texttt{BesselJ[4, 17.616 r] Cos[4}\theta\texttt{]} + 0.0811303\, e^{-43.3761\, t}\, \texttt{BesselJ[4, 20.8269 r] Cos[4}\theta\texttt{]}$$

Since we have the solution exactly at $t = 0$, let's also define

```
u[0, r_, θ_] := φ[r, θ]
```

The following command creates animation frames at each of $t = 0, 0.1, 0.2, \ldots, 1$. Six of the frames are shown below.

```
Do[ParametricPlot3D[{r Cos[θ], r Sin[θ], u[t, r, θ]},
   {r, 0, 1}, {θ, -π, π}, PlotRange → {0, 1.02},
   PlotPoints → {15, 45}, PlotLabel -> "t = " <> ToString[t],
   Ticks → {{-1, 0, 1}, {-1, 0, 1}, {.5}}], {t, 0, 1, .1}];
```

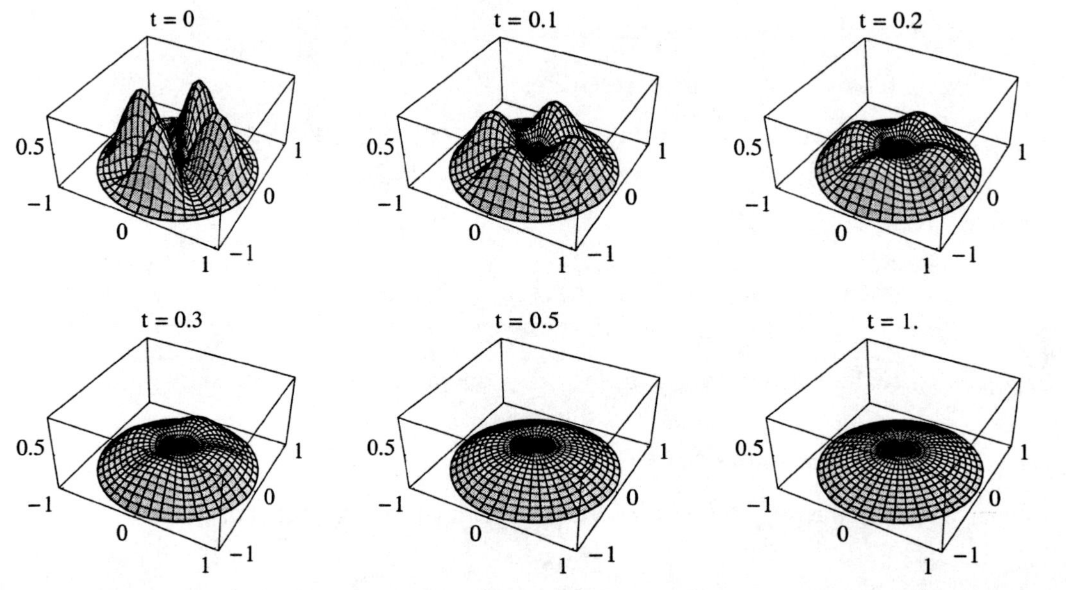

■ Neumann Boundary Condition

When $\alpha = 0$, the diffusion problem described at the beginning of this section becomes one involving a Neumann boundary condition:

$$\frac{\partial u}{\partial t} - k\,\Delta u = 0 \quad \text{for } t > 0,\ (x, y) \in \Omega,$$
$$\frac{\partial u}{\partial \mathbf{n}} = 0 \quad \text{for } t > 0,\ (x, y) \in \partial\Omega,$$
$$u(0, x, y) = \varphi(x, y) \quad \text{for } (x, y) \in \Omega.$$

Paralleling the previous examples involving the Dirichlet boundary condition, we will consider in the next two examples the cases in which Ω is a rectangle and a disk, respectively.

Example 11.3.3 (A Rectangular Domain)

Suppose that Ω is the rectangle in which $0 < x < L$ and $0 < y < H$. Along each vertical edge ($x = 0$ and $x = L$), the normal derivative is simply $\frac{\partial u}{\partial x}$, and along each horizontal edge ($y = 0$ and $y = H$), the normal derivative is simply $\frac{\partial u}{\partial y}$. Thus the diffusion problem under consideration is

$$\frac{\partial u}{\partial t} - k\,\Delta u = 0 \quad \text{for } t > 0,\ (x, y) \in \Omega,$$
$$\frac{\partial u}{\partial x} = 0 \quad \text{for } t > 0, x = 0\,, L,$$
$$\frac{\partial u}{\partial y} = 0 \quad \text{for } t > 0, y = 0\,, H,$$
$$u(0, x, y) = \varphi(x, y) \quad \text{for } (x, y) \in \Omega.$$

Let's assume that $L = 2, H = 1, k = 0.1$, and initial data are given (as in Example 11.3.1) by

$$\varphi(x, y) = \begin{cases} x(1-y), & \text{if } x \le 1; \\ 0, & \text{elsewhere in } \overline{\Omega}. \end{cases}$$

First we'll enter definitions for k and for the length and height of Ω.

```
k = 0.1; L = 2; H = 1;
```

Next we enter a definition for the coefficients in the double Fourier cosine series of φ:

```
a[0, 0] := 1/(L H) ∫_0^1 ∫_0^1 x (1 - y) dx dy

a[m_, 0] := 2/(L H) ∫_0^1 ∫_0^1 x (1 - y) Cos[(m π)/L x] dx dy /; m > 0

a[0, n_] := 2/(L H) ∫_0^1 ∫_0^1 x (1 - y) Cos[(n π)/H y] dx dy /; n > 0

a[m_, n_] :=
  4/(L H) ∫_0^1 ∫_0^1 x (1 - y) Cos[(m π)/L x] Cos[(n π)/H y] dx dy /; (m > 0 && n > 0)
```

Now we compute the sum of 63 terms of the solution's double Fourier sine series representation at time t.

```
u[t_, x_, y_] = Sum[Sum[a[m, n] Exp[-k (m^2/L^2 + n^2/H^2) π^2 t] Cos[(m π)/L x] Cos[(n π)/H y], {m, 0, 6}], {n, 0, 8}];
```

$$u[t\_, x\_, y\_] = \sum_{n=0}^{6}\sum_{m=0}^{8} a[m, n]\, \mathrm{Exp}\left[-k\left(\frac{m^2}{L^2}+\frac{n^2}{H^2}\right)\pi^2 t\right] \mathrm{Cos}\left[\frac{m\pi}{L}x\right] \mathrm{Cos}\left[\frac{n\pi}{H}y\right];$$

As usual, at $t = 0$ the solution is known exactly.

```
u[0, x_, y_] := If[x ≤ 1, x (1 - y), 0];
```

The following command creates animation frames at each of $t = 0, 0.05, 0.1, \ldots, 0.5$. Six of the frames are shown below. (Compare these frames with those shown in Example 11.3.1.)

```
Do[Plot3D[u[t, x, y], {x, 0, L}, {y, 0, H}, PlotRange → {0, 1},
   BoxRatios → {2, 1, 1}, Ticks → {{0, 1, 2}, {0, 1}, {.5, 1}},
   PlotPoints → {30, 15}, PlotLabel → "t = " <> ToString[t]],
  {t, 0, .5, .05}];
```

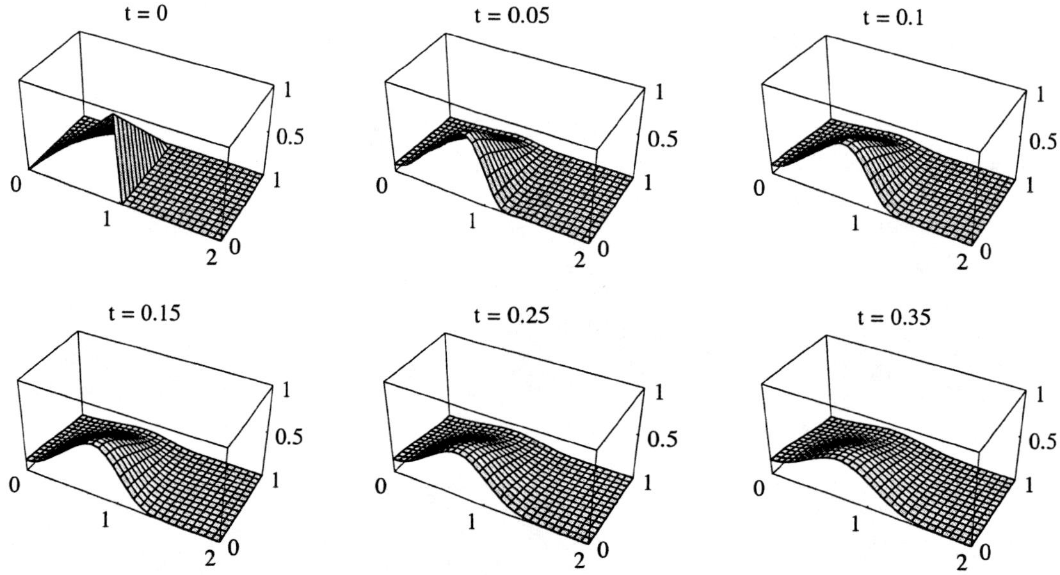

Example 11.3.4 (A Circular Domain)

Suppose that Ω is the open unit disk: $x^2 + y^2 < \rho$. The boundary $\partial\Omega$ is the circle $x^2 + y^2 = \rho$, which in polar coordinates is given by $r = \rho$. The normal derivative on $\partial\Omega$ is simply $\frac{\partial u}{\partial r}$. Thus the diffusion problem under consideration is

$$\frac{\partial u}{\partial t} - k\,\Delta u = 0 \quad \text{for } t > 0,\ 0 < r < \rho,\ -\infty < \theta < \infty,$$

$$\frac{\partial u}{\partial r} = 0 \quad \text{for } t > 0,\ r = \rho,\ -\infty < \theta < \infty,$$

$$u(0, r, \theta) = \varphi(r, \theta) \quad \text{for } 0 \le r \le \rho,\ -\infty < \theta < \infty.$$

(Also required are that u and $\frac{\partial u}{\partial r}$ be bounded for $0 \le r \le \rho, -\infty < \theta < \infty$.) Let us assume that ρ, k, and the initial data are given by

```
ρ := 1; k = 0.1; φ[r_, θ_] := r^4 Cos[2 θ]^4;
ParametricPlot3D[Evaluate[{r Cos[θ], r Sin[θ], φ[r, θ]}],
  {r, 0, 1}, {θ, -π, π}, PlotRange → All, PlotPoints → {15, 45},
  Ticks → {{-1, 0, 1}, {-1, 0, 1}, {.5}}];
```

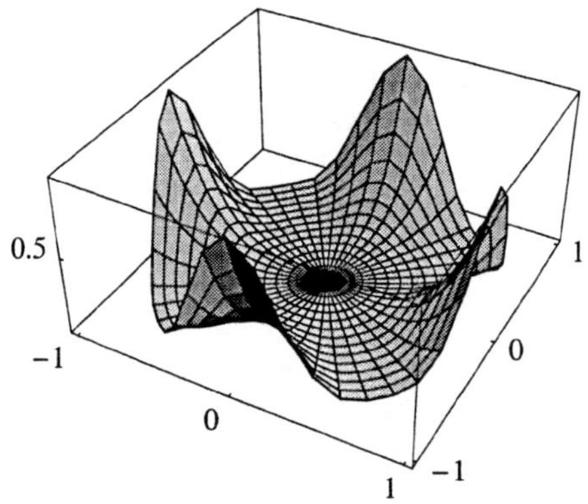

As in Example 11.3.2, the solution of the diffusion problem will be in terms of the *Bessel functions* $J_m(r)$, $m = 0, 1, 2, \ldots$. However in this case we will need the positive zeros of $J'_m(r)$. The following builds an 8×7 table.

```
<< NumericalMath`BesselZeros`;
JPrimezeros = Table[BesselJPrimeZeros[m, 7], {m, 0, 7}];
```

For convenience, let's define $\sigma_{m,n}$ to be the nth positive zero of J'_m divided by ρ, for $m = 0, 1, 2, \ldots$.

```
σ[m_, n_] := JPrimezeros[[m + 1, n]] / ρ
```

The coefficients in the Fourier-Bessel eigenfunction expansion of the initial data are as follows:

$$\texttt{a[0, 0] :=} \frac{\int_0^\rho \int_{-\pi}^{\pi} \varphi[\texttt{r}, \theta]\, \texttt{r}\, d\theta\, d\texttt{r}}{\int_0^1 \int_{-\pi}^{\pi} \texttt{r}\, d\theta\, d\texttt{r}}$$

$$\texttt{a[0, n\_] :=} \frac{\int_0^\rho \int_{-\pi}^{\pi} \varphi[\texttt{r}, \theta]\, \texttt{BesselJ[0, } \sigma\texttt{[0, n] r]}\, \texttt{r}\, d\theta\, d\texttt{r}}{\int_0^1 \int_{-\pi}^{\pi} \texttt{BesselJ[m, } \sigma\texttt{[m, n] r]}^2\, \texttt{r}\, d\theta\, d\texttt{r}}$$

$$\texttt{a[m\_, n\_] :=} \frac{\int_0^\rho \int_{-\pi}^{\pi} \varphi[\texttt{r}, \theta]\, \texttt{BesselJ[m, } \sigma\texttt{[m, n] r] Cos[m} \theta\texttt{]}\, \texttt{r}\, d\theta\, d\texttt{r}}{\int_0^1 \int_{-\pi}^{\pi} \texttt{BesselJ[m, } \sigma\texttt{[m, n] r]}^2\, \texttt{Cos[m}\theta\texttt{]}^2\, \texttt{r}\, d\theta\, d\texttt{r}} \ \texttt{/; m > 0}$$

$$\texttt{b[m\_, n\_] :=} \frac{\int_0^\rho \int_{-\pi}^{\pi} \varphi[\texttt{r}, \theta]\, \texttt{BesselJ[m, } \sigma\texttt{[m, n] r] Sin[m} \theta\texttt{]}\, \texttt{r}\, d\theta\, d\texttt{r}}{\int_0^\rho \int_{-\pi}^{\pi} \texttt{BesselJ[m, } \sigma\texttt{[m, n] r]}^2\, \texttt{Sin[m}\theta\texttt{]}^2\, \texttt{r}\, d\theta\, d\texttt{r}} \ \texttt{/; m > 0}$$

The (sum of 56 terms of the) solution is computed by entering

$$\texttt{u[t\_, r\_, } \theta\texttt{\_] = a[0, 0] +} \sum_{\texttt{m=0}}^{7} \sum_{\texttt{n=1}}^{7} e^{-\texttt{k}\, \sigma\texttt{[m,n]}^2\, \texttt{t}}\, \texttt{BesselJ[m, } \sigma\texttt{[m, n] r] (a[m, n] Cos[m} \theta\texttt{] + b[m, n] Sin[m} \theta\texttt{])}$$

To show the exact solution at $t = 0$, we also enter

```
u[0, r_, θ_] := φ[r, θ]
```

The following command creates animation frames at each of $t = 0, 0.05, 0.1, \ldots, 0.5$. Six of the frames follow.

```
Do[ParametricPlot3D[Evaluate[{r Cos[θ], r Sin[θ], u[t, r, θ]}],
   {r, 0, 1}, {θ, -π, π}, PlotRange → {-.05, 1},
   PlotPoints → {15, 45}, PlotLabel -> "t = " <> ToString[t],
   Ticks → {{-1, 0, 1}, {-1, 0, 1}, {.5}}], {t, 0, .5, .05}];
```

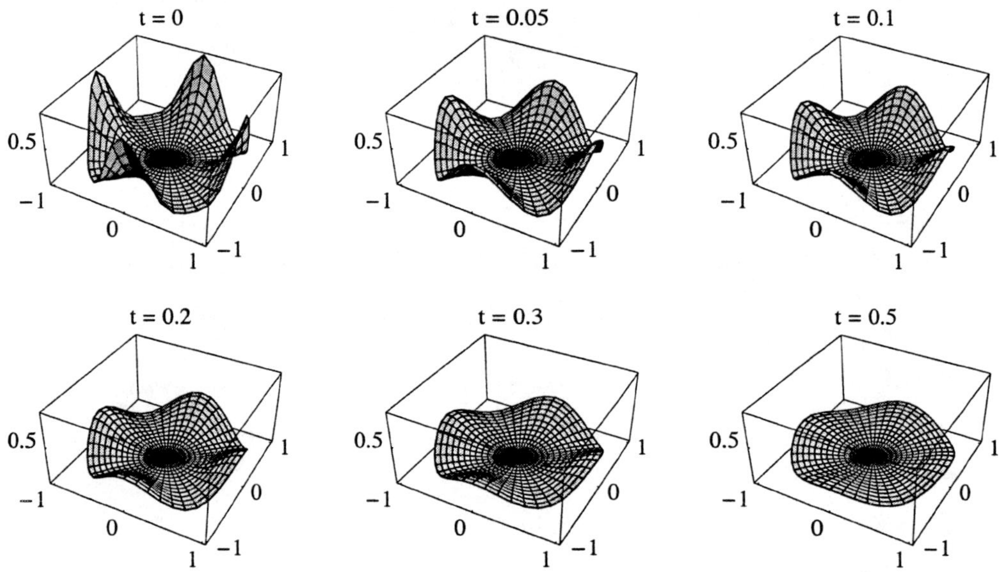

■ Exercises

In each of the cases described in Exercises 1–6, find and animate the solution of

$$\frac{\partial u}{\partial t} - 0.1\,\Delta u = 0 \quad \text{for } t > 0,\ (x, y) \in \Omega,$$

$$\alpha\, u + (1 - \alpha)\frac{\partial u}{\partial \mathbf{n}} = 0 \quad \text{for } t > 0,\ (x, y) \in \partial\Omega,$$

$$u(0, x, y) = \varphi(x, y) \quad \text{for } (x, y) \in \overline{\Omega} = \Omega \cup \partial\Omega.$$

The function h is the Heaviside `UnitStep` function.

1. $\Omega = \{(x, y) \mid 0 < x < 1,\ 0 < y < 1\},\ \alpha = 1,\ \varphi(x, y) = \sin(2\pi x)\,(\sin(\pi y) - \sin(3\pi y))$
2. $\Omega = \{(x, y) \mid 0 < x < 2,\ 0 < y < 1\},\ \alpha = 0,\ \varphi(x, y) = 2\,(4x - 3)\,h(x - \frac{3}{4})\cdot(3y - 2)\,h(y - \frac{2}{3})$
3. $\Omega = \{(x, y) \mid 0 < x < 5,\ 0 < y < 5\},\ \alpha = 1,\ \varphi(x, y) = 5\,(h(x - 2) - h(x - 3))\,(h(y - 2) - h(y - 3))$
4. $\Omega = \{(x, y) \mid 0 < x < 5,\ 0 < y < 3\},\ \alpha = 0,\ \varphi(x, y) = 5\,h(x - 4)\,h(y - 2)$

5 $\Omega = \{(r, \theta) \mid 0 \le r < 1\},\ \alpha = 0,\ \varphi(r, \theta) = 3\,(1 - r)^5$

6. $\Omega = \{(r, \theta) \mid 0 \le r < 1\},\ \alpha = 1,\ \varphi(r, \theta) = h(\theta) - h(\theta - \frac{\pi}{4})$ for $-\pi \le \theta \le \pi$

11.4 The 1-D Wave Equation

The one-dimensional wave equation on $(-\infty, \infty)$ is

$$\frac{\partial^2 u}{\partial t^2} - c^2 \frac{\partial^2 u}{\partial x^2} = 0 \quad \text{for } t > 0 \text{ and } -\infty < x < \infty.$$

Initial-value problems associated with the wave equation involve prescribed initial "displacement" and initial velocity:

$$u(0, x) = \varphi(x) \quad \text{for } -\infty \le x \le \infty,$$
$$\frac{\partial u}{\partial t}(0, x) = v(x) \quad \text{for } -\infty \le x \le \infty.$$

The **d'Alembert solution** of this initial-value problem is given by

$$u(t, x) = \frac{1}{2}\left(\varphi(x + ct) + \varphi(x - ct)\right) + \frac{1}{2c}\int_{x-ct}^{x+ct} v(s)\, ds.$$

One should verify this solution, as it is quite easy to do.

Example 11.4.1

Consider the problem described above with *wave speed* $c = 1$, zero initial velocity, and initial displacement given by

```
φ[x_] := If[1 < Abs[x] < 2, Sin[π x]^2, 0];
Plot[φ[x], {x, -3, 3},
  AspectRatio → Automatic, Ticks → {Automatic, {1}}];
```

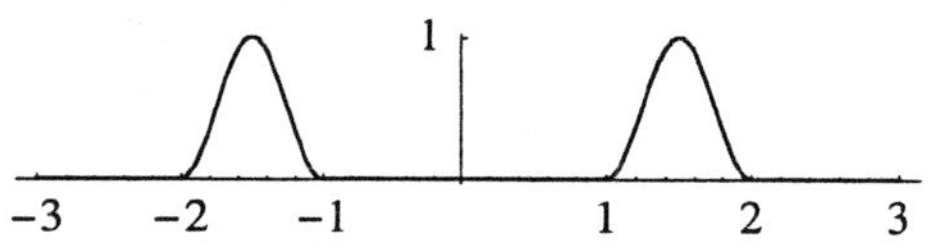

Since the initial velocity is zero, the solution is given simply by

```
u[t_, x_] := 1/2 (φ[x + t] + φ[x - t])
```

The following 3D plot and 2D animation illustrate how the initial displacement gives rise to four "travelling waves." Notice what happens when two of them "collide."

```
Plot3D[u[t, x], {t, 0, 3}, {x, -3, 3},
  PlotPoints → {30, 50}, PlotRange → All];
```

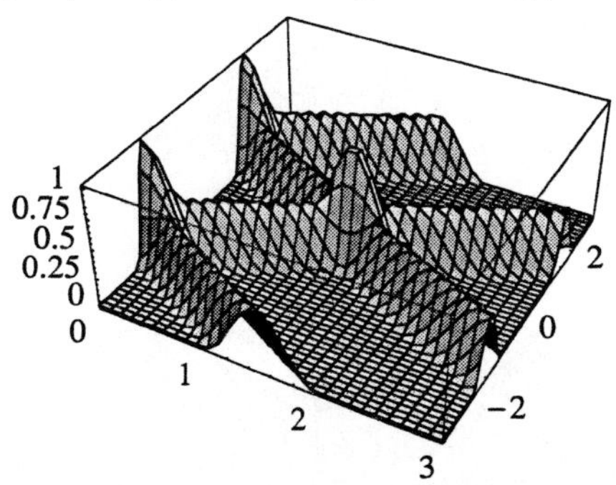

```
Do[Plot[u[t, x], {x, -4, 4}, AspectRatio → Automatic,
   PlotRange → {{-4, 4}, {0, 1.05}}, PlotStyle → Thickness[.005],
   AxesStyle → Hue[.67], PlotLabel -> "t = " <> ToString[t],
   Ticks → None, PlotPoints → 100], {t, 0, 5, .1}];
```

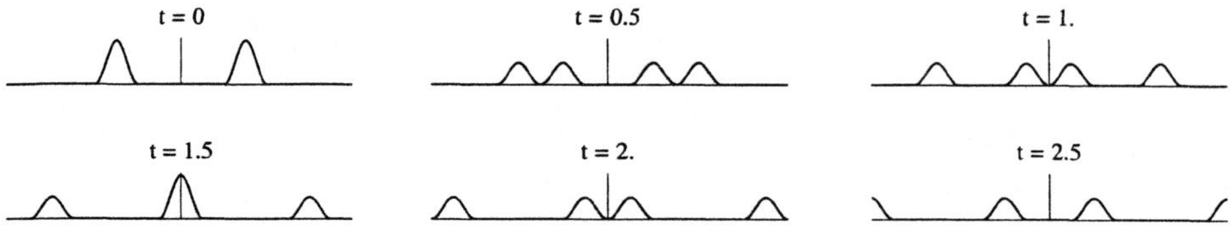

Example 11.4.2

Consider the problem described above with $c = 1$, zero displacement, and initial velocity

```
v[x_] := 2 Cos[π x]
```

Since the initial displacement is zero, the solution is given by

$$u[t\_, x\_] = \frac{1}{2}\int_{x-t}^{x+t} v[s]\, ds \; // \; Simplify$$

$$\frac{2\, Cos[\pi\, x]\; Sin[\pi\, t]}{\pi}$$

```
Plot3D[u[t, x], {t, 0, 3}, {x, -2, 2}, PlotPoints → 31];
```

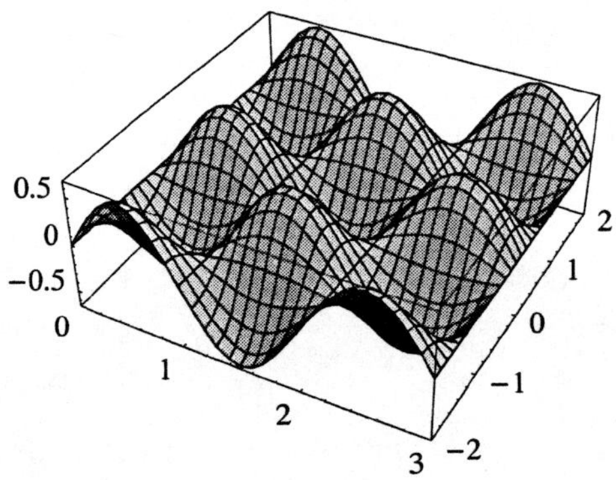

Notice how the wave motion is disguised as a stationary vibration (a "standing wave").

```
Plot[Evaluate[Table[u[t, x], {t, 0, 2, .125}]], {x, -2, 2},
  PlotStyle → Table[GrayLevel[.8 (1 - s)], {s, 0, 1, .125 / 2}],
  AspectRatio → Automatic];
```

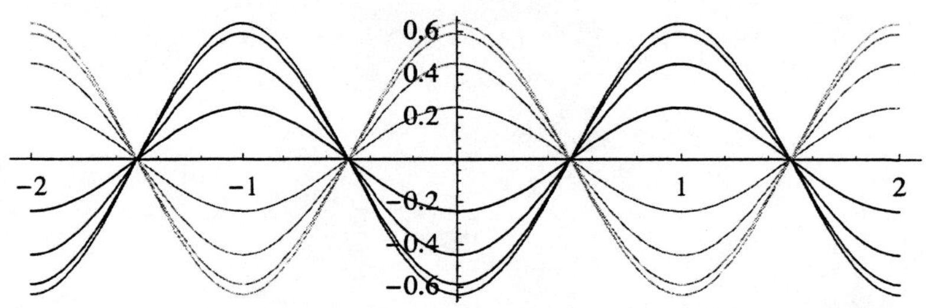

■ The Wave Equation on [0, *L*]

Let's consider now the wave equation,

$$\frac{\partial^2 u}{\partial t^2} - c^2 \frac{\partial^2 u}{\partial x^2} = 0 \quad \text{for } t > 0 \text{ and } 0 < x < L,$$

subject to initial conditions

$$u(0, x) = \varphi(x) \quad \text{for } 0 \le x \le L,$$

$$\frac{\partial u}{\partial t}(0, x) = 0 \quad \text{for } 0 \le x \le L,$$

and zero boundary conditions

$$u(t, 0) = u(t, L) = 0 \quad \text{for } t > 0.$$

A d'Alembert-type solution can be constructed as follows. First, let φ_o be the odd extension of φ to the interval $[-L, L]$:

$$\varphi_o(x) = \begin{cases} \varphi(x). & \text{if } 0 \le x \le L, \\ -\varphi(-x), & \text{if } -L \le x < 0. \end{cases}$$

Second, let $\tilde{\varphi}_o$ be the $2L$-periodic extension of φ_o to $(-\infty, \infty)$. Finally, the solution on $[0, L]$ is

$$u(t, x) = \frac{1}{2}\,(\tilde{\varphi}_o\,(x + c\,t) + \tilde{\varphi}_o\,(x - c\,t)).$$

Example 11.4.3

Consider the problem just described with $L = 2$, $c = 1$, and initial displacement given by

```
φ[x_] := .25 If[x < .5, Sin[2 π x], 0]
Plot[φ[x], {x, 0, 2},
  AspectRatio → Automatic, Ticks → {Automatic, {.25}}];
```

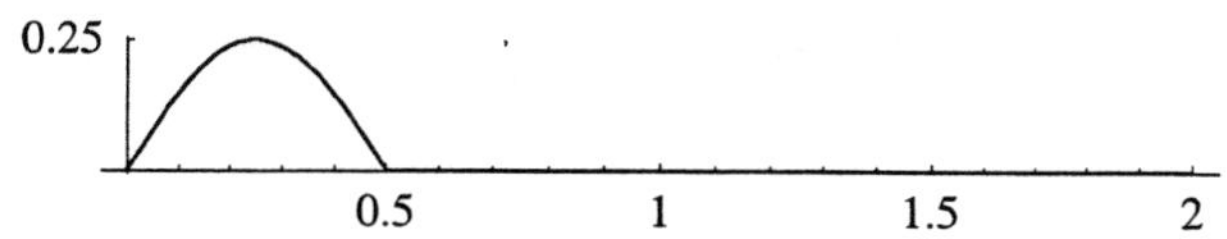

The odd extension of φ to $[-2, 2]$ is

```
oddφ[x_] := If[x < 0, -φ[-x], φ[x]]
Plot[oddφ[x], {x, -2, 2},
  AspectRatio → Automatic, Ticks → {Automatic, None}];
```

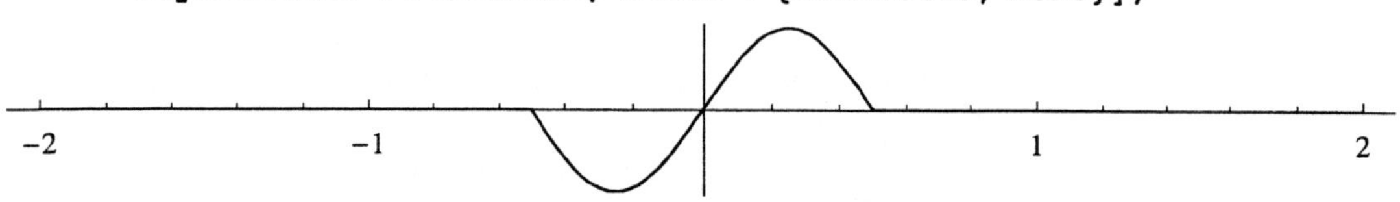

The 4-periodic extension to $(-\infty, \infty)$ is the composition of oddφ with the "shift function" $x - 2L\left\lfloor \frac{x+L}{2L} \right\rfloor$:

```
φ4[x_] := oddφ[x - 4 Floor[(x + 2)/4]]
```

```
Plot[φ4[x], {x, -5, 5}, AspectRatio → .1, Ticks → {Automatic, None}];
```

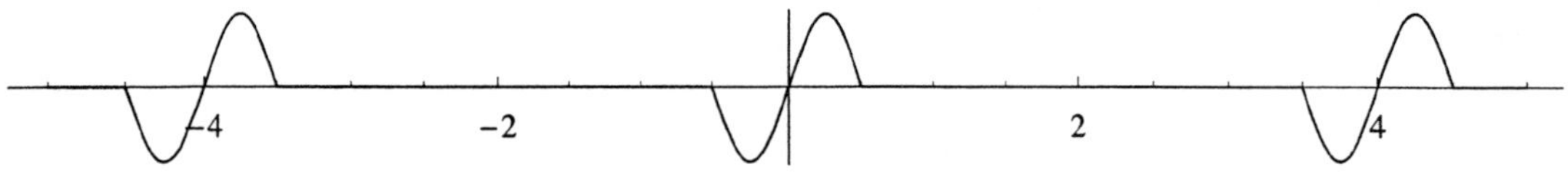

The solution is

```
u[t_, x_] := 1/2 (φ4[x + t] + φ4[x - t]);
Plot3D[u[t, x], {t, 0, 4}, {x, 0, 2},
  PlotPoints → 31, Ticks → None, AxesLabel → {t, x, u}];
```

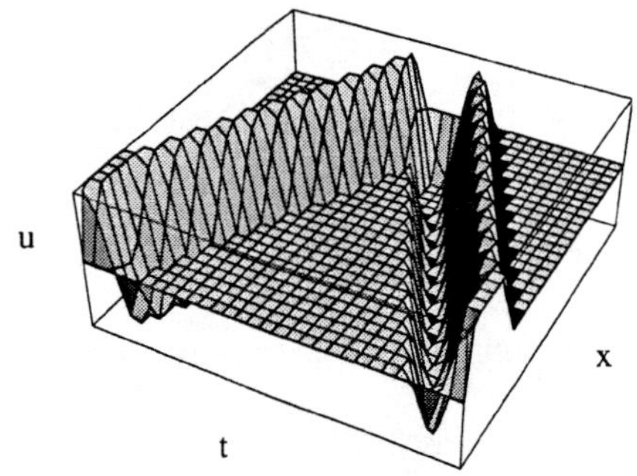

The following creates frames of an animation over half of one temporal cycle of the solution. Looping back and forth over the frames shows the wave motion in its entirety.

```
Do[Plot[u[t, x], {x, 0, 2},
   AspectRatio → Automatic, PlotRange → {{0, 2}, {-.3, .3}},
   PlotStyle → Thickness[.005], AxesStyle → Hue[.67],
   Ticks → None, PlotPoints → 100], {t, 0, 2, .05}];
```

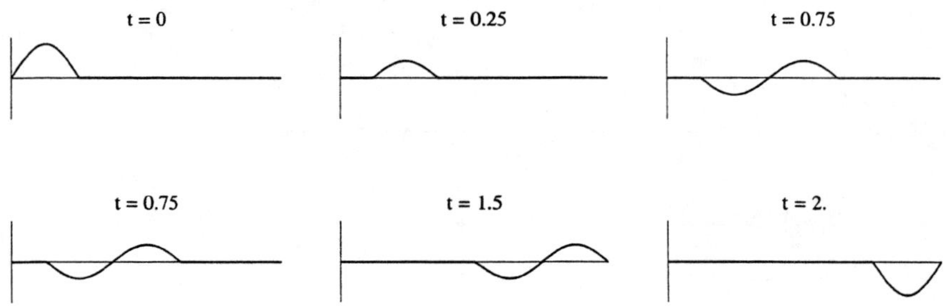

■ Fourier Series Solutions

Consider again the initial/boundary-value problem associated with the wave equation on $[0, L]$:

$$\frac{\partial^2 u}{\partial t^2} - c^2 \frac{\partial^2 u}{\partial x^2} = 0 \quad \text{for } t > 0 \text{ and } 0 < x < L,$$

$$u(0, x) = \varphi(x) \quad \text{for } 0 \le x \le L,$$

$$\frac{\partial u}{\partial t}(0, x) = v(x) \quad \text{for } 0 \le x \le L,$$

$$u(t, 0) = u(t, L) = 0 \quad \text{for } t > 0.$$

Assume that each of φ and v are given by Fourier sine series on $[0, L]$:

$$\varphi(x) = \sum_{n=1}^{\infty} a_n \sin(\tfrac{n\pi}{L}x), \text{ and } v(x) = \sum_{n=1}^{\infty} b_n \sin(\tfrac{n\pi}{L}x).$$

As a result, the solution is given by

$$u(t, x) = \sum_{n=1}^{\infty} (a_n \cos(\tfrac{cn\pi}{L}t) + \tfrac{L}{cn\pi} b_n \sin(\tfrac{cn\pi}{L}t)) \sin(\tfrac{n\pi}{L}x).$$

Example 11.4.4

Let's consider the problem described above with $c = 1$ and $L = 1$. Let's also assume that the initial velocity is zero and that the initial displacement is given by the first four terms of the Fourier sine series of x on $[0, 1]$.

$$\text{L := 1; c := 1; a[n\_] := }\frac{2}{1}\int_0^1 \text{x Sin[n }\pi\text{ x] dx;}$$

$$\varphi\text{[x\_] = }\sum_{n=1}^{4}\text{a[n] Sin}\left[\frac{n\,\pi\,x}{L}\right]\text{; Plot[}\varphi\text{[x], \{x, 0, 1\}];}$$

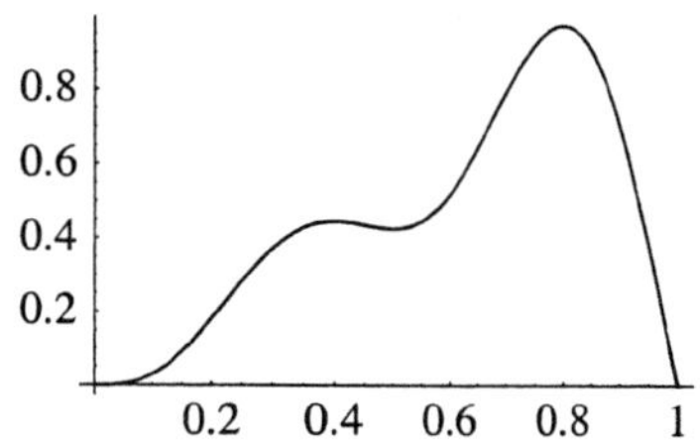

The solution is

$$\text{u[t\_, x\_] = }\sum_{n=1}^{4}\text{a[n] Cos}\left[\frac{c\,n\,\pi\,t}{L}\right]\text{ Sin}\left[\frac{n\,\pi\,x}{L}\right]$$

$$\frac{2\,\text{Cos}[\pi\,t]\,\text{Sin}[\pi\,x]}{\pi} - \frac{\text{Cos}[2\,\pi\,t]\,\text{Sin}[2\,\pi\,x]}{\pi} + \frac{2\,\text{Cos}[3\,\pi\,t]\,\text{Sin}[3\,\pi\,x]}{3\,\pi} - \frac{\text{Cos}[4\,\pi\,t]\,\text{Sin}[4\,\pi\,x]}{2\,\pi}$$

```
Plot3D[u[t, x], {t, 0, 2}, {x, 0, 1}, PlotPoints → 30];
```

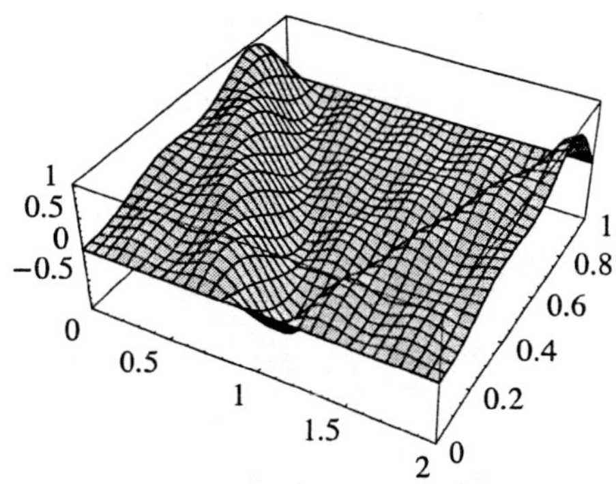

Enter the following to create an animation of the wave motion.

```
Do[Plot[u[t, x], {x, 0, 1}, PlotRange → {-1.1, 1.1},
   PlotStyle → Thickness[.005], Ticks → None], {t, 0, 1.95, .05}];
```

■ Exercises

Consider the wave equation on $(-\infty, \infty)$ with unit wave speed and zero initial velocity:

$$\frac{\partial^2 u}{\partial t^2} - \frac{\partial^2 u}{\partial x^2} = 0 \text{ for } t > 0 \text{ and } -\infty < x < \infty.$$
$$u(0, x) = \varphi(x) \quad \text{for } -\infty \le x \le \infty,$$
$$\frac{\partial u}{\partial t}(0, x) = 0 \quad \text{for } -\infty \le x \le \infty.$$

For the initial displacements in Exercises 1–5, create a surface plot of the d'Alembert solution, and create an animation corresponding to a well-chosen sequence of t values. (The function h is the Heaviside `UnitStep` function, and $\lfloor\ \rfloor$ denotes the `Floor` function.)

1. $\varphi(x) = h(x+1) - h(x-1)$
2. $\varphi(x) = x\,(h(x+1) - h(x-1))$
3. $\varphi(x) = \sqrt{1-x^2}\,(h(x+1) - h(x-1))$
4. $\varphi(x) = h(x+1) - 2\,h(x) + h(x-1)$
5. $\varphi(x) = -x\,(h(x+2) - 2\,h(x) + h(x-2))$

Consider the wave equation on $[0, 2]$ with unit wave speed, zero initial velocity, and zero boundary conditions:

$$\frac{\partial^2 u}{\partial t^2} - \frac{\partial^2 u}{\partial x^2} = 0 \text{ for } t > 0 \text{ and } 0 < x < 2,$$
$$u(0, x) = \varphi(x) \quad \text{for } 0 \le x \le 2,$$
$$\frac{\partial u}{\partial t}(0, x) = 0 \quad \text{for } 0 \le x \le 2,$$
$$u(t, 0) = u(t, 2) = 0 \quad \text{for } t > 0.$$

For the initial displacements in Exercises 6–10, create a surface plot of the d'Alembert solution, and create an animation corresponding to a well-chosen sequence of t values. (The function h is the Heaviside `UnitStep` function.)

6. $\varphi(x) = h(x - \frac{1}{2}) - h(x-1)$
7. $\varphi(x) = x\,(1 - h(x - \frac{1}{2}))$
8. $\varphi(x) = \sqrt{1-(x-1)^2}$
9. $\varphi(x) = 1 - |x-1|$
10. $\varphi(x) = 4\,x - \frac{1}{7}(32\,x - 8)\,h\left(x - \frac{1}{4}\right)$

Consider the wave equation on $[0, \pi]$ with unit wave speed, zero initial velocity, and zero boundary conditions:

$$\frac{\partial^2 u}{\partial t^2} - \frac{\partial^2 u}{\partial x^2} = 0 \text{ for } t > 0 \text{ and } 0 < x < \pi,$$
$$u(0, x) = \varphi(x) \quad \text{for } 0 \le x \le \pi,$$
$$\frac{\partial u}{\partial t}(0, x) = 0 \quad \text{for } 0 \le x \le \pi,$$
$$u(t, 0) = u(t, \pi) = 0 \quad \text{for } t > 0.$$

For the initial displacements in Exercises 11–15, create a surface plot of the Fourier series solution, and create an animation corresponding to a well-chosen sequence of t values.

11. $\varphi(x) = \frac{1}{\pi^2}\, x^2 \sin x$

12. $\varphi(x) = \sin x + \frac{1}{3} \sin 3x + \frac{1}{5} \sin 5x$

13. $\varphi(x) = \frac{1}{3}\,(\sin 5x + \sin 8x + \sin 13x)$

13. $\varphi(x) = \sin x + \frac{1}{5}\,(\sin 11x + \sin 18x)$

15. $\varphi(x) = \sin 3x$

11.5 The 2-D Wave Equation

In this section we will take a look at the two-dimensional wave equation,

$$\frac{\partial^2 u}{\partial t^2} - c^2 \Delta u = 0 \quad \text{for } t > 0,\ (x, y) \in \Omega,$$

with Dirichlet boundary condition,

$$u(t, x, y) = 0 \quad \text{for } t > 0,\ (x, y) \in \partial\Omega,$$

and given initial displacement and initial velocity,

$$u(0, x, y) = \varphi(x, y) \quad \text{for } (x, y) \in \overline{\Omega}.$$
$$\frac{\partial u}{\partial t}\,(0, x, y) = v(x, y) \quad \text{for } (x, y) \in \overline{\Omega}.$$

As usual, Ω is a two-dimensional domain with boundary $\partial\Omega$, and $\overline{\Omega} = \Omega \cup \partial\Omega$. (Also recall that $\Delta u = \frac{\partial^2 u}{\partial x^2} + \frac{\partial^2 u}{\partial y^2}$.) We assume that each of the functions φ and v are zero on $\partial\Omega$.

■ Rectangular Domain

Let Ω be the rectangle in which $0 < x < L$ and $0 < y < H$. We assume that each of φ and v are given by double Fourier sine series on $\overline{\Omega} = [0, L] \times [0, H]$:

$$\varphi(x) = \sum_{n=1}^{\infty} \sum_{m=1}^{\infty} a_{m,n} \sin(\tfrac{m\pi}{L} x) \sin(\tfrac{n\pi}{H} y), \text{ and } v(x) = \sum_{n=1}^{\infty} \sum_{m=1}^{\infty} b_{m,n} \sin(\tfrac{m\pi}{L} x) \sin(\tfrac{n\pi}{H} y).$$

As a result, the solution is the superposition $u = u_\varphi + u_v$ of

$$u_\varphi(t, x) = \sum_{n=1}^{\infty} \sum_{m=1}^{\infty} a_{m,n} \cos\Big(\sqrt{\tfrac{m^2}{L^2} + \tfrac{n^2}{H^2}}\ \pi c\, t\Big) \sin(\tfrac{m\pi}{L} x) \sin(\tfrac{n\pi}{H} y),$$

$$u_v(t, x) = \sum_{n=1}^{\infty} \sum_{m=1}^{\infty} \frac{b_{m,n}}{\sqrt{\frac{m^2}{L^2} + \frac{n^2}{H^2}}\ \pi c} \sin\Big(\sqrt{\tfrac{m^2}{L^2} + \tfrac{n^2}{H^2}}\ \pi c\, t\Big) \sin(\tfrac{m\pi}{L} x) \sin(\tfrac{n\pi}{H} y).$$

In the problem just described, suppose that $L = 3$, $H = 2$, and $c = 1$. Also suppose that the initial displacement and initial velocity are given respectively by

$$\varphi(x, y) = \sin(\tfrac{\pi}{3} x) \sin(\tfrac{\pi}{2} y) + \tfrac{1}{5} \sin(\tfrac{4\pi}{3} x) \sin(\tfrac{2\pi}{2} y) \quad \text{and} \quad v(x, y) = 0.$$

The solution is then

```
u[t_, x_, y_] = Cos[Sqrt[1/3^2 + 1/2^2] π t] Sin[π x/3] Sin[π y/2] +
   1/5 Cos[Sqrt[4^2/3^2 + 2^2/2^2] π t] Sin[4 π x/3] Sin[2 π y/2];
```

This can be verified by entering

```
∂t,t u[t, x, y] - (∂x,x u[t, x, y] + ∂y,y u[t, x, y]) // Simplify
0
```

The following creates animation frames at times $t = 0, 0.1, 0.2, \ldots, 5$. The first six frames are shown below.

```
Do[Plot3D[u[t, x, y], {x, 0, 3}, {y, 0, 2},
   PlotLabel → "t = " <> ToString[t], PlotRange → {-1, 1},
   BoxRatios → Automatic, Ticks → None], {t, 0, 5, .1}];
```

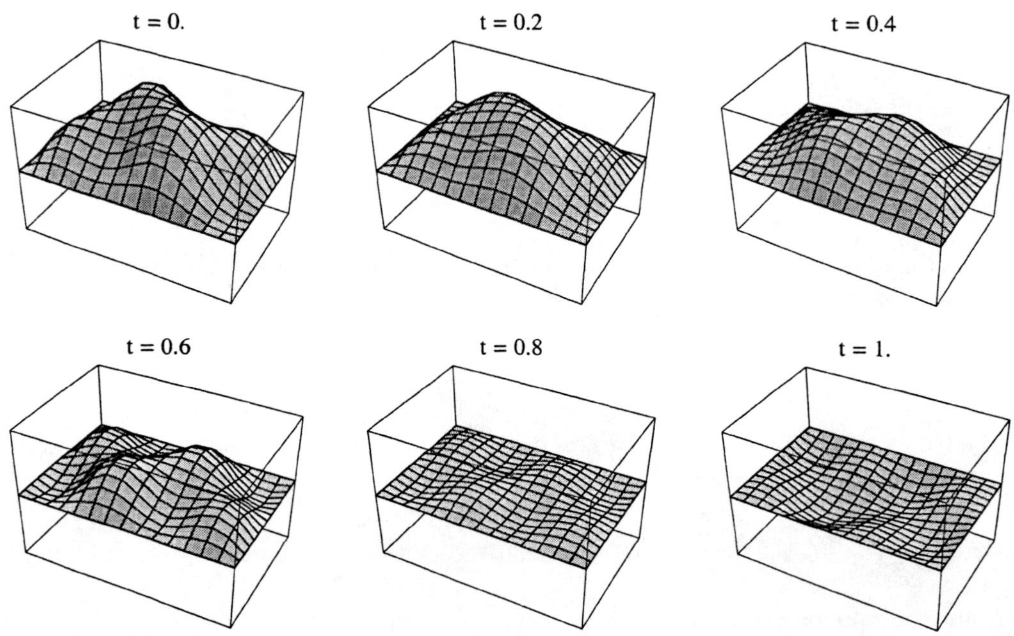

■ Circular Domain

Let Ω be the disk in which $r = \sqrt{x^2 + y^2} < \rho$. As in the case of the diffusion equation on a disk, we will require here Bessel functions and their zeros. We assume that each of φ and v are given by double Fourier-Bessel series on $\overline{\Omega}$:

$$\varphi(x) = \sum_{m=0}^{\infty} \sum_{n=1}^{\infty} J_m(\sigma_{m,n}\, r)\, (a_{m,n} \cos(m\theta) + b_{m,n} \sin(m\theta)),$$

$$v(x) = \sum_{m=0}^{\infty} \sum_{n=1}^{\infty} J_m(\sigma_{m,n}\, r)\, (\alpha_{m,n} \cos(m\theta) + \beta_{m,n} \sin(m\theta)).$$

As a result, the solution is $u = u_\varphi + u_v$, where

As a result, the solution is $u = u_\varphi + u_v$, where

$$u_\varphi(t, x) = \sum_{m=0}^{\infty} \sum_{n=1}^{\infty} \cos(\sigma_{m,n}\, c\, t)\, J_m(\sigma_{m,n}\, r)\, (a_{m,n} \cos(m\theta) + b_{m,n} \sin(m\theta)),$$

$$u_v(t, x) = \sum_{m=0}^{\infty} \sum_{n=1}^{\infty} \frac{1}{\sigma_{m,n}\, c} \sin(\sigma_{m,n}\, c\, t)\, J_m(\sigma_{m,n}\, r)\, (\alpha_{m,n} \cos(m\theta) + \beta_{m,n} \sin(m\theta)),$$

and $\rho\,\sigma_{m,n}$ is the nth positive zero of the Bessel function J_m.

Example 11.5.2

Consider the wave equation with wave speed $c = 1$ on the disk in which $r < \rho = 2$, and suppose that the initial displacement and velocity are

$$\varphi(r, \theta) = J_1(\sigma_{1,2}\, r) \cos\theta \quad \text{and} \quad v(r, \theta) = 3\, J_3(\sigma_{3,1}\, r) \cos 3\theta.$$

The solution is then

$$u(t, r, \theta) = \cos(\sigma_{1,2}\, t)\, J_1(\sigma_{1,2}\, r) \cos\theta + \frac{3}{\sigma_{3,1}} \sin(\sigma_{3,1}\, t)\, J_3(\sigma_{3,1}\, r) \cos 3\theta.$$

In order to create an animation of this solution, we first need to compute $\sigma_{1,2}$ and $\sigma_{3,1}$.

```
<< NumericalMath`BesselZeros`

      1
σ12 = — BesselJZeros[1, 2][[2]]
      2

3.50779

      1
σ31 = — BesselJZeros[3, 1][[1]]
      2

3.19008
```

Now we define the solution:

```
u[t_, r_, θ_] := Cos[σ12 t] BesselJ[1, σ12 r] Cos[θ] +
   3
  ——— Sin[σ31 t] BesselJ[3, σ31 r] Cos[3 θ]
  σ31
```

The following creates a 41-frame animation. A few of the frames are shown at the bottom of the next page.

```
Do[ParametricPlot3D[Evaluate[{r Cos[θ], r Sin[θ], u[t, r, θ]}], {r, 0, 2},
   {θ, 0, 2 π}, PlotLabel → "t = " <> ToString[t], PlotPoints → {10, 45},
   PlotRange → {-1, 1}, Ticks → None], {t, 0, 4, .1}];
```

■ Exercises

1. Let Ω be the square in which $0 < x < \pi$ and $0 < y < \pi$. Animate the solution of the wave equation with wave speed $c = 1$, a zero boundary condition, zero initial velocity, and initial displacement given by

$$\varphi(x, y) = \tfrac{1}{3} \sin x\, \sin y + \tfrac{1}{5} \sin 2x\, \sin y.$$

2. For the same scenario as in Exercise 1, use the following initial displacements to animate the first several fundamental modes of vibration:

(a) $\varphi(x, y) = \sin x \ \sin y$ (b) $\varphi(x, y) = \sin 2x \ \sin y$ (c) $\varphi(x, y) = \sin 3x \ \sin y$

(d) $\varphi(x, y) = \sin 2x \ \sin 2y$ (e) $\varphi(x, y) = \sin 3x \ \sin 2y$ (f) $\varphi(x, y) = \sin 3x \ \sin 3y$

3. Let Ω be the unit disk, and consider the wave equation, with wave speed $c = 1$, a zero boundary condition, and zero initial velocity. Use the following initial displacements to animate the first several fundamental modes of vibration:

(a) $\varphi(r, \theta) = J_0(\sigma_{0,1}\, r)$ (b) $\varphi(r, \theta) = J_0(\sigma_{0,2}\, r)$ (c) $\varphi(r, \theta) = J_0(\sigma_{0,3}\, r)$

(d) $\varphi(r, \theta) = J_1(\sigma_{1,1}\, r) \cos\theta$ (e) $\varphi(r, \theta) = J_1(\sigma_{1,2}\, r) \cos\theta$ (f) $\varphi(r, \theta) = J_1(\sigma_{1,3}\, r) \cos\theta$

(g) $\varphi(r, \theta) = J_2(\sigma_{2,1}\, r) \cos 2\theta$ (h) $\varphi(r, \theta) = J_2(\sigma_{2,2}\, r) \cos 2\theta$ (i) $\varphi(x, \theta) = J_2(\sigma_{2,3}\, r) \cos 2\theta$

(j) $\varphi(r, \theta) = J_3(\sigma_{3,1}\, r) \cos 3\theta$ (k) $\varphi(r, \theta) = J_3(\sigma_{3,2}\, r) \cos 3\theta$ (l) $\varphi(r, \theta) = J_3(\sigma_{3,3}\, r) \cos 3\,\theta$

4. Let Ω be the unit half-disk corresponding to $0 \le \theta \le \pi$, and consider the wave equation on Ω with wave speed $c = 1$, a zero boundary condition, and zero initial velocity. Animate the vibration resulting from the initial displacement

$$\varphi(r, \theta) = J_1(\sigma_{1,2}\, r) \sin\theta .$$

5. Let Ω be the unit quarter-disk corresponding to $0 \le \theta \le \frac{\pi}{2}$, and consider the wave equation on Ω with wave speed $c = 1$, a zero boundary condition, and zero initial velocity. Animate the vibration resulting from the initial displacement

$$\varphi(r, \theta) = J_2(\sigma_{2,1}\, r) \sin 2\,\theta .$$

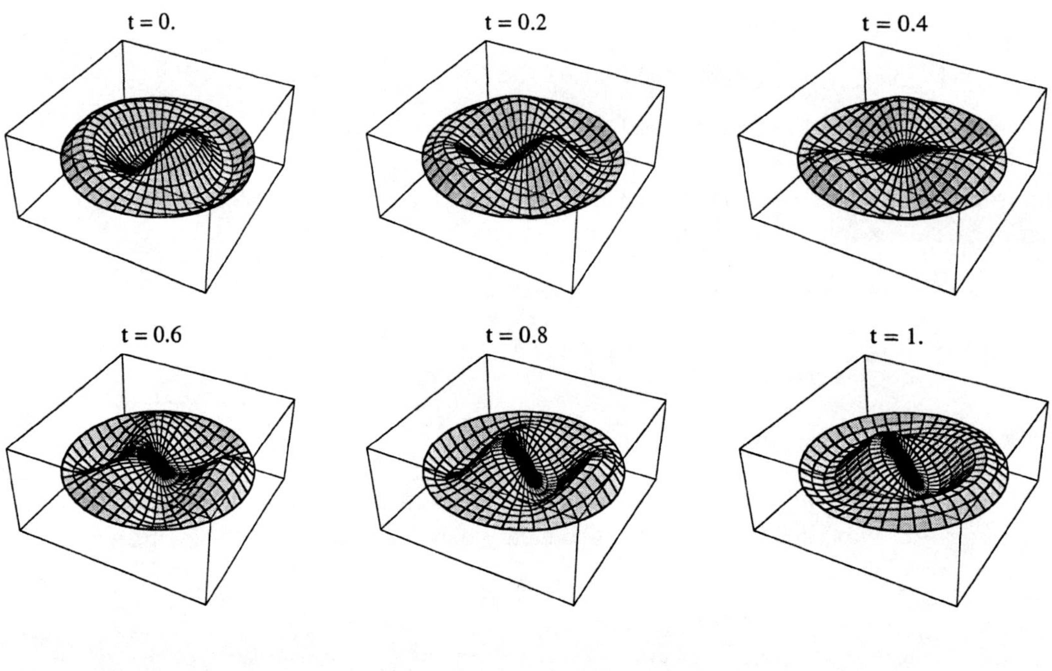

Appendix

The DiffEqs Packages

DiffEqs`DEGraphics`

This package provides functions for creating plots of solutions of differential equations.

`DEPlot[`*f*[*t*, *y*], {*t*, *tmin*, *tmax*}, {*y*, *ymin*, *ymax*}`]` creates a plot showing the direction field and solution curves for the differential equation $y' = f(t,y)$

`PhasePlot[`{*f*[*x*, *y*], *g*[*x*, *y*]}, {*t*, *tmin*, *tmax*}, {*x*, *xmin*, *xmax*}, {*y*, *ymin*, *ymax*}`]` creates a plot showing the direction field and solution curves for the autonomous system $x' = f(x,y)$, $y' = g(x,y)$.

`PhasePlot[`{*f*[*t*, *x*, *y*], *g*[*t*, *x*, *y*]}, {*t*, *tmin*, *tmax*}, {*x*, *xmin*, *xmax*}, {*y*, *ymin*, *ymax*},`]` creates a plot showing solution curves for the system $x' = f(t,x,y)$, $y' = g(t,x,y)$.

`NDPlot[`*eqns*, *fns*, {*t*, *tmin*, *tmax*}`]` uses `NDSolve` to compute a numerical solution of a system of (up to 3) first-order differential equations and plots the solution. Returns {*solution*, `-Graphics-`}.

`PoincareTimeSection[`{*f*[*t*, *x*, *y*], *g*[*t*, *x*, *y*]}, {t, t_0, *tmax*, Δt}, {x, x_0}, {y, y_0}`]` creates a Poincaré time section plot of the solution of $x' = f(t,x,y)$, $x(t_0) = x_0$, $y' = g(t,x,y)$, $y(t_0) = y_0$.

`ViewProjections[`{*f*[*t*], *g*[*t*], *h*[*t*]}, {*t*, *tmin*, *tmax*}, {*x*, *y*, *z*}`]` creates a GraphicsArray of projections of the curve $(f(t), g(t), h(t))$ onto the xy, xz, and yz coordinate planes.

`PlotImplicit[`*F*[*t*, *y*], {*t*, *tmin*, *tmax*}, {*y*, *ymin*, *ymax*}`]` creates a contour plot of the function F.

`TimeStatePlot[`{*x*[*t*], *y*[*t*]}, {*t*, *tmin*, *tmax*}, {*y*, *ymin*, *ymax*}`]` creates a 3D plot of the "time-state trajectory" $(t, x(t), y(t))$ with projections onto the tx, ty, and xy planes.

Plotting solutions of differential equations.

■ DEPlot

The following options can be used in `DEPlot`.

option name	*default value*	
`InitialPoints`	`Automatic`	Initial points for generated curves.
`MaxCurves`	`{1}`	"The number of curves to generate from each initial point.
`Gap`	`0.1`	Controls the spacing between automatically generated starting points.

Curve options for `DEPlot`.

`MaxCurves` may be a single number or a list of numbers corresponding to the list of initial points.

option name	*default value*	
`Arrows`	`{11, 9}`	The number of arrows to draw in each direction.
`ArrowColor`	`RGBColor[0, .6, .6]`	Color of the arrows.
`ScaleArrows`	`1`	Controls the arrows' length.

Arrow style options for `DEPlot`.

option name	*default value*	
`Isoclines`	`False`	Whether to draw isoclines. May also be *a* list of slope values.
`ShowStartPoints`	`False`	Whether to show the initial point from which each curve is drawn.
`ShowEquilibriumPoints`	`False`	Whether to show equilibrium points (on the *y* axis).
`ShowField`	`True`	Whether to show the direction field.
`UseVectorField`	`False`	Whether to show the vector field rather than the direction field.

Display options for `DEPlot`.

option name	*default value*	
`StepSize`	`0.1`	The stepsize for the numerical method.
`Method`	`RungeKutta`	The numerical method to use : `Euler`, `Heun`, `RungeKutta`, or `Automatic`

Numerical options for `DEPlot`.

`Method→Automatic` causes `DEPlot` to use `NDSolve` successively over intervals $[t, t+h]$, where $h =$ `StepSize`. (Thus `StepSize` does not affect the accuracy in this case—only the smoothness of the curve drawn.) `Method→Automatic` is generally slower at a given stepsize, but for difficult (i.e., stiff) equations gives much better results.

■ PhasePlot

Options for `PhasePlot` are the same as for `DEPlot`, except that `Isoclines` is replaced by the following.

option name	*default value*	
`ShowNullclines`	`False`	Whether to draw the nullclines of the system.

Options for `PhasePlot`.

■ NDPlot

Any of the options for `NDSolve`, `Plot`, `ParametricPlot`, or `ParametricPlot3D` can be used in `NDPlot`.

■ PoincareTimeSection

Options for `PoincareTimeSection` are the same as for `ListPlot`.

■ ViewProjections

Options for `ViewProjections` are the same as for `ParametricPlot` with the addition of `GraphicsSpacing`.

■ TimeStatePlot

`TimeStatePlot` uses `ParametricPlot3D` and `Shadow` from the `Graphics3D` package. The following options can be used.

option name	*default value*
`ShadowColors`	`{RGBColor[1, 0, 0], RGBColor[0, 1, 0], RGBColor[0, 0, 1]}`
`ShadowPositions`	`{-.67,.8,-.67}`

Options for `TimeStatePlot`.

■ PlotImplicit

`PlotImplicit` uses `ContourPlot` and `DensityPlot` and accepts any of their options. In addition, the following option can be used.

option name	*default value*	
`ShowDensity`	`False`	Whether to include a density plot under the curves.

Options for `PlotImplicit`.

DiffEqs`DETools`

This package provides the following functions for solving differential equations.

`FOLDSolve[`*eqn*, *y*[*t*], *t*`]`
solves a first-order linear differential equation for $y(t)$.

`BernoulliDSolve[`*eqn*, *y*[*t*], *t*`]`
solves a Bernoulli-type differential equation for $y(t)$.

`RiccatiDSolve[`*eqn*, *y*[*t*], *t*, *z*[*t*]`]`
solves a Riccati-type differential equation for $y(t)$, given a known solution $z[t]$.

`SeparableDSolve[`*f*[*x*] *dx* == *g*[*y*] *dy*`]`
returns an implicit solution of a separable differential equation.

`SeriesDSolve[`*eqn*, *y*[*t*], {*t*, t_0, *n*} `]`
returns the first *n* terms of a series expansion about t_0 of the solution of a first- or second-order differential equation.

Finding solutions of differential equations.

The following option can be used in `FOLDSolve`, `BernoulliDSolve`, `RiccatiDSolve`, and `SeparableDSolve`.

option name	*default value*
`ConstantName`	`C[1]`

Option for `FOLDSolve`, `BernoulliDSolve`, `RiccatiDSolve`, and `SeparableDSolve`.

`SeparableDSolve` can also be given the following option.

option name	*default value*
`InitialPoint`	`None`

Option for `SeparableDSolve`.

`SeriesDSolve` has no options.

DiffEqs`MatrixTools`

This package provides functions for manipulating the columns of a matrix.

`TakeColumn[`*matrix*, *n*`]` returns as a list the *n*th column of *matrix*.

`DropColumn[`*matrix*, *n*`]` returns *matrix* with its *n*th column dropped.

`DropColumn[`*matrix*, {*m*, *n*}`]` drops columns *m* through *n*.

`ReplaceColumn[`*matrix*, *newcol*, *n*`]` replaces the *n*th column of *matrix* with *newcol*.

`AppendColumn[`*matrix*, *newcol*`]` appends *newcol* to *matrix*.

`PrependColumn[`*matrix*, *newcol*`]` prepends *newcol* to *matrix*.

`InsertColumn[`*matrix*, *n*`]` inserts *newcol* into *matrix* as the *n*th column.

Manipulating the columns of a matrix.

In all instances, *newcol* must be supplied as a list.

DiffEqs`FourierTools`

This package provides functions for computing partial sums of Fourier series.

`FourierSum[`*f*[*x*], {*x*, −*L*, *L*}, *m*`]`
returns the first *m* terms of the Fourier series of *f*[*x*] on [−*L*, *L*].

`NFourierSum[`$f[x]$`, `$\{x, -L, L\}$`, `m`]`
returns the first m terms of the Fourier series of $f[x]$ on $[-L, L]$ with numerical coefficients.

`FourierCosSum[`$f[x]$`, `$\{x, 0, L\}$`, `m`]`
returns the first m terms of the Fourier cosine series of $f[x]$ on $[0, L]$.

`NFourierCosSum[`$f[x]$`, `$\{x, 0, L\}$`, `m`]`
returns the first m terms of the Fourier cosine series of $f[x]$ on $[0, L]$ with numerical coefficients.

`FourierSinSum[`$f[x]$`, `$\{x, 0, L\}$`, `m`]`
returns the first m terms of the Fourier sine series of $f[x]$ on $[0, L]$.

`NFourierSinSum[`$f[x]$`, `$\{x, 0, L\}$`, `m`]`
returns the first m terms of the Fourier sine series of $f[x]$ on $[0, L]$ with numerical coefficients.

Computing terms of Fourier series.

Index

E

F

M

N

O

P

T

U

V

W

X

Y

Z

! @#...

$